Protected Area Management - Recent Advances

Edited by Mohd Nazip Suratman

Published in London, United Kingdom

IntechOpen

Supporting open minds since 2005

Protected Area Management - Recent Advances
http://dx.doi.org/10.5772/intechopen.95152
Edited by Mohd Nazip Suratman

Contributors
Cristiane Gomes Barreto, Jose Augusto Drummond, Sandra Daniela Manzano Guzmán, Nurul Izza Ab Ghani, Wardah Arifin, Ahmad Ismail, Adelakun Kehinde Moruff, Shafiu Kilishi Halidu, Azeez Olalekan Ibrahim, Olorunfemi Boye Oyediran, Dillip Kumar Das, Binh V. Ngo, Ya-Fu Lee, Tsholofelo Lori, Joanna Korzeniowska, Javier Ferreiro da Costa, Pablo Ramil-Rego, Manuel A. Rodríguez Guitián, Luis Gómez-Orellana, Hugo López Castro, Carlos Oreiro Rey, José Antonio Fernández Bouzas, José Eduardo Ezaquias, Daniel T. Dalton, Michael Jungmeier, Kathrin Pascher, Vanessa Berger, Klaus Steinbauer, Djumanto, Muhammad Erdi Lazuardi, Imam Musthofa Zainudin, Sukma Ridarwati, Jean-Eudes Beuret, Anne Cadoret, Mirjam de Koning, Oliver Avramoski, Boycen Mudzengi, Thembinkosi Keith Gumede, Antonia Thandi Nzama, Paraskevi E. Mpeza, Teresa Mouga, Ronald Kupar Lyngdoh Tron, Duwaki Rangad, Wankitlang Shangpliang, Baiakmenlang Manners, Iasyllok Rynjah, Mohd Nazip Suratman

First published in London, United Kingdom, 2022 by IntechOpen
IntechOpen is the global imprint of INTECHOPEN LIMITED, registered in England and Wales, registration number: 11086078, 5 Princes Gate Court, London, SW7 2QJ, United Kingdom
Printed in Croatia

British Library Cataloguing-in-Publication Data
A catalogue record for this book is available from the British Library

Additional hard and PDF copies can be obtained from orders@intechopen.com

Protected Area Management - Recent Advances
Edited by Mohd Nazip Suratman
p. cm.
Print ISBN 978-1-83969-812-5
Online ISBN 978-1-83969-813-2
eBook (PDF) ISBN 978-1-83969-814-9

For EU product safety concerns:
IN TECH d.o.o., Prolaz Marije Krucifikse Kozulić 3, 51000 Rijeka, Croatia, info@intechopen.com or visit our website at intechopen.com.

Meet the editor

Mohd Nazip Suratman is Professor of Forestry, Faculty of Applied Sciences, Universiti Teknologi MARA (UiTM), Malaysia. He obtained an MSc in Forest Biology from the University of Nebraska-Lincoln, USA. He was then awarded a prestigious Canadian Commonwealth Scholarship to pursue a Ph.D. in Forest Resources Management at the University of British Columbia, Canada. He has been very active in conducting research, obtaining national and international grants, and publishing hundreds of articles in books, journals, and conference proceedings. He was a recipient of UiTM's Best Researcher award in 2015 and the Top Talent Award (Science) in 2021. His research interests include forest carbon sequestration, forest modelling, climate change, applied remote sensing, and biodiversity conservation.

Contents

Preface

Globally, protected areas and national parks are becoming increasingly important for biodiversity conservation and protection. They are home to thousands of species of flora and fauna that depend upon each other and are linked directly and indirectly to numerous ecosystems. Biodiversity is an intricate interaction on which civilization depends on various medicinal, economical, spiritual, and recreational resources. Unfortunately, climate change and the destruction of global biodiversity have taken a toll on the environment. To achieve a symbiosis between species conservation and ecotourism, a management approach to the usage of protected areas should be adopted. Based on the vital roles of protected areas and their importance for conservation, this book conveys important messages about what it takes for biodiversity to survive amidst degradation. Protected Area Management – Recent Advances is a platform for future planning and policy-making. The multidisciplinary approach in understanding such complex interaction results in a greater understanding of its function for conservatory decision-making. Chapters present an extensive array of best practices in protected area management along with research findings.

Mohd Nazip Suratman
Faculty of Applied Sciences,
Universiti Teknologi MARA (UiTM),
Shah Alam, Malaysia

Section 1

Introduction

Chapter 1

Introductory Chapter: Tropical Biodiversity Research in Protected Area of Pahang National Park, Malaysia

Mohd Nazip Suratman

1. Introduction

The tropical rainforests are earth's most complex biome that harbour a greater diversity of life than any other terrestrial habitat. The tropical rainforests of Kuala Keniam in Pahang National Park, Malaysia is no exception. They are home to biodiversity, which consists of thousands of species of flora and fauna that depend upon each other and are linked directly or indirectly with ecosystems. Kuala Keniam forests are one of the biodiversity hotspots in Pahang National Park located at the mouth of and along the Keniam River. A variety of species of plants, bacteria, insects, fungi, birds, mammals, reptiles, fish, invertebrates and amphibians live together with non-living elements like water, soil and air to make a functioning ecosystem [1]. This amazing spectrum of life makes Kuala Keniam forests a living laboratory of scientific research, and is regarded as one the most exciting habitat by researchers, scientists, botanists and naturalists within Pahang National Park.

Historically, Pahang National Park was declared as a National Park by the British Administrators with the name King George V National Park under Enactment 1939, to commemorate the installation of King George V in England [2]. The Department of Wildlife and National Parks (acronym: DWNP in English and PERHILITAN in Malay) was established in 1972 by the Government of Malaysia under the Wildlife Protection Act, 1972. The department is responsible to strengthen wildlife conservation programmes through management, enforcement, enrichment and research of wildlife [3]. In addition, it maintains the integrity of protected areas and enhances knowledge, awareness and public participation towards wildlife conservation. In 2007, the Universiti Teknologi MARA (UiTM), Malaysia in collaboration with DWNP established a field research station in a protected area of Pahang National Park known as UiTM-PERHILITAN Research Station.

This introductory chapter aims to document the great richness of biodiversity in the tropical rainforests of Pahang National Park, Malaysia, and share the findings from a scientific expedition undertaken by UiTM in its effort to carry out a multidisciplinary research programme focusing on the aspects of biodiversity in the area.

2. Establishment of UiTM-PERHILITAN research station

UiTM-PERHILITAN Research Station (hereafter is referred to as Kuala Keniam forests) was established with a primary focus to provide a great natural laboratory to conduct scientific investigations of the biodiversity of tropical rainforest as well as

Figure 1.
Accommodation facilities for researchers and visitors in Kuala Keniam forests, Pahang National Park.

ecosystem processes and interactions between them within the National Park landscape and on ways of managing them. Physical facilities at Kuala Keniam forests are designed to accommodate programmes of research, education and services (**Figure 1**). Currently available facilities include a multi-purpose building, boats, chalets, bathrooms, toilets, prayer room and base camp which can accommodate 50–70 people at a particular time. The interpretive trails and forest arboretum in Kuala Keniam forests are used for teaching purposes in the fields of dendrology, forest botany, environmental sciences, park and recreation management and wildlife management. Transect lines and permanent field plots from research activities were established to provide hands-on training in forest ecology, silviculture and forest inventory.

Initial research was funded through a top-down Fundamental Research Grant Scheme (FRGS) awarded by the Ministry of Higher Education, Malaysia [4]. UiTM teamed with DWNP in a joint Memorandum of Understanding (MoU) to collaborate with three main objectives. Firstly, to carry out joint programmes of scientific research, conservation, management of biological resources. Secondly, to strengthen the capacity of the DWNP in terms of training, attachment and networking and to provide the necessary framework to develop expertise in the research, conservation and management of biological resources. Finally, to provide consultancy related to research, conservation and management of biological resources.

3. Topography, climate and vegetation of Kuala Keniam forests

The total area of Pahang National Park is 4,343km^2 which covers the three states, namely Pahang, Kelantan and Terengganu [2]. It is considered one of the oldest tropical rainforests in the world of more than 130 million years old. Geographically, the park lies between 80 and 2,187 m above sea level with Mount Tahan is the highest peak. The weather is characterised by permanent high temperatures ranging from 20°C at night and 35°C in the daytime with high relative humidity (above 80%) [5, 6]. Annual rainfall is approximately 2,260 mm with the highest rainfall occurring

Figure 2.
Riparian vegetations along Tahan River, Pahang National Park.

in October to November with about 312 mm of rainfalls [7]. The lowest rainfall occurs in March with only about 50 mm. Keniam River is one of the major rivers flowing into the Tembeling River with the presence of *Diptercarpus oblongifolius* (Keruing neram) trees along with the riparian habitats (**Figure 2**).

The great richness of this tropical rain forest is illustrated by a hectare plot that contains more than 280 tree species, with the Shannon index of diversity (*H*') ranging from 1.9 to 2.5 [5]. The rainforest consists of tall evergreen trees that attain heights up to 50 m. It is typically very damp and rich in herbaceous, shrubs, epiphytes, lianas and tree species from the family of Dipterocarpaceae [8]. *Elasteriospermum tapos* (Perah) is dominant and frequently occurs in Kuala Keniam and its vicinity [6]. Other trees include *Intsia palembanica* (Merbau), *Koompassia malaccensis* (Tualang), *Knema patentinervia* (Penarahan), *Aporosa prainiana* (Sebasah), *Macaranga lowii* (Balik angin) and *Koompassia malaccensis* (Kempas) [5, 6]. *Dipterocarpus oblongifolius* (Keruing neram), *Pometia pinnata* (Kasai daun besar) and *Pterocambium javanicum* (Melembu) trees commonly occur along the river banks together with other riparian vegetation such *Milletia hemsleyana* (Jada), *Neonauclea subdita* (Mengkai) and *Dysoxylum angustofolium* (Maris) [1]. The understory trees and shrub community consist of the families Rubiaceae, Myrtaceae and Euphorbiaceae with *Rennelia* spp., *Syzygium* spp., *Mallotus* spp. being the common genera [5]. Common tree crown epiphytes include *Asplenium nidus* (bird nest fern) and *Platycerium coronarium* (staghorn fern).

4. Biodiversity in Malaysia

Located within the tropical belt of the world, Malaysia is richly endowed with a luxuriant and diverse natural environment that offers unlimited opportunities for exploration and discovery of biological diversity of tropical rainforests. The Convention on Biological Diversity in 1993, rated Malaysia as one of the world's 12th mega diversity countries. In terms of flora diversity, the tropical rainforests of the country houses over 15,000 flowering plants, 1,159 ferns and fern allies, 513 palms, 3,000 orchids, 432 mosses and 700 fungi. There is also a greater diversity of fauna in the country. For vertebrates, there are about 300 species of wild mammals, 700–750 species of birds, 350 species of amphibians and more than 300 species of freshwater fishes [9].

5. Species richness of Kuala Keniam forests

5.1 Flowering plants

Some of the common flowering plants of Kuala Keniam are the umbel-shaped inflorescences of *Trevesia burckii* (Tapak hantu), which are found on a spiked stem (**Figure 3**), the striking large orange-red flowers of *Spathodea campanulate* (African

Figure 3.
A common flowering plant of Trevesia burckii (Tapak hantu) in Kuala Keniam forests, Pahang National Park.

tulips) and the sweet pink flowers of *Cassia nodosa* (Bebusuk) [1]. Attractive combination of orange and pink flowerheads of a shrub, *Lantana camara* (Bunga tahi ayam), clusters of white flowers of *Ixora* spp.

5.2 Medicinal plants

Zingiber spectabile (Tepus tanah) (**Figure 4**), *Labisia pumilla* (Kacip Fatimah) and *Eurycoma longifolia* (Tongkat Ali) are a few examples of common medicinal plants thriving in Kuala Keniam forests. Many more of the lesser-known plants are no less valuable with virtues waiting to be discovered. *Psycotria malayana*, *Knema glauca* and *Horsfeldia tomentosa*, to name a few, deserve attention from interested researchers searching for bioactive molecules to be turned into yet another important medicinal agent. *Rennellia elliptica* (Segemuk) has been dubbed as the Malaysian ginseng. Decoction of the roots of this plant is drunk for general health booster and also claimed to be anti-diabetic and aphrodisiac.

5.3 Mosses

Belong to the group of Bryophytes, mosses are commonly recognised as seedless nonvascular plants, which are tiny and small, but some may be as large as 35cm tall. In Kuala Keniam forests, *Calymperes moluccense* is found mainly on the tree base, *Pelekium velatum* well spread on rotten bark and *Syrrhopodon spiculosus* is found on bark of living trees in moist and shaded lowland areas. Mosses do not have direct economic value; however, they are important in many respects. They play an important role in the water balance of ecosystems in the forests by storing large amounts of water. They also prevent erosion by fixing the uppermost layer of soil, and also providing microhabitats for small animals, germination of seeds or food for certain beetles.

5.4 Ferns and allies

At Kuala Keniam forest, ferns and their allies are widespread and can be found on forest floors, tree branches and trunks, rocks and also in the drains. Ferns can be recognised by the groups of sporangia called sori on the back of the leaves (rarely on the margin). Examples of the fern allies found at Kuala Keniam forests are *Selaginella willdenowii*, *S. wallichii* and *Schizaea dichotoma*. Some ferns are good to prevent soil erosion such as *Dicranopteris* spp. Some of them can be used for food, as a biological fertilizer, and also grown in horticulture as landscape plants, for

Figure 4.
A species of true ginger Zingiber spectabile (Tepus tanah) has long been used in traditional medicine.

cut foliage and as houseplants, especially *Asplenium nidus* (birds-nest fern) and *Platycerium coronarium* (staghorn fern).

5.5 Rattans and bamboo

Rattans are classified into the family of Arecaceae or Palmae and the subfamily of Calamoideae. In Kuala Keniam forests, Abdul Hamid and Suratman [10] found thirteen species of rattans with the three most dominant species are *Daemonorops didymorphylla* (Rotan jernang), *Calamus caesius* (Rotan sega) and *C. luridus* (Rotan kerai). Bamboo belongs to the *Gramineae* family and the Bambusoideae subfamily. It is considered to be one of the most fast-growing plants on earth [11]. Certain species of bamboo can grow at the rate of 900mm per day. Asari and Suratman [12] found five bamboo species grow in Kuala Keniam forests. These are *Schizostachyum grande* (Buluh semeliang), *S. latifolium* (Buluh nipis), *S. brachycladum* (Buluh lemang), *Bambusa vulgaris* (Buluh aur) and *Gigantochloa scortechinii* (Buluh semantan).

5.6 Fungi

Fungi play important roles in a rainforest ecosystem. They are the parasites, saprophytes or symbionts. As parasites, they invade living plants, cause diseases and eventually bring deaths to the plants. As saprophytes, they grow on dead plants, break organic materials and return the nutrients back to the soil. As symbionts, they form mycorrhizae with plant roots and form lichens with algae. In addition, some fungi can be food for animals in the forest. Kuala Keniam forests have a rich diversity of fungal flora, which is dominated by various kinds and forms of mushrooms and polypore or bracket fungi. For instance, the bird's nest fungus, coral fungus, long net stinkhorn (**Figure 5**), cup fungus, jelly fungus and puffball are thriving in the area.

Figure 5.
Phallus indusiatus (long net stinkhorn) is one of many fungi species that grows in the forest floor of Kuala Keniam forests.

6. Insects of Kuala Keniam forests

Despite being relatively small in size compared to many animal groups, insects are well-adapted to their environments. In order to make sure of their survival, insects have strategies by applying camouflage and mimicry, which refer to adaptations some animals use as protection from predators. Some insects display warning colours to scare off predators. They use camouflage and mimicry as methods of

Figure 6.
Shield bug (Hemiptera) uses colours as protection from predators.

hiding from predators or ambushing prey (**Figure 6**). For example, pray mantis (Mantidae), katydids (Orthoptera: Tettigoniidae), grasshoppers (Orthoptera: Acrididae) are insects that have developed camouflaged bodies. Instead of using camouflage to blend in with their habitat, some insects use mimicry to distinguish themselves from other insects or resemble certain marks. By doing so, the insects benefit by deterring the predators.

7. Rivers of Pahang National Parks

A few kilometres upstream of Tahan River, a spectacular view greets the visitors as one can observe the convergence between two rivers that has two different colours; Tahan River, a coffee-coloured river, and Tenor River with greenish colour due to its mineral contents. The dark colour results from the tannins released from the decaying leaves and organic materials of the vegetation in the area. However, in the monsoon season, this river loses its clarity and colour due to the sedimentation from the runoff of the surrounding forest. Another tributary is the Keniam River, where it is considered as a clearwater river as the water only receives minimal loading of suspended matter. Rapids and ancient rock beds are common, hence, the water has a higher pH than other tributaries.

8. Recreational opportunities

A canopy walkway in Kuala Tahan offers a close-up view of activity from the rainforest canopy. Observation hides (bumbun) are another great way to observe wildlife, especially at night to catch opportunities to observe animals in their natural habitat. Cave explorations to Luas Cave and Kepayang Cave are available from Kuala Keniam research station.

Along the Tembeling River from Kuala Tahan to Kuala Keniam (approximately 25 km), there are seven sets of rapids, which provide an exciting ride when the river is in full flood. Main activities are river rapid shooting and rafting. There are two rivers in Kuala Keniam forests (i.e., Keniam and Perkai Rivers), where wildlife viewing is possible from the water. A distinctive plant community lines the riverbanks as a home to riparian flora and fauna.

9. Wildlife, birds and fishes of Kuala Keniam

Abundant food plants, a variety of habitat types and major drainage systems provide all necessary living requirements for wildlife. *Elephas maximus* (Elephant) herds are distributed in Ulu Atok, Jenut Kumbang and Trenggan of Pahang National Park. Other mammals found in Kuala Keniam forests are *Tapirus indicus* (Malayan tapir), *Cervus unicolor* (Sambar deer), *Sus scrofa* (Wild boars), *Muntiacus muntjac* (Barking deer) and *Tragulus* spp. (Mouse deer). The forest habitat in Kuala Keniam forests contains about over 20 species of birds which include *Pycnonotus plumosus* (Olive-winged bulbul), *Arachnothera longirostra* (Little spider hunter) and *Hirundo tahitica* (Pacific swallow). In addition, there are also many fish species found in Keniam River. *Hampala maerdepidota* (Sebarau), *Mystacholeucus marginatus* (Sia), *Labiobarbus* spp. (Kawan) and *Tor tombroides* (Kelah) are the most widely distributed species.

10. Conclusions

As highlighted in the chapter, tropical rainforests of Pahang National Park are ecologically diverse and exceedingly rich, however, many species remain uncovered. Kuala Keniam forests provide a spectacular example of a living laboratory and the ultimate in biodiversity. The abundance and diversity of nature in these forests are phenomenal, making it one of the world's most complex and rich ecosystems. Therefore, there is a need to enhance efforts in research as much of the country's biodiversity has yet to be scientifically investigated. Opportunities for multidisciplinary research should be fully explored and enhanced to strengthen our scientific base and at the same time build up researchers' competencies through collaborative programmes nationally and internationally working towards shared goals.

Acknowledgements

A special appreciation is extended to Y. Bhg. Dato' Professor Dr. Rahmat Mohamad who was instrumental to the establishment of the UiTM-PERHILITAN Research Station. The author is grateful for the strategic partnership and cooperation of the Department of Wildlife and National Parks, Peninsular Malaysia. The author wishes to thank the Ministry of Higher Education Malaysia for providing research funding and the Universiti Teknologi MARA, Malaysia for providing logistic and financial support.

Author details

Mohd Nazip Suratman
Faculty of Applied Sciences and Institute for Biodiversity and Sustainable Development, Universiti Teknologi MARA (UiTM), Shah Alam, Malaysia

*Address all correspondence to: nazip@uitm.edu.my

References

[1] Suratman MN, Mohamad R, Abdullah Z, Mohamad Amin MA, editors. Kuala Keniam: Nature's Treasure Trove in the Heartland. Shah Alam, Malaysia: Universiti Teknologi MARA; 2011. p. 156

[2] UNESCO. National Park (Taman Negara) of Peninsular Malaysia. 2014. Available from: https://whc.unesco.org/en/tentativelists/5927/ [Accessed: December 20, 2021]

[3] DWNP. Department of Wildlife and National Parks, Peninsular Malaysia, Ministry of Energy and Natural Resources, Malaysia. 2022. Available from: https://www.wildlife.gov.my/index.php/en/ [Accessed: December 20, 2021]

[4] Mohamad R, Suratman MN, Abdullah Z, editors. Abstracts: UiTM-PERHILITAN Taman Negara Research Projects. Malaysia: UiTM Press; 2010. p. 54

[5] Suratman MN, Kusin M, SAK Y, Salleh K, Ahmad M, Bahari SA. Tree species composition and forest stand structure of Taman Negara Pahang: A case study from Kuala Keniam forest. In: Proceedings of Regional Seminar on Science, Technology and Social Science; 2010. pp. 395-410

[6] Suratman MN. Tree species diversity and forest stand structure of Pahang National Park, Malaysia. In: Lameed GA, editor. Biodiversity Enrichment in a Diverse World. Rijeka: InTech; 2012. pp. 473-492

[7] Zani NF, Suratman MN. Estimation of above ground biomass of Keniam Forests, Taman Negara Pahang. In: Proceedings International Symposium on Business, Engineering and Industrial Application (ISBEIA2011). U.S.A.: Institute of Electrical and Electronics Engineers (IEEE), Curran Associates, Inc.; 2010. pp. 80-83

[8] Md Sabri MD, Suratman MN, Daim MS, Kassim AR, Khamis S. Influence of ecotourism activities on forest structure and light intensity of forests in Taman Negara Pahang. In: Proceedings IEEE Symposium on Business, Engineering and Industrial Applications (ISBEIA2012). New York: IEEE; 2012. pp. 322-325

[9] NRE. National Policy of Biological Diversity 2016-2025. Malaysia: Ministry of Natural Resources and Environment; 2012. p. 114

[10] Abd Hamid NH, Suratman MN. Assessment of rattan abundance and species composition at Kuala Keniam, Taman Negara Pahang. In: Proceedings of IEEE International Conference on Science and Social Research. New York: IEEE; 2010. pp. 1053-1057

[11] Aminuddin M, Abd Latif M. Bamboo in Malaysia: Past, present and future research in bamboo in the Asia Pacific. In: Paper presented at the Fourth International Bamboo Workshop, Chiangmai, Thailand. Bangkok: FAO; 1991

[12] Asari N, Suratman MN. Distribution and species diversity of bamboo at Kuala Keniam, Pahang National Park. In: Proceedings of IEEE International Conference on Science and Social Research. New York: IEEE; 2010. pp. 1031-1035

Section 2

Biodiversity and Genetic Resources

Chapter 2

Conservation Genetics for Managing Biodiversity

Nurul Izza Ab Ghani, Wardah Arifin and Ahmad Ismail

Abstract

Conservation genetics is a field derived from a combination of evolution, ecology, behaviour, and genetics. It is an applied discipline of crisis-oriented science of biodiversity resource management that is highlighted when the world realizes the increasing anthropogenic impact and natural populations are declining towards species extinction. It helps to understand and explain the importance of evolutionary factors — mutations, non-random mating, gene flow, genetic drift, and natural selection — for the survival of populations/species that justify the need for prudent biodiversity management. The four justifications for maintaining prudent biodiversity are the economic value of bioresources, ecosystem services, esthetics, and rights of living organisms to exist ensure functioning community and ecosystem services. Hence, conservation genetics must be an essential part of policies and programs in wildlife and biodiversity management.

Keywords: biodiversity, conservation, evolutionary factors, genetic management, genetic variability

1. Introduction

The need to preserve wildlife arises because the earth's biological diversity is rapidly depleted as a direct or indirect result of human action. To date, a number of unknown but many species have become extinct, meanwhile many other species have reduced population sizes and putting them at risk of extinction. IUCN [1] reported that more than 38,500 species are threatened with extinction — highlighting 26% of mammals, 14% of birds, 41% of amphibians, 37% of sharks and rays, 28% of selected crustaceans, 34% of conifers, and 33% of reef corals. Hence, many species are now required human intervention to ensure their survival through effective management and conservation of biodiversity resources. But a statistically robust Population Viability Analysis (PVA) has yet to be developed to assess the ecological and genetic risks faced by the Essential Evolutionary Unit (ESU) which is the unit of biodiversity that is of concern to conservation geneticists. Though, International Union for Conservation of Nature (IUCN) has recognized the need to manage and conserve biodiversity resources at three levels; genetic diversity, species diversity, and ecosystem diversity. Genetic information is involved in all of these three levels. Thus, geneticists (specifically known as conservation geneticists) are playing an increasingly important role in the management and conservation of biodiversity resources — identifying and monitoring the genetic variability that directly relates to evolutionary factors of biodiversity units.

Conservation geneticists deal with evolutionary factors causing rarity; endangerment and extinction of threatened population and species, and genetic management to minimize impacts of evolutionary factors in threatened population and species, as well as resolving taxonomic uncertainties in threatened species, understanding the biology of threatened population and species through their genomic, and wildlife forensics. All of these are important research courses in conservation genetics with the ultimate goal to manage biodiversity resources with utmost care through preserving and maintaining the ability of populations and species to evolve. Thus, reducing the extinction risk of population and species, while ensuring a functioning community and ecosystem services. All research courses in conservation genetics can be disentangled by using molecular genetics methods through the use of various molecular markers. The common molecular markers which have been used are single-locus markers (allozymes), DNA minisatellite fingerprints, random amplified polymorphic DNA (RAPD), mitochondrial DNA (mtDNA) sequences, chloroplast DNA (cpDNA) sequences, genic sequences such as Major histocompatibility complex (MHC) genes, and nuclear DNA (nDNA) sequences including microsatellites and single nucleotide polymorphism (SNP). To date, conservation geneticists also have started to use whole-genome sequence which offers a more powerful assessment to disentangle evolutionary factors and their implications towards population/species rarity and survival to manage biodiversity.

Yet, efforts to implement conservation genetics for managing biodiversity have been done for very few threatened species. Therefore, the aim of this chapter is to briefly highlight the importance of assimilating conservation genetics to manage biodiversity with a review of the relevant literature. This chapter is comprised of three parts. The first part introduces readers to the genetic management of biodiversity units that are seldom been misinterpreted. The second part points out the essence of genetic variability in managing biodiversity due to its importance for determining future population/species evolution. The final section hopes to engage readers with an appreciation of research courses in conservation genetics by briefly describing evolutionary factors influencing genetic variability of threatened populations/species including mutations, non-random mating, gene flow, genetic drift, and natural selection.

2. Genetic management of biodiversity unit

Poorly planned conservation management plans can significantly cause local adaptation damage (overcoming depression) and reduce the viability of the population, especially the threatened population. PVA is a methodology that has been used by conservation geneticists to assess the ecological and genetic risks faced by wildlife or captive population, and thus appropriate conservation management plan can be developed. PVA refers to a group of mathematical models that are useful for predicting the probability of population extinction at some point in the future. The early PVA models considered demographic data (growth rate, current population size, and birth rate) and environmental stochastic data. But Gilpin and Soulé [2] has been further enhanced the ability of the PVA model in predicting the extinction of a species by including genetic factors. Genetic factors including mutation, genetic drift, non-random mating, gene flow, and natural selection have significantly influenced genetic variability. It is clearly expressed through its effects on demographic factors that influence population dynamics, especially in small isolated/threatened populations. This shows that genetic factors contribute to extinction probabilities through a very complex manner of interactions affecting the genetic variability and fitness of a population [3, 4]. Unfortunately, little is understood regarding genetic factors' linkage to ecological factors. Thus, statistically strong PVAs have not yet been developed sufficiently to provide comprehensive biodiversity management.

The biodiversity unit of concern by conservation geneticists in PVA is ESUs. ESUs represent genetically differentiated populations whereby depicting deep phylogenetic subdivisions typically within a species (i.e., subspecies) or occasionally as entire species in the case of local endemics or distinct population segments (DPS - Endangered Species Act 1973). ESUs are classified based on genetic criteria; both genetic diversity and multilocus genetic similarity using multilocus mtDNA or nDNA (preferably microsatellites) variation. mtDNA shows evidence for significant long-term genetic divergence and reciprocal monophyly. Whereas microsatellites show evidence for the significant recent divergence of allele frequencies at nuclear loci. A refine ESUs in wildlife conservation include pedigree analysis. Pedigree analysis has been used to understand the established kinship and individual founder contributions, to determine genetically desirable and undesirable individuals as well as their descendants, to elucidate population structure and mating system, and to designate appropriate individuals for translocation or reintroduction. Hence, pedigree management programs based on mean kinship or minimal founder contributions are commonly used to minimize inbreeding in local subpopulations and metapopulations. Delineating refine ESUs is important when considering long-term conservation actions especially translocations and captive breeding programs. Translocation between ESUs should be avoided in order to successfully replenish the diversity and viability of severely declining and nearly monomorphic populations with severe inbreeding depression (low heterozygosity, low fertility (e.g., poor sperm and ovum quality and cryptorchidism), and low disease susceptibility). Whereas captive breeding programs between ESUs may lead to reduce genetic variability and increase populations' susceptibility to extinction.

3. Genetic variability as the heart of managing biodiversity

Conservation of the genetic variability within a species is necessary as a part of global efforts to manage and conserve biodiversity. High levels of genetic variability in most natural populations of plants and animals are determinants of future population/species evolution. Genetic variability which is determined by genetic diversity can be interpreted at several levels including karyotypic variation (usually low within a species), allozyme variation (usually high within a species), and DNA sequence variation (maybe very high within a species at nongenic region e.g., short repeat sequences (microsatellites/SSR) of nuclear DNA, and maybe low within a species at genic region).

Genetic diversity can be assessed by determining kinship lineage and home range within and between a particular species/population of wildlife by using DNA analyses. Through DNA analyses, crucial information including identification of parentage, distant relatives, founders of a population, unidentified individuals, and population structure (mating system, sex ratio, estimate past population size and patterns of variability over periods time) can be correctly done to ensure genetic effective population size (Ne) is present in a particular wildlife population/species. DNA analysis expressed as genetic distance allows interpopulation comparisons to uncover spatial structuring and historical patterns of gene flow within a species. The absolute values of genetic distances which can be calculated from dissimilarities in genetic diversity vary between species, and they are increased over geological time. Therefore, accurate ESUs for effective conservation management purposes can be justified. Widely use DNA analyses by conservation geneticists are allozymes, DNA minisatellite fingerprints, RAPD, mtDNA sequences, cpDNA sequences, genic sequences such as MHC genes, and nDNA sequences including microsatellites and SNP. The recent DNA analysis used by conservation geneticists involves the investigation of a whole-genome that is typically challenged with a huge amount of DNA base. Both nuclear and mitochondrial sequence data still provide the most informative characterizing variability at or above the level of populations. Whereas for characterizing variation within populations, polymorphic nuclear microsatellite loci and SNP are ideal markers. The various DNA analyses provide different resolutions of pedigree, population, and species-level answers and all methods are correct. Most importantly, DNA analyses can be performed for wildlife populations without requiring plants to be disturbed and animals to be seen and disturbed, as well as for museum and fossil specimens (e.g., dodo, moa, thylacine, and quagga). This can be done by using non-invasive (shed tissues, faeces, urines, and scent markings) and non-destructive (toe, tail and ear clips, and fish scales) samples. Nevertheless, DNA of some types of non-invasive and non-destructive samples may deteriorate rapidly, and hence be very difficult to work with, but it is possible with extra technical care and patience.

DNA sequence variation at the genic region is the focus of conservation geneticists. This is due to in natural populations, much of genetic variability at genic region have been discovered are appeared to be selectively neutral or near-neutral in their effects on the phenotypes (i.e., cryptic variations). Hence, the individuals carrying these allelic variants/genetic diversities appear phenotypically normal. In addition, some cryptic variations have shown circumstantial evidence that they are beneficial — provide long-term population perseverance and evolvability [5]. However, their relationship between genetic variability and individual fitness is not well understood. In a world whose change is unpredictable, alleles that are selectively neutral for thousands of generations can suddenly become a saviour for the individual who carries it. Experiments and field observations on several species have shown that there is a positive relationship between genetic variability at the genic region and individual adaptability or evolvability in important ecological aspects and significant phenotypes. The phenotypes are including body size, symmetry of body parts, growth rate, size at maturity, fecundity, hatching date, predator avoidance behaviour (e.g., escape speed, defence method, aggression, etc.), and health as measured by parasite load. Hence, conservation genetics have been putting efforts to understand genetic variability at these phenotypes through understanding the genetic diversity to explain the cause of rarity, endangerment, and extinction of a genetically deteriorate species/population. For example, cheetah (*Acinonyx jubatus*) with a low level of genetic diversity has been proved to have reduced genetic variability and hence has increased susceptibility to diseases [6, 7]. Genetic variability in these phenotypes; quantitative trait loci (QTL) are controlled by several to many genes (i.e., oligogenic and polygenic) that work additively in dominance/recessive relationships or epistatically, and their expression profiles are usually induced by environmental factors as consequences of local adaptation known as phenotypic plasticity (i.e., an adaptive mechanism). According to Fisher's fundamental theorem of natural selection, additive genetic variation (i.e., innate genetic variability; heritability (h^2)) in QTL fitness is positively related to a population's ability to respond to natural selection (i.e., evolutionary success; the ability of a species to persist despite changes in climate and environment as well as exposure to new challenges including new competitors, diseases and predators). Therefore, the heritability of such phenotypes is of conservation geneticists' interest. High heritabilities of a QTL on a trait demonstrate that a population has a great potential for evolution. Whereas low heritabilities demonstrate a more limited ability of a population to respond to environmental change. Unfortunately, such heritability is difficult to measure because it requires pedigree studies over several generations or long-term manipulative experiments such as laboratory-raised plants and animals. Heritability is the ratio of the variance of a genetically inherited proportion of a trait (additive genetic variance, VA; a component in genetic variance (VG)) which response to directional selection) to the total phenotypic variance (VP) measured in a particular population and time. Estimating VA is complicated by the need to estimate environment variance (VE) as well as other genetic components in VG that are nonadditive genetic variances including dominance (VD) and epistasis (VI). However, QTL analyses using studbook records for captive populations of plants and animals, and the comparison of laboratory-raised offspring to their parents in the wild have allowed conservation geneticists to predict a reliable population's risk of extinction. This provides conservation biologists with important information on how biodiversity can be best protected against climate change and anthropogenic impacts.

On the other hand, management of genetic diversity at a large number of neutral polymorphic sites (nongenic region) has provided a useful scientific assistant to clarify for setting a species/population recovery priorities and protection. Whereby it allows more explicit estimates of Ne, migration rate, populations dynamics, and population structure (units of management). It also permits better assessment of introgression concerning management against the breeding of hybrid organisms and closely related individuals. Thus, de-extinction that is bringing back extinct wildlife species and reintroducing them to their previously inhabited landscapes with optimum Ne can be successfully done. Asexually reproducing species including clonal plants, hermaphrodite invertebrates, fish, and lizards, as well as threatened species are mostly genetically invariant in their nongenic region although they may exhibit a great ecological success [8]. Therefore, they are more prone to become extinct when their environment changes than their closely related sexually reproducing species. This has been proved in several threatened species — e.g., cheetah and ice-breeding seals whereby they are ecologically successful in the wild because of their innate genetic variability despite low absolute levels of genetic diversity; both genic and nongenic genetic diversity and being classified as threatened wildlife [8, 9].

Evolution is largely dependent on genetic variability; both genic and nongenic genetic diversity, whereby the conservation and survival of species significantly depend on the conservation of their innate genetic variability [5]. Different types of genetic variability will respond differently to evolutionary factors, population collapse, and habitat fragmentation. Hence, genetic variability is an important biological factor to determine the presence of genetic diversity or it lost, understand the causes of the loss and make recommendations to overcome its ultimate effects in wildlife conservation.

4. Evolutionary factors influencing genetic variability

4.1 Mutation

Mutations encompass a wide range of phenomena; from a change of a single base pair in the genetic code to an inadvertent doubling of the number of chromosomes. Many mutations are deleterious or lethal, some are near neutral and a small number may be beneficial (usually exist as rare alleles). A large number of mutations are completely invisible in the phenotype and can only be detected with various genetic techniques. Hence, mutations are of concern for conservation geneticists in a couple of circumstances. First, mutations in small, remnant, or isolated populations with deleterious effects. Second, whether the emergence of new mutations will replace variability lost due to population extinction and genetic erosion. Mutation rates are usually in the order of one per 105 cell divisions with time for the accumulation of new variants in a population taking tens of thousands of years.

Deleterious alleles (alleles that are accountable for genetic defects such as albinism) are usually very rare and have a frequency less than 0.0001. In a large population, natural selection purges very rare alleles of deleterious mutations from the population almost immediately. However, in a small, remnant, or isolated population, purging for such deleterious alleles in the context of the conservation of threatened species breeding program should be controlled or eliminated instantly artificially because natural selection is inefficient. Even though extinction due to the presence of deleterious mutations is almost unknown, but their contribution to the extinction process should not be ignored. Theoretically, the accumulation of deleterious mutations can significantly induce inbreeding depression and genetic erosion of fitness [10]. Deleterious alleles if not eliminated in a population, will gradually increase in frequency and become a serious problem when the frequency exceeds 0.05 or 1/(2Ne). Fortunately, this process took hundreds of generations.

On the other hand, conservation geneticists are often being demanded to save rare alleles including mildly deleterious alleles in threatened populations as they may be important for the population's adaptation towards environmental changes. The maintenance of desirable rare alleles including mildly deleterious alleles require very large population sizes and is simply not possible in most captive management programs. The risk of extinction due to fixation of rare alleles including mildly deleterious mutations of equal importance to environmental stochastics and can reduce the long-term viability of populations with Ne of less than a few thousand. An optimum Ne = 10,000 is required to ensure genetic and demographic factors act synergistically for avoiding inbreeding depression and for suppressing genetic erosion of fitness [11]. Small populations (Ne < 500) can decline fitness rapidly with the accumulation of mildly deleterious mutations, called mutational meltdown [11–13]. However, many threatened species currently have insufficient individuals to ensure long-term viability if Ne = 10,000 is strictly required.

Therefore, conservation geneticists are often left with conflict to design conservation plans that will further maintain rare alleles including mildly deleterious alleles, and eliminate deleterious alleles in threatened populations without jeopardizing populations' fitness. If the purpose of a conservation program is to return captive populations to the wild, then managers should maximize the genetic variability of rare alleles including mildly deleterious mutants. On the other hand, if the population cannot be returned to the wild and must be sustained in captivity for many generations, managers should either purge or rigorously control deleterious mutations and maintain rare alleles including mildly deleterious mutants as they are identified. For example, the homozygous recessive rare allele of White tigers (*Panthera tigris*) show no severe physiological defects but are needed to be strictly controlled in the captive populations and curbed from transmission to the wild populations to maintain the wild tiger populations [14].

4.2 Non-random mating

The ideal population genetic theory is based on random mating. It is widely accepted that random mating in sexual reproduction species evolved in part because of chromosomal crossing over and recombination facilitated by outbreeding. Most plants and animals species have effective immunological and behavioural mechanisms to favour outbreeding. These include asynchronous maturation of male and female gametes, sex-biased dispersal of the juvenile from their natal population, complex courtship behaviours, and the evolution of diverse self-incompatibility systems. Though, such mating behaviour is rarely observed in the nature of non-random mating species. The three extreme modes of non-random mating species are self-fertilized hermaphroditic, obligate outbreeding dioecious, and females preferentially mate (also known as selective breeding).

The most extreme consequence of non-random mating is the rise of inbreeding. Inbreeding refers to the mating of close relatives — mattings between father and daughter, brother and sister, or first cousins. Many species of plants and animals have evolved mechanisms to minimize close inbreeding. Species differ greatly in their tolerance to inbreeding; for example, some trees and dioecious plants are obligate outcrosses. In wild populations, the occurrence of gradual inbreeding allows natural selection to purge the first generation but the partially recessive near-neutral mutations continue to increase in frequency and significance. Inbreeding results in increased homozygosity of recessive partially deleterious mutants and by chance, in small isolated populations, these alleles can become fixed. In the simplest genetic example of a trait under the control of this recessive allele, there is an increased risk that the offspring of two related healthy but heterozygous individuals will inherit the harmful allele from each parent and die. Although the risk, in this case, is only one in four, this is a very strong fitness difference in which natural selection will act. Generalizing from this simplest single-locus example, geneticists discuss inbreeding depression as an overall manifestation of the genomic effects of mating between close relatives. These effects may involve outright genetic disease (congenital abnormalities) but are more often subtle and appear as decreased growth rate, behavioural abnormalities, and reduced fertility and fecundity. Inbreeding is rare in typically outbreeding populations but becomes a serious problem in small isolated populations. In small isolated populations and fragmented populations, inbreeding depressions can intimidate population viability. Animal and plant breeders have learned this lesson from their centuries of experience with artificial selection, and therefore they limit inbreeding rates to less than 2% per generation. The genetic underpinnings of inbreeding depression (i.e., reduced viability and fecundity) are best studied and understood in inbred strains of laboratory-reared Drosophila and Mice, in which recessive lethal mutations and mildly deleterious mutations arise due to non-random mating [5].

There is abundant evidence that isolated wildlife populations suffer inbreeding depressions. Inbreeding depression can be avoided in the short term if $Ne > 50$ [12]. The inbreeding coefficient (F) increases by $1/2Ne$ per generation and centuries of animal breeding experience show that a 1% increase in F per generation is tolerable. Thus, $Ne = 50$ is necessary to avoid inbreeding depression [12]. Jamieson and Allendorf [12] further concluded that $Ne > 500$ was necessary to enable a population to continue to evolve in the long term. Although this 500 number has been revised upwards, the theory behind the 50 number is still accepted [15]. But it is important to realize that its derivation was based on controlled laboratory experiments; larger Ne ($Ne = 10{,}000$) are required in nature, where environmental fluctuations are more severe and stressful.

4.3 Gene flow

One of the fundamental agents in evolution that interest conservation geneticists are the dispersal of genes (i.e., gene flow) between populations of a species. Gene flow can be either active or passive, often gender-biased and limited to certain phases of the life cycle. It may be accelerated under certain climatic conditions that occur at frequencies of many years or irregular intervals of many years apart. Gene flow is typically can be estimated from allele frequency data and presented in terms of the number of successful establishment migrants per generation in the new population. In theory, one migrant per generation between two populations will ensure the two populations remain genetically homogeneous and related, as well as reduce inbreeding depression. In the future, overcoming genetically depauperate populations. Whereas lack of gene flow allows interpopulation differentiation. Hence, understanding historical patterns and rates of gene flow in a conserved population are crucial. Particularly if previously continuous populations become fragmented, the patterns of historical dispersal and gene flow may be disrupted with potentially serious consequences for population viability. For example, if young female orangutans can no longer migrate and confine from their natal social group due to habitat destruction in the surrounding area, their isolated natal populations will experience significantly increased inbreeding. On the other hand, if previously fragmented populations with each population have the unique genetic basis for adapting to local conditions become interacted, gene flow can erode the genetic differences between populations. Consequently, the two populations become one and some unique genes/alleles may be lost (see genetic drift).

In nature, widespread interspecific gene flow may occur between members of two different but related species (i.e., semispecies) or between very distantly related conspecific individuals in hybrid zones and produce hybrids. Hybrids are commonly sterile, or partial sterile in one sex or have high neonate mortality or have genetic disorders, and rarely are fertile. However, if fertile interspecific hybrids (also known as introgressive hybridization) exist, it causes a dilemma in conservation management. Because their occurrence reduces the value of the taxon. But at the same time, it is interesting because they show that the evolution of many groups of species involves both lineage splitting and lineage anastomosis. Hybridization is more common observe in plants than in animals; therefore, not surprisingly in plants, there are many examples of rare species being hybridized with the more common sympatric congeners (genetic assimilation) and become extinct (e.g., [16]).

4.4 Genetic drift

Genetic drift is referring to the loss of alleles from a population by chance due to a sudden reduction in Ne. This results in loss of fitness unless there is a rapid and continuous recovery. Often in nature, genetic drift happens almost clocklike regularly [5, 7] and followed by a rapid population recovery is referred to as a demographic bottleneck. They can have an immediate impact on variability at molecular genetic loci as genetic drift snatch the innate variation in a population. The evidence of a demographic bottleneck may persist for hundreds of thousands to millions of generations in low levels of variation in the loci of allozyme and molecular genetic markers. On the other hand, a demographic bottleneck can result in a short-term increase in population variation because epistatic variation (due to interactions among genes controlling a trait) is transformed into additive variation. However, whether it is beneficial or harmful to population viability is unknown.

The rate at which alleles are lost from a population by genetic drift can be statistically estimated. Sewall Wright theoretical model showed analytically how the rate of allele loss varies with population size, and concluded that census population size (N) is not important but rather the Ne. Ne is almost always less than N under some populations. Ne taking into account the fact that closely related individuals will share alleles with the same lineage, unequal numbers of males and females, increased variances in family size, and temporal fluctuations. Ne can be defined and estimated in a variety of ways using temporal ecological data, DNA sequences, and a variety of methods to estimate migration rates. Some estimation methods have theoretical value but little operational utility. Even so, by estimating Ne the effects of different population management strategies can be evaluated.

In many threatened populations, Ne is only 10–30, and at such levels, genetic variation becomes significant for the viability of the population.

Very low genetic variability has been known in many sexually reproductive species whose currently large populations have recovered from one or recurrent demographic bottleneck or extinction. Meanwhile, in a large continuously distributed population (metapopulation) with frequent extirpation and recolonization of subpopulations, reduce metapopulation Ne orders of magnitude below than N can mimic the genetic effects of a demographic bottleneck. In small isolated populations with the absence of factors driving genetic variation (mutation and gene flow), the impacts of demographic bottlenecks are severe. Whereby demographic bottleneck reduces genetic variation (loss of heterozygosity), leading to increased homozygosity and loss of evolutionary adaptability to change (genetic variability or selectively neutral variation). The genetic variability is expected to be lost ½Ne per generation and mostly lost within 2Ne generations. Ne of 10 is predicted to lose heterozygotes five times faster than Ne of 100. This is because 50% of heterozygosity in Ne = 10 will be lost in approximately 20 generations. Therefore, in theory, small isolated populations have a higher rate of loss of heterozygosity and faster loss of variability by genetic drift than large populations and metapopulations).

4.5 Natural selection

In nature, differences in the survival and reproduction of some genotypes over others as the major agents of microevolutionary changes are known as natural selection. Natural selection attracts the interest of conservation geneticists for two reasons. First, human activities can radically alter selection coefficients in both natural and control populations. Such evolutionary changes of human influence are referred to as artificial selection whether intentional or not. This can be seen in many commercially exploited wildlife species, whereby has resulted in rapid behavioural and natural history changes and consequently reduced fitness. Examples include reduced body size in the game and commercial fish and the impact of hunting only horned or tusked male mammals on social behaviour.

Second, the major challenges to assist wildlife species adapt to ongoing global climate change. In the past, in the absence of humans, natural selection favoured individuals adapted to change and many species shifted their ranges towards accommodating major changes. The rate of directional selection that a population can control in response to some environmental change is in part, is determined by its inherent variability. Unfortunately, in the 21st century, environmental change and destruction such as those associated with global warming are happening too quickly for many species to respond to it. Hence, effective conservation management is necessary to ensure that many species survive.

5. Conclusions

The importance of incorporating conservation genetics in managing biodiversity is undeniable. This is because the understanding of the relationship between evolutionary factors including mutations, non-random mating, gene flow, genetic drift, and natural selection in population/species survival is very important in the current situations where many natural populations are declining towards species extinctions. Therefore, with the relevant literature review in this chapter, it is hoped to provide brief explanations of the importance of assimilating conservation genetics to manage biodiversity. Especially to those who are less aware of the scope of genetic conservation studies.

Acknowledgements

Special thanks are given to Rabi'atul Adawiyah Fauzi and Nur Aisyah Abdul Hamid for sharing their literature review collections to complete this book chapter. Thank you are addressed to Genome Wide Association Laboratory members and anonymous reviewers for helping us to improve the early version of this book chapter. This research was supported by the Ministry of Education (MOE) through (FRGS/1/2018/STG03/UPM/01/1), (FRGS/1/2020/WAB11/UPM/02/2) and (TRGS/1/2020/UPM/01/4).

Conflict of interest

The authors declare no conflict of interest.

Appendices and nomenclature

N	Census population size
cpDNA	Chloroplast DNA
DPS	Distinct population segments
Ne	Effective population size
ESU	Essential Evolutionary Unit
IUCN	International Union for Conservation of Nature
MHC	Major histocompatibility complex
mtDNA	Mitochondrial DNA
nDNA	Nuclear DNA
PVA	Population Viability Analysis
RAPD	Random amplified polymorphic DNA
allozymes	Single-locus markers
SNP	Single nucleotide polymorphism
QTL	Quantitative trait loci

Author details

Nurul Izza Ab Ghani[1,2*], Wardah Arifin[1] and Ahmad Ismail[1,3]

1 Faculty of Science, Department of Biology, Universiti Putra Malaysia, Serdang, Selangor, Malaysia

2 Laboratory of Sustainable Animal Production and Biodiversity, Institute of Tropical Agriculture and Food Security (ITAFoS), Universiti Putra Malaysia, UPM, Serdang, Selangor, Malaysia

3 Malaysian Nature Society (MNS), Bukit Persekutuan, Kuala Lumpur, Wilayah Persekutuan Kuala Lumpur, Malaysia

*Address all correspondence to: nurul_izza@upm.edu.my

References

[1] IUCN. The IUCN Red List of Threatened Species. Version 2021-2. Cambridge: International Union for Conservation of Nature and Natural Resources; 2021. Available from: https://www.iucnredlist.org. [Accessed: 20-8-2021]

[2] Gilpin ME, Soulé ME. Minimum viable populations: Processes of species extinction. In: Soulé ME, editor. Conservation Biology: The Science of Scarcity and Diversity. Sunderland: Sinauer Associates; 1986. pp. 19-34

[3] Akçakaya HR, Sjögren-Gulve P. Population viability analysis in conservation planning: An overview. Ecological Bulletins. 2000;**48**:9-21. http://www.jstor.org/stable/20113245

[4] Beissinger SR, McCullough DR, editors. Population Viability Analysis. Chicago: The University of Chicago Press; 2001. p. 593

[5] Woodruff DS. Populations, Species, and Conservation Genetics. Encyclopedia of Biodiversity. 2001;**4**: 811-829. DOI: 10.1016/B0-12-226865-2/00355-2

[6] O'Brien SJ, Roelke ME, Marker L, Newman A, Winkler CA, Meltzer D, et al. Genetic basis for species vulnerability in the Cheetah. Science. 1985;**227**:1428-1434. DOI: 10.1126/science.2983425

[7] O'Brien SJ. Conservation genetics of the Felidae. In: Avise JC, Hamrick JL, editors. Conservation Genetics: Case Histories from Nature. New York: Chapman & Hall; 1996. pp. 50-74

[8] Avise JC, Hamrick JL, editors. Conservation Genetics: Case Histories from Nature. New York: Springer; 1996. p. 512

[9] Stoffel MA, Humble E, Paijmans AJ, Acevedo-Whitehouse K, Chilvers BL, Dickerson B, et al. Demographic histories and genetic diversity across Pinnipeds are shaped by human exploitation, Ecology and Life-history. Nature Communications. 2018;**9**(4836): 1-12. DOI: 10.1038/s41467-018-06695-z

[10] Bosse M, Megens HJ, Derks M, de Cara Á, Groenen M. Deleterious alleles in the context of domestication, inbreeding, and selection. Evolutionary Applications. 2018;**12**(1):6-17. DOI: 10.1111/eva.12691

[11] Frankham R, Brook BW, Bradshaw CJA, Traill LW, Spielman D. 50/500 rule and minimum viable populations: Response to Jamieson and Allendorf. Trends in Ecology and Evolution. 2013;**28**(4):187-188. DOI: 10.1016/j.tree.2013.01.002

[12] Jamieson IG, Allendorf FW. How does the 50/500 rule apply to MVPs? Trends in Ecology and Evolution. 2012;**27**(10):578-584. DOI: 10.1016/j.tree.2012.07.001

[13] Jamieson IG, Allendorf FW, et al. A school of red herring: Reply to Frankham. Trends in Ecology and Evolution. 2013;**28**(4):188-189. DOI: 10.1016/j.tree.2013.01.012

[14] Xu X, Dong G-X, Hu X-S, Miao L, Zhang X-L, Zhang D-L, et al. The genetic basis of White Tigers. Current Biology. 2013;**23**(11):1031-1035. DOI: 10.1016/j.cub.2013.04.054.

[15] Wang T, Fujiwara M, Gao X, Leu H. Minimum Viable Population Size and Population Growth Rate of Freshwater Fishes and their Relationships with Life History Traits. Scientific Reports. 2019;**9**(3612):1-8. DOI: 10.1038/s41598-019-40340-z

[16] Faucon M-P, Tshilong BM, Van Rossum F, Meerts P, Decocq G, Mahy G. Ecology and Hybridization Potential of Two Sympatric Metallophytes, the Narrow Endemic Crepidorhopalon perennis (Linderniaceae) and its More Widespread Congener C. Tenuis. BIOTROPICA. 2012;**44**(4):454-462. DOI: 10.1111/j.1744-7429.2011.00845.x

Chapter 3

Conserving Endemic Plant Species in Oceanic Island's Protected Areas

Teresa Mouga

Abstract

Oceanic islands are known for their high levels of plant diversity, due to disjunct geographical distribution that leads to speciation. The main factors contributing to genetic speciation includes the creation of a barrier within a previously widely distributed taxon and the limited dispersal of seeds, which favours genetic differentiation and, thus, fosters rapid speciation. Plant survival and population fitness vary according to environmental factors and to human interference. This chapter depicts the importance of oceanic islands as biodiversity hotspots, discusses the threats to which endemic plants on islands are exposed, namely climate change, invasive alien species, urbanisation, touristic activities, fire, changes in agriculture practices and collecting pressure. The best practices worldwide to protect endemic plant species in protected areas are also addressed, namely the implementation of prevention and mitigation actions, the programs executed to protect endemic species, and management plans to avoid future threats.

Keywords: small islands, vegetation, invasive alien species, climate change, endemism, conservation

1. Introduction

Oceanic islands are those that never had a connection to continental land masses, being generally composed of volcanic rock, reef limestone or both. Those of volcanic origin are formed over oceanic plates, being a product of volcanism or tectonic uplift. These islands tend to be steep and relatively high for their area and, over time, become highly sheared due to erosion processes. Usually, they lack native mammals and amphibians, but a fair number of birds and insects, as some reptiles are usually present [1]. Not classified as "real" islands, atolls and reef are marine habitats islands, built up by small coelenterate animals (corals) that secrete a calcareous exoskeleton. These form an annular reef rim surrounding a central lagoon, with the rim being more or less occupied by calcareous sand or coral shingle and rubble [2], such as the reef islands of the Maldives, the Solomon Islands, the Bahamas, the Tarawa atoll in Kiribati, and many other islands and atolls in the Pacific Ocean. Coral islands tend to be very low-lying and flat; some only raised a few meters above sea level [1]. As defined by Paulay [3] all these are considered oceanic islands (**Figure 1a** and **b**).

Oceanic island are mainly small islands, which are defined as those which present less than 10,000 km^2 [4]. The largest oceanic island on Earth is Iceland, with more than 100 thousand square kilometres, but all the other oceanic island

Figure 1.
Examples of oceanic islands: left: S. Miguel, one of the nine islands of the Azores volcanic archipelago; right: an atoll at the Maldives, one of the 1192 coral islands that are grouped in 26 atolls.

are much smaller, being New Britain (Papua New Guinea), Grande Terre (New Caledonia), Negros (Philippines), and Hawaii (USA), the other large oceanic islands.

Besides these, there are millions of small islands and islets. **Table 1** states these small oceanic islands and oceanic archipelagos throughout the globe, being referred the main island of the archipelago (if any), the size, and the location. To avoid being over-exhaustive in this analysis, only the main oceanic island of each archipelago is presented, in addition to the isolated islands.

In contrast to oceanic islands, continental islands were joined to continental land in the past, namely during the Quaternary ice ages, and becoming separated owing to sea level rise or to tectonic events, and still sit on the continental shelf. As such such, terrestrial mammals and amphibians are usually present [1, 3]. Most of the larger islands are of continental origin, such as Greenland, New Guinea, Borneo, Madagascar, Baffin Island, Sumatra, Honshu, Victoria Island, or Great Britain.

Oceanic islands are usually smaller, younger, more isolated from the continent, more isolated from the nearest neighbour island and present less plant species than continental islands [5]. Their climate has, evidently, a strong oceanic influence, with the low islands being much drier and the high islands presenting heavy orographic rainfall. Most oceanic islands have freshwater reservoirs, both volcanic and atolls, which depend on rainfall percolating through the island. Small islets, however, may lack such lens, being therefore adverse for plant growth [1].

When a new island emerges, an ecological succession begins with the species that were able to reach the land colonising the island but subjected to island isolation. High dispersal capabilities are more likely to overcome distance, which determines that plants, birds, and insects, for example, are much more common on islands than other taxa with lower dispersal capacity. Of the newly arrived species, only a few will be able to survive and establish new populations. As a result, islands have fewer species than mainland habitats. Island populations are small, exhibit low genetic variability and are isolated from the predators and competitors with which they initially evolved [6]. These small islands are also known to present high levels of endemism, mainly due to disjunct geographical distribution and limited dispersal of seeds. These favour genetic differentiation, which, in turn fosters endemism [7–9]. These endemisms have small population distribution, and present low competitive ability [6].

The isolation and small size of the oceanic islands makes them very vulnerable, highly susceptible to threats such as climate change, natural catastrophes, coastal erosion, seawater intrusion, and overexploitation of natural resources [10]. They are also very vulnerable to invasive alien species, that compete with the native taxa,

Main island	Is. area [km²]	Archipelago (Ac)	Ac area [km²]	Country	Ocean	Coordinates
Iceland	102,775			Iceland	Arctic	64°08′N 21°56′W
Santorini	73	Cyclades	2,572	Greece	Mediterranean	36° 23′ N 25° 27′ E
Lipari	37	Aeolian Islands	115	Italy	Mediterranean	38°28′N 14°57′E
S. Miguel	759	Azores	2,351	Portugal	North Atlantic	37°44′28″N 25°40′50″W
Tenerife	2,034	Canary Islands	7,493	Spain	Northeast Atlantic	28°28′N 16°15′W
Santiago	991	Cape Verde	4,033	Cape Verde	Northeast Atlantic	14°55′N 23°31′W
Madeira	740	Madeira	801	Portugal	Northeast Atlantic	32°39′N 16°55′W
Bermuda	53	The Somers Isles		UK	Northwest Atlantic	32°18′N 64°47′W
New Providence	207	Bahamas	13,878	Bahamas	West Indies, Atlantic	25°4′N 77°20′W
Fernando Noronha	18	Atlantic Islands	26	Brazil	Southwest Atlantic	3°51′13″S 32°25′25″W
Montague	120	South Sandwich Islands	3,903	UK	South Atlantic	58°25′S 26°23′W
Tristan da Cunha	96	Tristan da Cunha Islands	207	UK	South Atlantic	37°4′S 12°19′W
Heard	368	Heard Isl. & McDonald Isls.	371	Australia	Atlantic (Antarctic)	53°06′S 73°31′E
La Grande Terre	6,675	Kerguelen Islands	7215	France	Atlantic (Subantarctic)	49°20′55″S 70°13′09″E
Île de la Possession	67	Crozet Islands	352	France	Atlantic (Subantarctic)	46°24′S 51°46′E
Bouvet	49			Norway	Atlantic (Subantarctic)	54°25′S 3°22′E
Guadeloupe	1,628	Antilles volcanic arc	14,364	France	Caribbean Sea, Atlantic	16°02′04″N 61°41′56″W
Grande Comoro	1,147	Comoros Islands	1,861	Comoros	Western Indian	11.699°S 43.256°E
La Réunion	2,511			France	Western Indian	21°06′52″S 55°31′57″E
Malé	8	Maldives	300	Maldives	Indian	4°10′31″N 73°30′32″E
Diego Garcia	30			UK	Indian	7°18′48″S 72°24′40″E
St. Paul	6			France	Indian	38°43′S 77°13′E
Unimak	4,070	Aleutians	17,670	Russia/USA	North Pacific	52°02′N 174°02′W
Iturup	3,139	Kuril Islands	10,503	Japan, Russia	Northwest Pacific Ocean	45°02′N 147°37′E

Main island	Is. area [km^2]	Archipelago (Ac)	Ac area [km^2]	Country	Ocean	Coordinates
Hokkaido	378	Japanese	83,424	Japan	Northwest Pacific	43°4′N 141°21′E
Tidore	1,550	Moluccas Islands	74,505	Indonesia	Western Pacific	0°41′N 127°24′E
Negros	13,350	Visayas	71,503	Philippines	Western Pacific	10°40′35″N 122°57′03″E
New Britain	36,520	Bismarck	49,700	Papua-New Guinea	Western Pacific	5°44′S 150°44′E
Bougainville Island	9,318	Solomon Islands	28,400	Papua-New Guinea	Melanesia, Pacific	6°14′40″S 155°23′02″E
Espiritu Santo	3,955	New Hebrides	12,189	Vanuatu	Melanesia, South Pacific	17°44′S 168°19′E
Grande Terre	16,372	New Caledonia	18,576	France	Melanesia, South Pacific	22°16′S 166°28′E
Tarawa	31	Kiribati	811	Kiribati	Micronesia, Pacific	1°28′N 173°2′E
Majuro	10	Marshall Islands	181	Marshall Islands	Micronesia, Pacific	7°7′N 171°4′E
Guam	540	Mariana Islands	1,036	USA	Micronesia, Pacific	16°37′N 145°37′E
Hawaii	10,432	Hawaiian	28,311	USA	Polynesia, Pacific	19°34′N 155°30′W
Savai'i	1,694	Samoa	2,842	Samoa	Polynesia, Pacific	13°50′S 171°45′W
Vaitupu	6	Ellice Islands	26	Tuvalu	Polynesia, Pacific	07°28′S 178°41′E
Nuku Hiva	339	Marquesas Islands	1,049	France	Polynesia, Pacific	8°52′S 140°08′W
Tahiti	1,044	Society islands	1,590	Tahiti	Polynesia, Pacific	17°40′S 149°25′W
Tongatapu	260	Tonga Islands	750	Tonga	Polynesia, Pacific	21°08′S 175°12′W
Rarotonga	67	Cook Islands	237	Cook Islands	Polynesia, Pacific	21.235°S 159.778°W
Tubual	45	Austral islands	152	France	Polynesia, Pacific	23°22′12″S 149°28′48″W
Henderson Island	37	Pitcairn Islands	47	UK	Polynesia, Pacific	24°22′01″S 128°18′57″W
Isabela	4,586	Galápagos Islands	7,880	Equator	East Pacific	0°30′S 90°30′W
Socorro	132	Revillagigedo	158	Mexico	East Pacific	18°50′N 112°50′W
San Ambrosio	3	San Félix Islands	5.36	Chile	East Pacific	26°20′37″S 79°53′28″W
Easter Island	164			Chile	East Pacific	27°7′S 109°22′W
Selkirk	50	Juan Fernández Islands	100	Chile	East Pacific	33°45′04″S 80°47′00″W

Main island	Is. area [km²]	Archipelago (Ac)	Ac area [km²]	Country	Ocean	Coordinates
Auckland	443	Auckland islands	626	New Zealand	South Pacific	50.7°S 166.1°E

Table 1.
Main oceanic islands, including the archipelago, the country, the ocean, the island and the archipelago total area, and the coordinates of the main city.

causing severe ecological and economic problems. Besides, deforestation is frequently a major problem, both for agriculture and for timber, and tourism is causing additional infrastructural and pollution pressure [6]. Biodiversity conservation and sustainability are accordingly major concerns in relation to the oceanic island, to preclude the degradation and destruction of the natural heritage.

The effort of creating protected areas is the first key step to the conservation of threatened natural and cultural heritages. This step should be followed by a successful management of the protected area, which considers both the conservation of ecosystems and the socio-economic development of island inhabitants and considers the specificities of island territories.

This chapter discusses the importance of the oceanic islands, of its endemic plants, the threats they are currently facing, and the conservation measures being implemented to protect these important ecosystems.

2. Biodiversity and endemism of oceanic islands

There are around 374,000 plants species on earth [11], but their distribution is uneven, with the tropical environments presenting larger numbers than other environments. This is a result of ecoevolutionary drivers which include the climatic stability over the past million years associated with time, energy availability, and biotic interactions [12]. Consequently, some areas of the globe have been recognised as global biodiversity hotspots as they exhibit exceptionally high species richness and high endemism levels [13]. Mittermeier et al. [14] have defined 35 biodiversity hotspots, many of which are oceanic and continental island archipelagos.

The colonisation of the small oceanic islands depends on geographical and environmental drivers, being inversely related to the distance to other lands [15]. Another important factor is the dispersal ability of the organisms. The geographical range of a *taxon* depends on its ability to disperse its pollen and its seeds. In the case of an island, this dispersal can occur through anemocory (wind dispersal), endozoochory (in the gut of animals), epizoochory (attached to the exoskeleton, fur, feathers of scales of animals) or thalassochory (floating in the water) [16].

When a plant species is able to reach a new territory, it depends on its ability to adapt to the physical and chemical characteristics of the island, and to other biotic factors such as competition, herbivory, parasitism, and symbiosis [3]. The few taxa that survived and adapted to the new environment may therefore evolve into new species. Due to the time these adaptive processes take, island age is an important factor for the biodiversity of oceanic islands, as older islands have a higher probability of successful colonisation. They also had more time for selection processes to act on the first colonisers, so that natural selection takes place, thus constituting a favourable factor for speciation. Because of their evolutionary processes, oceanic islands are poor in the number of species for their size, but present a remarkable high ratio of endemism, and the ecosystems exhibit much higher biodiversity than terrestrial ecosystems for the same area [1, 6]. E.g., the East Melanesian Islands,

Figure 2.
Endemic plant species from oceanic islands. Top left: Hibiscus arnottianus, *from Hawaii; top right:* Brachycereus nesioticus *from Galapagos; bottom left:* Bikkia tetrandra *from Mariana; bottom right:* Viola paradoxa *from Madeira.*

comprising the Solomon islands, Vanuatu and Papua New Guinea, include around 8,000 plant species of which about 3,000 are endemic, the Atlantic islands of Macaronesia are the third richest hotspot in the world in terms of its plant biodiversity (25,000 species); 5,330 species of native vascular plants are native to Polynesia-Micronesia, of which more than 3,070 are endemic, Japan has more than 5,600 plant species of which roughly a third are endemic [17]. Hawaii archipelago also has about 1180 native vascular plants, of which 1000 are angiosperms. Of these, about 900 are endemic (**Figure 2**) [18].

These endemic species, however, present restricted geographical range, specialised environmental niche, limited dispersal ability and reduced size population and distribution [19]. The islands with high large proportion of endemic plants are mainly the high volcanic islands, while most the low islands are species poor. The smaller the island is, the more isolated, and the less the topographic relief, the poorer the island. This is due to the reduced variety of habitats and the broad mix of the typically sea-dispersed strand species that dominate their floras [1].

3. Natural and anthropogenic disturbance

All habitats are exposed to an ecological succession and to natural disturbances, namely volcanic eruptions, or tropical cyclones, that significantly alter the animal and plant populations. As defined by Pickett, disturbance is "a change in the minimal structure of an object caused by a factor external to the level of interest" [20].

Oceanic islands are also subject to numerous disruptive events such as hurricanes, high winds, heavy rains, high pressure systems, earthquakes, volcanic eruptions, tsunamis, extreme tides, the introduction of exotic species and human activities. These have mechanical, physiological, or biotic impacts that can last for years. In fact, because most oceanic islands are small and located in harsh environments, these disturbance events tend to have more severe consequences on oceanic islands than on continental land masses [1].

In addition to these natural disturbances, humans have had a profound impact on biodiversity, altering the composition and functioning of ecosystems. These events are of the utmost importance for the survival of wild habitats and the viability of populations.

After a disturbance event, when the number of individuals falls below a specific threshold, the species loses genetic diversity, which reduces its ability to adapt to change and therefore increases the risk of extinction. Island endemic species are usually very localised and have small numbers of individuals, which makes them highly vulnerable to disturbance and therefore to extinction [21].

3.1 Biological invasions

With human settlement on oceanic islands new species were introduced as farm stock, crops, for fibres or furs, domestic animals, pets, sports, or solely as ornamentals [22, 23]. Other species, however, were introduced due to military operations, international trade, and globalisation, either ship cargoes, ballast water, shipwrecks, which unintentionally transported these exotic species to the island, whether plants or animals (**Figure 3**) [24]. More recent invasions drivers are climate change, land-use change providing new habits, pollution, and the positive interaction among non-native species, a process known as invasion meltdown [25, 26].

Figure 3.
*Invasive alien mammals: top right: mouse (*Mus muscullus*) native to south Asia is invasive worldwide; top centre: rabbit (*Oryctolagus cuniculus*) native to Europe; top right: feral goat (*Capra hircus*); bottom left: wild boar (*Sus scrofa*) native to Eurasia and Africa; bottom centre: red deer (*Cervus elaphus*) native to Europe; bottom right: grey-squirrel (*Sciurus carolinensis*) native to America.*

An introduced species is a species that (1) owing to human activity colonises a new area where it was not previously present, (2) is remotely dispersed with a wide geographic discontinuity, and (3) becomes naturalised by perpetuation of new generations without human intervention [27]. Luckily, most introduced species do not become established, due to mortality during translocation, unsuitable environmental conditions and biotic resistance exerted by the host community [28].

Nevertheless, once established, it can become a new invasive alien species (IAS) when it has an undesirable effect on the native ecosystems. The ecological and economic impact of IAS may be after the invader is well established and have wide range, and then the damage may be extremely severe. IAS are responsible for altering the ecosystem functioning, modifying native species richness and abundance, and increasing the risk of extinction, breaking down biogeographic realms, affecting the genetic biological diversity, changing the phylogenetic diversity across communities, and modifying the trophic networks, as well as disturbing human health and/or socioeconomic values at the individual, population, or community level [25, 29–31]. "Habitat transformers" species, which cause changes in ecosystem nutrient cycling at microbial or higher plant levels [32] and "ecosystem engineer" species, which are landscape modifier species [33], are particularly dangerous for they are strongly competitive IAS with the ability to alter environmental conditions, being a major contributor to species diversity loss. As such, IAS alter the composition of plant and animal communities, and also interfere with other ecosystem processes such as nutrient cycling, hydrological cycles, and primary productivity [34].

Accordingly, IAS may have severe negative impacts on oceanic islands because these ecosystems are species-poor and have few highly competitive species [30]. IAS impacts on islands are intensified through the interaction with other global change threats, including over-exploitation of natural resources, agricultural intensification, urban development, and climate change, exacerbating some invasions, and facilitating others, escalating the impact and the extent of IAS [35]. Currently, IAS may be the main cause for ecological disintegration globally, and thus the early detection, rapid action in eradication and good planning is of utmost importance, mainly on islands or other limited habitats [23].

3.2 Climate change

Climate change poses serious risks for human and natural systems. Species are shifting their geographic ranges and altering the numbers of individuals in their populations, variations in seasonal activities, migration patterns and interactions between different species are also occurring in response to ongoing climate change. The impact from recent climate-related extremes, such as heat waves, floods, droughts, cyclones, and fires, reveal significant vulnerability and risk of many ecosystems, some irreversible. To make matters worse, carbon stored in the terrestrial biosphere in peatlands, permafrost, and forests, among others, may be lost to the atmosphere, exacerbating ecosystem degradation. Furthermore, the sea level rise projected for the 21st century and beyond will have an enormous impact on coastal systems, islands, and low-lying areas, which will suffer adverse impacts such as submergence, flooding and coastal erosion. These impacts will be extremely severe on low-lying developing countries and small island states [36].

Due to climate change, the intensity and frequency of wildfires is also increasing [37]. Besides the noticeable economic impact, heat dramatically disturbs soil surface, often causes a decrease in diversity and abundance of soil biota, and strongly increases the risk of erosion by wind and water [38]. These effects depend upon fire severity, and some fire regimes are beneficial to ecosystems. These are controlled by environmental factors such as amount, nature, and moisture of live and dead fuel, air temperature and humidity, wind speed, and topography of the site [39, 40]. Due to climate change, induced wildfires are becoming more frequent and are more aggressive and, thus, have frequently severe negative impact on the vegetation and on sensitive species.

Islands are particularly vulnerable to climate change disturbance, owing to the vulnerability of island endemic plants, due to habitat loss and interactions with introduced species [41]. The IAS may benefit from climatic change, as they are opportunistic, very competitive species, thus less vulnerable due to their adaptability to new environments [42]. Manes et al. [41] study stated a 100% risk of extinction for island ecosystem due to climate change and a risk of extinction 3 and 10 times higher for endemic than native and introduced species, respectively.

As such it is expected a decline of endemic plants in oceanic islands, a degradation of mangroves, wetlands, and seagrass around small islands, a degradation of groundwater and freshwater ecosystems due to saline intrusion, a spread of warm water species into the Mediterranean, namely IAS, among many other negative impacts attributed to climatic change [36].

Steffen et al. [43] postulate that the Anthropocene' era is rapidly approaching levels of human-induced greenhouse gases that are approaching critical levels. When reaching an irreversible threshold, the devastating consequences will be irreparable for the distributions of species and in the composition of biological communities. Many of these impacts may already be permanent.

3.3 Tourism and recreation

Disregarding the impact of the pandemic Tourism and Leisure are among the fastest growing economic activities of recent decades [44]. Yet, touristic activities are well known by their negative consequences, being responsible, for instance, for greenhouse gas emissions [45], high patterns of visitor consumption and waste generation [46], for plant damage, including vegetation removal and changes in land cover and land use [47], tourists trampling and spreading weeds and pathogens, and altering fire regimes [17, 48]. Tourists also often pick flowers, threatening the more charismatic species [49]. Tourism, thus, have negative impact in the wildlife, health, physiology, reproduction rate, and behaviour of the wild species [45, 50–53], prompting the decline of sensitive plants, while favouring the growth of resistant species, frequently opportunistic and exotic ones [49].

Thus, tourism is frequently an unsustainable activity not complying with the UNWTO definition of sustainable tourism as "tourism that takes full account of its current and future economic, social and environmental impacts, addressing the needs of visitors, the industry, the environment and host communities" [54].

The presence of tourists in Protected Areas is especially sensitive, for the number of visitors in a protected area increase the number of exotic species on site, since visitors increase propagule pressure and disturbance [28]. More disturbed habitats create open space that may allow IAS to establish and, thus, offer invaders an edge against native species [24].

Yet, due to the dependence on a healthy and safe environment, a social change seems to be arising within tourists and policymakers, increasingly seeking more environmentally friendly practices and tourism activities, through the development of nature-based tourism and ecotourism [55, 56]. In fact, more sustainable tourism activities are increasingly supporting wildlife conservation and local populations welfare are becoming a reality in many countries with pristine ecosystems and charismatic species [57–59].

3.4 Agriculture, and deforestation

Agriculture is intensifying at global level, and this trend will continue in the next years, to meet the growing human population needs. This agriculture expansion will bring ecosystem simplification, loss of ecosystem services, and species extinctions [60]. The agricultural spreading could have major impacts on biodiversity hotspots, as these are areas where there is significant population growth, often poor and with a low development index, where there is an increasing pressure to produce food and promote economic growth through the commercial use of natural resources [61]. In fact, many tropical protected areas, are suffering forest loss through agricultural intrusion, often to grow palm trees for biofuels, being a cheap source of oil [62].

Forest loss has also been occurring through legal or illegal logging, conversion to small-scale agriculture, and larger-scale commercial plantations, namely in the Amazon, Africa, and Asia, but also in small tropical islands, such as New Britain [63]. At the community level, large trees contribute extensively to ecosystem functioning and provide key habitats for biodiversity [64]. Logging is known to degrade forest structure, creating gaps, removing soil, and fostering the proliferation of IAS [65].

3.5 Urbanisation

Human population has more than doubled since 1950 and for the next half century there should be a continued rapid growth in the least developed regions [66]. This massive growth in human population has serious consequences for natural habitats, with increasing pollution, the spread of IAS, carbon emissions and the consumption and destruction of natural resources, resulting in the change of many of the last remaining wild spaces on the planet [67]. Therefore, fewer world ecosystems are away from human pressure, and many are experiencing biodiversity loss and ecosystem degradation due to the construction of infrastructures, for vehicles, for the industry, for hydraulic and harbour set-ups, hydroelectric infrastructures, among others, with severe impacts on many ecosystems and species. Roads, for example, open new opportunities for habitat fragmentation, fires, logging, and land speculation [68, 69]. The rapid proliferation of roads will also strongly influence the footprint of agriculture. Thus, wild regions, parks and protected areas, relics of intact habitat within biodiversity hotspots, such as islands, are among the environments where roads and other infrastructure should be limited, allowing the conservation of such habitats and species [68, 70].

Besides the roads, the building of infrastructures for urban expansion, tourism, or for other economic activities, has, evidently, direct impact in the vegetation clearance, to open the area. However, beyond the immediate impact on the vegetation, such infrastructures have a long-term impact, due to habitat fragmentation, the changes caused on the soil hydrology, pollution runoff, and as already mentioned, as a corridor for the introduction of pathogens and IAS [71, 72].

4. Conservation measures

Protected areas (PAs) are the main pillar of conservation activities and are therefore the first integrated approach for the conservation of biodiversity and ecosystem services worldwide [73]. Acknowledging the worldwide recognition of the importance of the PAs as a tool for the economic, social, and scientific importance, and for their role in environmental well-being, the total PA has increased tenfold from 1959 until 2016, from roughly 2 Mkm^2 to almost 20 Mkm^2, corresponding to 202,467 total PAs. In 2014 around 17% of the world island biomes were protected, mainly temperate (23%) and polar ecosystems (17.5%), while less than 13% of tropical islands were protected, where endemism is higher [74]. Also, the Intergovernmental Science-Policy Platform on Biodiversity and Ecosystem Services report [36] and the recent Global Biodiversity Outlook [75] noted some interesting progress in the conservation and sustainable use of biodiversity in PAs.

Although more recent reports do not include data on islands, between 2014 and 2020 the protected land and inland water ecosystems increased from 15.4% to 16.64% (with a total of 22.5 million km^2 and 248,113 protected areas), and the protected coastal waters and the ocean increased from about 4.5% to 7.74% (28.1 million km^2 and 17,828 protected areas) [76, 77]. This growth falls within the conservation efforts tackled by the Aichi Biodiversity Targets under the Convention on Biological Diversity (CBD). Still, despite the progress in conservation and sustainable use of biodiversity, the Strategic Goal 11 has been tightly missed: "by 2020, at least 17% of terrestrial and inland water areas and 10% of coastal and marine areas, especially areas of particular importance for biodiversity and ecosystem services" [78].

While many of the endemic species' populations are within protected areas, often these are not enough to fully protect them, because, among other problems, management quality is not satisfactory, and thus biodiversity loss has persistently continued [79]. Therefore, it seems that the extensive conservation efforts are not being successful and new approaches are needed.

Current conservation strategies are still largely based on the assumption that we live in a dynamic but slowly changing world. Such an assumption needs to be revised considering the rapid rate of climate change already experienced in recent years, which is expected to continue at this pace if not increasing, over the coming decades, forcing researchers and managers to rethink and recalibrate the conservation responses [80]. On the other hand, conservationist classical approaches are based mainly on *ex-situ* conservation of endangered species, and reintroduction measures from which they have been lost, while restoring degraded or lost ecosystems [81]. When it comes to conservation of plants, and endemic species in particular, the scenario seems even more ineffective, with plants becoming increasingly rare around the world. Successful plant conservation includes research on the species distribution and rarity. Then an efficient management plan to tackle conservation efforts, prioritisation of measures, stakeholders' interests, and training capacity are important to mitigate threats facing threatened species. To implement such plans, policy and funding are foundation stones to support continued capacity of conservation. Ultimately, the last but not the least, a deep education plan for the public, so to understand and support the importance of plants and the need for their conservation is of utmost importance to achieve efficient conservation. These are not simple or isolated actions. Coordination of plant conservation efforts is also needed to ensure that resources and expertise are used in a strategic, efficient, and effective manner [82].

4.1 Data collection

Due to lack of knowledge and interest, plants are often under protected by policy, their conservation efforts are underfunded, and their importance is under cherished. To overcome such lack of information, an Important Plant Areas (IPAs) criteria system was defined, offering a pragmatic and scientifically rigorous mean of delivering these datasets, assisting the informed decision making and conservation prioritisation [83]. This database generates essential data for other databases such as the IUCN Key Biodiversity Areas (KBAs) programme [77] producing a worldwide network of relevant information. The database, however, is still rather limited, for many countries have not yet made available the data on the distribution, rarity and threat status of plant species and their habitats, mainly in the tropical areas.

The IPA criteria, for the first time, recognises the socio-economically valuable plant species providing essential goods, such as the importance of plants as a food source, medicines, timber, fuel, materials for clothing, ornamental, social and cultural traditions, besides the vital ecosystem services [83].

The identification of the biodiversity hotspots and endemism centres, along with the assessments summarised by the IUCN red list categorisation [84] and creating global, national, and regional lists of threatened species, are, likewise, valuable tools in conservation prioritisation and planning [85]. Most countries have national agencies responsible for gathering information on native ecosystems, habitats, endemic species, PAs, in regional or national databases, fundamental information for the implementation of conservation actions.

The improvement of biological and ecological knowledge will allow to better target conservation measures.

4.2 Legal protection

Besides the legal protection at regional and national levels, there are several international cooperation treaties to tackle the threats on wildlife and nature protection. The following are some of the most important, within the plant conservation:

1. Ramsar Convention on Wetlands (1971) which promotes de wise use of wetlands, encouraging the research, training, and management of these ecosystems [86].

2. The Convention for the Protection of the World Cultural and Natural Heritage (1972), aiming to ensure the identification, protection, conservation, presentation, and transmission to future generations of the cultural, and natural heritage [87].

3. Convention on International Trade in Endangered Species of Wild Fauna and Flora (CITES) (1973), seeking to regulate the international trade in endangered animals and plants, and in products derived from them [88].

4. Convention on Biological Diversity (1992), which aims at the conservation of biological diversity, the sustainable use of its components, and the fair and equitable sharing of the benefits of utilising the genetic resource. It also set ambitious goals to restore and safeguard ecosystems, promote sustainability, halt biodiversity loss, combat desertification, among others [74].

5. International Plant Protection Convention (1951) aims to protecting the world's plant resources from the spread and introduction of pests and promoting safe trade [89].

Although each of these international treaties stand on its own, regarding their objectives and commitments, they are inter-linked between their goals and complement each other. Each convention governing body set out specific mandates for cooperation between the biodiversity-related conventions, providing a framework for joint action of biodiversity and a foundation for sustainable development [74].

4.3 Creation and management of protected areas

The PA creation, as stated, is probably the number one national and international conservation policy. They are regarded as the primary defence against biodiversity loss, as long as they are well maintained and managed [6, 67, 81]. The Aichi Biodiversity Targets are a strong showcase of the political priority given to the creation of protected areas at the international level. The following are key messages to achieve the Aichi Targets for APs [77]:

1. Ensuring a more sustainable future [...] will require greater recognition of the important role that PAs play in underpinning sustainable development.

2. Making PAs a key part of national and local responses to address harmful incentives to biodiversity (Target 3), biological invasions (Target 9), anthropogenic impacts and climate change challenges (Targets 10, 15) will help to halt biodiversity loss (Targets 5 and 12), [...].

Complying with these guidelines, IUCN developed a set of educational tools for teaching about PAs and governance aiming to produce a "well-implemented legal frameworks [to create and maintain] effective and sustainable PAs, which provide fundamental infrastructure for conservation of biological diversity and ecosystem services" [90]. These guidelines are helping to create and implement efficient management plans, making them an effective tool to guide managers and other stakeholders in the decision-making process towards achieving the conservation goals.

However, PAs coverage and management plans are not enough to ensure the PA conservation success. Presently, not all the important biodiversity hotspots occur inside the PAs [91–94], because the PA area is at times inadequately defined in terms of extent, ecological representation, and key biodiversity areas [95]. Another major bottleneck is that many PA are inadequately managed and, therefore, do not fulfil their goal of providing a safe and secure site for the species, populations, and ecosystems to thrive.

While biodiversity conservation is the primary objective of a PA, successful management must also address the funding and training requirements of conservation actions, as well as ensuring the sustainability and socio-economic development of local communities [6]. Balancing conservation interests and human well-being is often the most difficult challenge to successfully manage a PA. Therefore, local populations ought to be involved at all stages of the PA management planning, notably in defining the mission, vision, and goals of the PA [6].

Besides all these challenges, in the present days, the greatest threat to PAs is, probably, climate change. How far protected areas will continue to be effective in protecting biodiversity under projected climate change scenarios is still uncertain, but it is expected that some PAs will virtually cease to function, with massive species loss and shift, others may survive relatively undisturbed, while others may even experience an increase in species, leading to changes in the species assemblages [81].

When it comes to island PAs, the intrinsic characteristics of island species and ecosystems cause a particular vulnerability due to the small population sizes, low habitat availability, and isolated evolution [96, 97]. Strong local anthropogenic pressure added to the impacts of climate change increase the threats to island ecosystems and plants. Due to the high degree of endemism in island floras, there is a particularly high potential for biodiversity loss in these ecosystems. Climate change impacts on oceanic island, though, are not evenly distributed, with the greatest vulnerabilities to be expected on smaller islands with low elevation and uniform topography, which will experience higher disruptions rates associated with ecosystems co-modification and co-extinction [98]. Thus, islands PAs are much more vulnerable than other land ecosystems, and management plans must take this into account.

4.4 Control of invasive species

In oceanic islands, as stated, biological invasions can lead to severe large-scale ecosystem alterations. Thus, the eradication of IAS has been a common management practice in island PAs, being widely recommended [23, 96, 99–104].

Eradication of IAS in general is a complex and controversial management action. On islands it is attainable in the early stages of invasion [35], but later it is largely restricted to a few invasive mammals such as rabbits and rats [105] and then, for most species, permanent pest control is the only option.

Most of the already mentioned measures must be applied to the control of IAS. First, the knowledge of the IAS present is fundamental. There are many IAS listed around the world, a study that has been undertaken during the past 50 years or so. The Invasive Species Specialist Group developed a global invasive species database [106], and many countries have regional and national databases, although there is still work to do on this subject.

Coordination between countries and trans-national management plans are required to allow the development of joint actions across geographical areas that go beyond each country's frontiers. To this purpose the Aichi Target 9 established "By 2020, invasive alien species and pathways are identified and prioritized, priority species are controlled or eradicated, and measures are in place to manage pathways to prevent their introduction and establishment" [78].

This target addressed the following implementation measures:

1. Improved border controls and quarantine [...].

2. Development of early warning mechanisms, rapid response measures and management plans.

3. Prioritise control and eradication efforts to those species and pathways which will have the greatest impact on biodiversity and/or which are the most resource effective to address.

4. A special reference is made for the island's ecosystems, due to the acute impact of invasive alien species on island ecosystems.

The Invasive Species Specialist Group also developed a Toolkit for the economic analysis of Invasive species [107] which addresses the causes and the impacts of IAS, the related costs and benefits, the valuation of ecosystem impacts and the actions to address IAS.

Besides the information, the international and national legislations, the definition of biosecurity programs is also important, identifying IAS that pose a high risk of causing damage, and establishing measures to protect natural resources and citizens. Currently, biosafety on plant IAS is governed internationally by the International Plant Protection Convention, which establishes harmonised guidelines and standards between countries to limit the spread of IAS while promoting free trade [25].

Addressing IAS control in islands is less difficult than in continental land masses since it may be possible to prevent the entry of these IAS at the border in the management plan. Yet, it is a complex operation. The engagement of the community (citizen science) is of utmost importance, to allow early identification of new invasions. Engaging volunteers in surveillance and monitoring is also a low-cost, large-scale, and a long-term option, for those countries that are not able to implement integrated IAS surveillance programs [25].

Established populations of IAS have traditionally been managed by mechanical or physical control, chemical control, and biological control, all with successes and failures, but with increasing efficiency [108]. New management and innovative eradication technologies have been implemented in recent years, based on molecular genetics, notably the use of gene-silencing for the control of invasive populations that affect plants [109], or gene-editing technology, together with transgenes, which is a whole new technological approach that can help in the control and management of IAS [110].

4.5 Conservation and restauration

As defined by article 8 of the CBD, *in-situ* conservation is "the conservation of ecosystems and natural habitats and the maintenance and recovery of viable populations of species in their natural surroundings [...]" [74]. This definition includes the conservation of natural and semi-natural ecosystems in various types of PAs, aiming to conserve the ecosystem biodiversity, the landscape, to provide habitat for targeted organisms, such as endemic species. It also involves the conservation of targeted species in their natural habitat or ecosystem through conservation or management plans, the definition of recovery programmes for threatened, rare or endangered wild species and the restoration, and the recovery, or rehabilitation of habitats [111].

The *in-situ* conservation action is often complemented with *ex-situ* conservation actions, such as the cultivation in botanical gardens, the maintenance of seeds in seedbanks, arboreta collections, clone banks, cryopreservation, seed production, or other activities, while removed from many of their natural ecological processes, and being managed by humans [112]. The *ex-situ* conservation has enabled research into the causes of the primary threats, such as habitat loss, IAS, and exploitation, while also enabling conservation training and education activities. Different *ex-situ* activities allow the restauration of threatened wild populations, which can be used for population restauration (reinforcement or reintroduction) or conservation introduction, improve the demographic or genetic viability of wild plant populations by reducing the impact of anthropogenic or stochastic threats on these populations [112].

The use of *in-situ* and *ex-situ* conservation action has been an integrated approach increasingly used in the management of islands PAs, namely, to conserve endemic species [105, 111, 113–116]. The Hawai'i islands alone, e.g., have 14 state, federal, non-profitable and international institutions involved in *ex-situ* and *in-situ* conservation programmes, which are responsible for research in plant conservation, native ecosystems, managing wild plants, tissue culture, seed bank maintenance, species populations recovery, besides data management, defining strategy, priorities and planning, outreach, and training, among other activities [117]. A good example is the *ex-situ* conservation of the Hawaiian Vulcan palm (*Brighamia insignis*) which currently survives mainly in gardens.

Inter-situ conservation is a mixture of the *in-situ* and *ex-situ* conservation practices, creating a new community or ecosystem that is partly managed and partly wild. This conservation strategy is used when a threatened species had to be removed from its original range due to threats, and, thus, is conserved in a new location where those threats could be mitigated or are absent [118]. A step forward in conservation measures is "conservation-oriented restoration"

[119], which aims to conserve biodiversity in partially degraded habitats, either for assisted establishment or assisted colonisation. The concept aims to create partially new ecosystems with species compositions that differ from their historical analogues. This restoration aims to conserve endangered species and their habitats, rather than to improve the well-being of local communities by improving ecological services. The concept makes ecological restoration an integral part of conservation planning and implementation and uses threatened plant species in habitat restoration. Another interesting approach within the restauration measures are the Nature based Solutions (NbS), defined as "actions to protect, manage, and restore natural or modified ecosystems, that address societal challenges effectively and adaptively, simultaneously providing human well-being and biodiversity benefits" [120]. This rather new concept aims to work with the ecosystems and the native species within these ecosystems, using them to adapt and mitigate climate change. NbS are categorised into five main approaches [121]:

1. Ecosystem restoration approaches, including ecological restauration.

2. Issue-specific ecosystem-related approaches, including ecosystem-based adaptation, and ecosystem-based disaster risk reduction.

3. Ecosystem-based management approaches, such as integrated coastal zone management.

4. Ecosystem protection approaches, including protected area management.

5. Natural and green infrastructure-related approaches.

Accordingly, many NbS being implemented in PAs fall within the species and ecosystems conservation measures, as well as within the management tools that must be adopted when PAs are involved. In small oceanic islands, NbS can provide significant human wellbeing and biodiversity benefits, linking ecological, climate, and human wellbeing issues in an integrated, ocean-focused, and climate-responsive manner [122, 123].

5. Conclusions

The conservation of endemic plants in protected areas of oceanic islands is a vast, complex, and challenging topic, which has received the attention of many researchers in the past. These plants grow in small population due to low habitat availability, and isolated evolution. Therefore, the islands' ecosystems and their endemic plants are very vulnerable to current threats, such as climate change and the introduction of invasive alien species, but also to pollution, habitat fragmentation, fire, and other anthropogenic threats.

The conservation measures implemented so far are not consensual and many have not been successful, although important steps have been taken. The study and definition of major biodiversity hotspots, the establishment of thousands of protected areas, the creation of databases with information on relevant habitats and species, and the implementation of many *in-situ* and *ex-situ* conservation projects, with their pros and cons, are some of the cornerstones of conservation knowledge and management.

New scientific approaches are appearing in conservation, namely the Nature Based Solutions, the conservation-oriented restoration, the gene-editing technology together with transgenes, which are already showing promising results in plant conservation.

Despite the scientific efforts, the importance of efficient management of protected areas and of the political priority given to conservation should be stressed. Without them, all scientific achievements are irrelevant.

Acknowledgements

This study had the support of Fundação para a Ciência e Tecnologia (FCT), through the strategic project UIDB/04292/2020 granted to MARE. I acknowledge the Saxifraga foundation (http://www.freenatureimages.eu/) for providing the free photo of *Viola paradoxa*. All the other photos were taken from royalty free sites.

Author details

Teresa Mouga
MARE – Marine and Environmental Sciences Centre, Polytechnic Institute of Leiria, Leiria, Portugal

*Address all correspondence to: mougat@ipleiria.pt

References

[1] Whittaker RJ, Fernández-Palacios JM. Island biogeography: ecology, evolution, and conservation. Oxford, U.K.: Oxford University Press; 2007.

[2] Woodroffe CD, McLean RF, Smithers SG, Lawson EM. Atoll reef-island formation and response to sea-level change: West Island, Cocos (Keeling) Islands. Mar Geol. 1999;160(1-2):85-104. Doi: 10.1016/S0025-3227(99)00009-2

[3] Paulay G. Biodiversity on Oceanic Islands: Its Origin and Extinction. Am Zool. 1994;34:134-144.

[4] Beller W, D'Ayala P, Hein P, editors. Sustainable development and environmental management of small islands. Vol. 5, Series man and the biosphere. Paris, France: The Parthenon Publishing Group; 1990.

[5] Irl SDH, Anthelme F, Harter DEV, Jentsch A, Lotter E, Steinbauer MJ, et al. Patterns of island treeline elevation - a global perspective. Ecography. 2016;39(5):427-436. Doi: 10.1111/ecog.01266

[6] H. Calado, C. Fonseca, M. Vergílio, A. Costa, F. Moniz, A. Gil, et al. Small Islands Conservation and Protected Areas. J Integr Coast Zo Manag. 2014;14(2):167-174. doi: 10.5894/rgci523

[7] Givnish TJ. Ecology of plant speciation. Taxon. 2010 59(5): 1326-66.

[8] Crawford DJ, Archibald JK. Island floras as model systems for studies of plant speciation: Prospects and challenges. J Syst Evol. 2017;55(1):1-15. Doi: 10.1111/jse.12234

[9] Thompson JD. Population differentiation in Mediterranean plants: insights into colonization history and the evolution and conservation of endemic species. Heredity. 1999;82: 229-236.

[10] Rietbergen S, Hammond T, Sayegh C, Hesselink F, Mooney K. Island voices-island choices Developing strategies for living with rapid ecosystem change in small islands. Ecosystem Management Series n. 6. Gland, Switzerland; 2008.

[11] Christenhusz MJM, Byng JW. The number of known plants species in the world and its annual increase. Phytotaxa. 2016;261(3):201-17. Doi: 10.11646/phytotaxa.261.3.1

[12] Fine PVA. Ecological and Evolutionary Drivers of Geographic Variation in Species Diversity. Annu Rev Ecol Evol Syst. 2015;46:369-392. Doi:10.1146/annurev-ecolsys-112414-054102

[13] Myers N, Mittermeier RA, Mittermeier CG, Fonseca GAB da, Kent J. Biodiversity hotspots for conservation priorities. Nature. 2000;403:853-858.

[14] Mittermeier RA, Turner WR, Larsen FW, Brooks TM, Gascon C. Global Biodiversity Conservation: The Critical Role of Hotspots. In: Zachos FE, Habel JC, editors. Choice Reviews Online. Biodiversity hotspots: distribution and protection of conservation priority areas. Berlin Heidelberg: Springer-Verlag; 2011. p. 3-22.

[15] MacArthur R, Wilson E. The theory of island biogeography. Princeton University Press; 1967. Doi: 10.1515/9781400881376/html

[16] Heleno R, Vargas P, Heleno RH. How do islands become green? Glob Ecol Biogeogr. 2015;24:518-526. Doi: 10.1111/geb.12273

[17] Buckley RC. Conservation Tourism. Oxfordshire, U.K.: CAB International; 2010.

[18] Funk VA, Wagner WL. Biogeographic Patterns in the Hawaiian islands. In: Funk VA, Wagner WL, editors. Hawaiian Biogeography - Evolution on a Hot spot Archipelago. Washingtoon D.C.: Smithsonian Series in Comparative Evolutionary Biology, Smithsonian Institution Press; 1995. p. 379-419.

[19] Staude IR, Navarro LM, Pereira HM. Range size predicts the risk of local extinction from habitat loss. Glob Ecol Biogeogr. 2020;29(1):16-25. Doi: 10.1111/GEB.13003

[20] Pickett STA, Kolasa J, Armesto JJ, Collins SL. The Ecological Concept of Disturbance and Its Expression at Various Hierarchical Levels. Oikos. 1989 Feb;54(2):129.

[21] Montmollin B de, Strahm W, (Eds). The Top 50 Mediterranean Island Plants: Wild plants at the brink of extinction, and what is needed to save them. IUCN. Montmollin B, Strahm W, editors. The Top 50 Mediterranean Island Plants: Wild plants at the brink of extinction; and what is needed to save them. Gland, Switzerland and Cambridge, UK; 2005. 110 p.

[22] Brockie RE, Loope LL, Usher MB, Hamann O. Biological invasions of island nature reserves. Biol Conserv. 1988;44(1-2):9-36.

[23] Veitch CR, Clout MN, (edt.). Turning the Tide: The Eradication of Invasive Species. Veitch CR, Clout MN, editors. Gland, Switzerland; Cambridge, UK: IUCN SSC Invasive Species Speciatlist Group.; 2002. 414 p.

[24] Dalmazzone S, Giaccaria S. Economic drivers of biological invasions: A worldwide, bio-geographic analysis. Ecol Econ. 2014;105:154-165. Doi: 10.1016/J.ECOLECON.2014.05.008

[25] Pyšek P, Hulme PE, Simberloff D, Bacher S, Blackburn TM, Carlton JT, et al. Scientists' warning on invasive alien species. Biol Rev. 2020;95(6):1511-1534. Doi: 10.1111/brv.12627

[26] Braga RR, Gómez-Aparicio L, Heger T, Vitule JRS, Jeschke JM. Structuring evidence for invasional meltdown: broad support but with biases and gaps. Biol Invasions 2017;20(4):923-936. Doi: 10.1007/s10530-017-1582-2

[27] Boudouresque C-F, Verlaque M. Assessing scale and impact of ship-transported alien fauna in the Mediterranean Sea. CIESM Work Monogr no 20 Alien Mar Org Introd by ships Mediterr Black Seas. 2002:53-61.

[28] Lonsdale WM. Global patterns of plant invasions and the concept of invasibility. CONCEPTS Synth. 1999;80(5):1522-1536.

[29] Boudouresque C, Verlaque M. Biological pollution in the Mediterranean Sea: invasive versus introduced macrophytes. Mar Pollut Bull. 2002;44(1):32-38.

[30] Jäger H, Kowarik I, Tye A. Destruction without extinction: Long-term impacts of an invasive tree species on Galápagos highland vegetation. J Ecol. 2009 Nov;97(6):1252-1263. Doi: 10.1111/J.1365-2745.2009.01578.X

[31] S. Brunel, Brundu G, Fried G. Eradication and control of invasive alien plants in the Mediterranean Basin: towards better coordination to enhance existing initiatives. Bull OEPP. 2013;43(2):290-308. Doi: 10.1111/epp.12041

[32] Vardarman J, Berchová-Bímová K, Pěknicová J. The role of protected area zoning in invasive plant management. Biodivers Conserv 2018 278. 2018;27(8):1811-29. Doi: 10.1007/S10531-018-1508-Z

[33] Nummi P, Pöysä H. Population and community level responses in Anas-species to patch disturbance caused by an ecosystem engineer, the beaver. Ecography (Cop). 1997;20(6): 580-584.

[34] Berio Fortini Christina Leopold Kim S Perkins Oliver A Chadwick Stephanie G Yelenik James D Jacobi Kai LR, Bishaw Makani Gregg ena I, Berio Fortini L, Leopold CR, Perkins KS, Chadwick OA, et al. Landscape level effects of invasive plants and animals on water infiltration through Hawaiian tropical forests. Biol Invasions. 2021; Doi: 10.1007/s10530-021-02494-8

[35] Russell JC, Meyer J-Y, Holmes ND, Pagad S. Invasive alien species on islands: impacts, distribution, interactions and management. Environ Conserv [Internet]. 2017;44(4):359-370. Doi: 10.1017/S0376892917000297

[36] IPCC. IPCC Summary for policymakers - Global Warming of 1.5°C. An IPCC Special Report on the Impacts of Global Warming of 1.5°C Above Pre-industrial Levels and Related Global Greenhouse Gas Emission Pathways, in the Context of Strengthening the Global Response to the Threat of Climate Change, Sustainable Development, and Efforts to Eradicate Poverty. Josef Settele. Geneva, Switzerland: Alistair Woodward; 2018.

[37] Chitale V, Behera MD. How will forest fires impact the distribution of endemic plants in the Himalayan biodiversity hotspot? Biodivers Conserv 2019 288. 2019;28(8):2259-73. Doi: 10.1007/s10531-019-01733-8

[38] Acea MJ, Diz N, Prieto-Fernández A. Microbial populations in heated soils inoculated with cyanobacteria. Biol Fertil Soils. 2001;33(2):118-125. Doi: 10.1007/s003740000298

[39] Certini G. Effects of fire on properties of forest soils: a review. Oecologia. 2005;143(1):1-10. Doi: 10.1007/s00442-004-1788-8

[40] Mataix-Solera J, Guerrero C, García-Orenes F, Bárcenas GM, Torres MP. Forest Fire Effects on Soil Microbiology. In: Cerdà A, Robichaud PR (Eds) Fire Effects on Soils and Restoration Strategies Science Publishers, Enfield, NH, USA. 2009. p. 133-175.

[41] Manes S, Costello MJ, Beckett H, Debnath A, Devenish-Nelson E, Grey KA, et al. Endemism increases species' climate change risk in areas of global biodiversity importance. Biol Conserv. 2021 May 1;257:109070. Doi: 10.1016/J.BIOCON.2021.109070

[42] Oduor AMO, Leimu R, van Kleunen M. Invasive plant species are locally adapted just as frequently and at least as strongly as native plant species. J Ecol. 2016 Jul 1;104(4):957-968. Doi: 10.1111/1365-2745.12578

[43] Steffen W, Rockström J, Richardson K, Lenton TM, Folke C, Liverman D, et al. Trajectories of the Earth System in the Anthropocene. Proc Natl Acad Sci. 2018;115(33):8252-8259. Doi: 10.1073/PNAS.1810141115

[44] UNWTO. International Tourism Highlights, 2019 Edition [Internet]. [cited 2019 Aug 30]. Available from: www.e-unwto.org/doi/book/10.18111/9789284421152

[45] Lenzen M, Sun Y-Y, Faturay F, Ting Y-P, Geschke A, Malik A. The carbon footprint of global tourism. Nat Clim Chang. 2018;8(6):522-528. Doi: 10.1038/s41558-018-0141-x

[46] Niäiä M, Ivanovič S, Drpič D. Challenges to sustainable development in Island tourism. South East Eur J Econ Bus. 2010;5(2):43-53. Doi: 10.2478/v10033-010-0014-3

[47] López-Pujol J, Orellana MR, Bosch M, Simon J, Blanché C. Effects of Habitat Fragmentation on Allozyme Diversity and Conservation Status of the Coastal Sand Dune Plant Stachys maritima (Lamiaceae) in the Iberian Peninsula. Plant Biol. 2003;5(05):504-512. Doi: 10.1055/s-2003-44787

[48] Ballantyne R, Packer J, Sutherland LA. Visitors' memories of wildlife tourism: Implications for the design of powerful interpretive experiences. Vol. 32, Tourism Manag. 2011;3284):770-79. Doi: 10.1016/j.tourman.2010.06.012

[49] Ballantyne M, Pickering C. Ecotourism as a threatening process for wild orchids. J Ecotourism. 2012;11(1):34-47. Doi: 10.1080/14724049.2011.628398

[50] Hall CM. Policy learning and policy failure in sustainable tourism governance: from first- and second-order to third-order change? J Sustain Tour. 2011;19(4-5):649-671. Doi:10.1080/09669582.2011.555555

[51] Das M, Chatterjee B. Ecotourism: A panacea or a predicament? Tour Manag Perspect. 2015;14:3-16. Doi: 10.1016/J.TMP.2015.01.002

[52] Jitpakdee R, Thapa GB. Sustainability Analysis of Ecotourism on Yao Noi Island, Thailand. Asia Pacific J Tour Res. 2012;17(3):301-325. Doi: 10.1080/10941665.2011.628328

[53] Kroeker-Maus D. The Protected Area as Enclave: Towards New Geographies of Tourism and Conservation. Geogr Compass. 2014;8(11):796-807. Doi. 10.1111/gec3.12179

[54] UNWTO. Definition | Sustainable Development of Tourism [Internet]. 2012 [cited 2021 Sep 06]. Available from: https://sdt.unwto.org/content/about-us-5

[55] Budeanu A, Miller G, Moscardo G, Ooi C-G. Sustainable tourism, progress, challenges and opportunities: an introduction. J Clean Prod. 2016;111:285-294. Doi: 10.1016/J.JCLEPRO.2015.10.027

[56] Buckley R. Sustainable tourism: Research and reality. Ann Tour Res. 2012;39(2):528-546. Doi: 10.1016/j.annals.2012.02.003

[57] Griffin LP, Brownscombe JW, Gagné TO, Wilson ADM, Cooke SJ, Danylchuk AJ. Individual-level behavioral responses of immature green turtles to snorkeler disturbance. Oecologia. 2017;183(3):909-917. Doi: 10.1007/s00442-016-3804-1

[58] Mendes S, Martins J, Mouga T. Ecotourism based on the observation of sea turtles-A sustainable solution for the touristic promotion of São Tomé and Príncipe. Cogent Social Sciences. 2019; 5(1). Doi: 10.1080/23311886.2019.1696001

[59] Tisdell C, Wilson C. Ecotourism for the survival of sea turtles and other wildlife. Biodivers Conserv. 2002;11(9):1521-1538. Doi: 10.1023/A:1016833300425

[60] Tilman D, Fargione J, Wolff B, D'Antonio C, Dobson A, Howarth R, et al. Forecasting agriculturally driven global environmental change. Science. 2001;292(5515):281-284. Doi: 10.1126/SCIENCE.1057544

[61] Laurance WF, Sayer J, Cassman KG. Agricultural expansion and its impacts on tropical nature. Trends Ecol Evol. 2014;29(2):107-116. Doi: 10.1016/J.TREE.2013.12.001

[62] MacKinnon K. Sound Investments: Protected Areas as Natural Solutions to Climate Change and Biodiversity Conservation. In: Baillie J, Joppa L, Robinson JG, editors. Protected areas are they safeguarding biodiversity. West

Sussex, UK: Wiley Blackwell; 2016. p. 49-65.

[63] Buchanan GM, Butchart SHM, Dutson G, Pilgrim JD, Steininger MK, Bishop KD, et al. Using remote sensing to inform conservation status assessment: Estimates of recent deforestation rates on New Britain and the impacts upon endemic birds. Biol Conserv. 2008;141(1):56-66. Doi: 10.1016/J.BIOCON.2007.08.023

[64] Katovai E, Katovai DD, Campbell M, Laurance SG, Edwards W, Laurance WF. Structural Recovery of Logged Forests in the Solomon Islands: Implications for Conservation and Management. Trop Conserv Sci 2021; 14:1-13. Doi: 10.1177/19400829 211028125

[65] Harrison RD, Swinfield T. Restoration of logged humid tropical forests: An experimental programme at harapan rainforest, Indonesia. Trop Conserv Sci 2015;8(1):4-16. Doi: 10.1177/194008291500800103

[66] Bongaarts J. Human population growth and the demographic transition. Phil Trans R Soc B 2009; 364: 2985-2990. doi:10.1098/rstb.2009.0137

[67] Baillie JEM, Joppa L, Robinson JG. Introduction: Do Protected Areas Safeguard Biodiversity? In: Joppa LN, Baillie JEM, John G. Robinson, editors. Protected Areas: Are They Safeguarding Biodiversity? Oxford, U.K.: John Wiley & Sons, Ltd; 2016.

[68] Laurance WF, Goosem M, Laurance SGW. Impacts of roads and linear clearings on tropical forests. Trends Ecol Evol. 2009;24(12):659-669. Doi: 10.1016/J.TREE.2009.06. 009

[69] Balmford A, Chen H, Phalan B, Wang M, O'Connell C, Tayleur C, et al. Getting Road Expansion on the Right Track: A Framework for Smart Infrastructure Planning in the Mekong. PLOS Biol. 2016;14(12):e2000266. Doi: 10.1371/journal.pbio.2000266

[70] Caro T, Dobson A, Marshall AJ, Peres CA. Compromise solutions between conservation and road building in the tropics. Curr Biol. 2014;24(16): R722–R725. Doi: 10.1016/J. CUB.2014.07.007

[71] Pickering CM, Hill W. Impacts of recreation and tourism on plant biodiversity and vegetation in protected areas in Australia. J Environ Manage. 2007;85(4):791-800. Doi: 10.1016/J. JENVMAN.2006.11.021

[72] Sherley G. Invasive species in the Pacific: A technical review and draft regional strategy. Apia, South Pacific Regional Environment Programme, Samoa; 2000.

[73] Foxcroft LC, Pysek P, Richardson DM, Pergl J, Hulme PE. The Bottom Line: Impacts of Alien Plant Invasions in Protected Areas. In: Foxcroft LC, Pysek P, Richardson DM, Genovesi P, editors. Plant Invasions in Protected Areas - Patterns, Problems and Challenges. Invading N. Springer; 2013. p. 19-41.

[74] CBD. The Convention on Biological Diversity. [Internet]. Secretariat of the Convention on Biological Diversity 2021[cited 2021 Aug 28]. Available from: https://www.cbd.int/ecosystem/principles.shtml

[75] Bongham HC, Lewis E, Tayleur J, Cunnigham C, Kingston N, Burgess ND, et al. Coverage | Protected Planet Report 2020 [Internet]. UNEP-WCMC. 2021 [cited 2021 Aug 26]. Available from: https://livereport.protectedplanet.net/chapter-3

[76] Juffe-Bignoli D, Burgess ND, Bingham H, Belle EMS, de Lima MG, Deguignet M, et al. Protected Planet Report 2014 - Tracking progress towards

global targets for protected areas. Cambridge, UK; 2014.

[77] UNEP-WCMC, IUCN. Protected Planet Report 2016.How Protected Areas contribute to achieving Global Targets for Biodiversity [Internet]. Cambridge UK and Gland, Switzerland.: UNEP-WCMC and IUCN; 2016. Available from: https://wdpa.s3.amazonaws.com/Protected_Planet_Reports/2445 Global Protected Planet 2016_WEB.pdf

[78] CBD. Aichi Biodiversity Targets [Internet]. Convention on Biological Diversity. 2021 [cited 2021 Sep 6]. Available from: https://www.cbd.int/sp/targets/

[79] Moreno-Saiz JC, Albertos B, Ruiz-Molero E, Mateo RG. The European Union can afford greater ambition in the conservation of its threatened plants. Biol Conserv. 2021 Sep 1;261:109231. Doi: 10.1016/J.BIOCON.2021.109231

[80] Heywood VH. Plant conservation in the Anthropocene - Challenges and future prospects. Plant Divers. 2017;39:314-30. Doi: 0.1016/j.pld.2017.10.004

[81] Heywood VH. Conserving plants within and beyond protected areas – still problematic and future uncertain. Plant Divers. 2019;41(2):36-49. Doi: 10.1016/J.PLD.2018.10.001

[82] Havens K, Kramer AT, Jr. EOG. Getting Plant Conservation Right (or Not): The Case of the United States. Int J Plant Sci. 2015;175(1):3-10. Doi: 10.1086/674103

[83] Darbyshire I, Anderson S, Asatryan A, Byfield A, Cheek M, Clubbe C, et al. Important Plant Areas: revised selection criteria for a global approach to plant conservation. Biodivers Conserv 2017 268. 2017;26(8):1767-800. Doi: 10.1007/s10531-017-1336-6

[84] IUCN. IUCN Red List of Threatened Species [Internet]. 2021 [cited 2021 Sep 7]. Available from: https://www.iucnredlist.org/resources/categories-and-criteria

[85] Kougioumoutzis K, Kokkoris IP, Panitsa M, Kallimanis A, Strid A, Dimopoulos P. Plant endemism centres and biodiversity hotspots in Greece. Biology. 2021;10(2):1-27. Doi: 10.3390/biology10020072

[86] Ramsar Convention. Convention on Wetlands | Ramsar [Internet]. 2021 [cited 2021 Sep 7]. Available from: https://www.ramsar.org/about-the-convention-on-wetlands-0

[87] UNESCO. UNESCO World Heritage Centre - Convention Concerning the Protection of the World Cultural and Natural Heritage [Internet]. 2021 [cited 2021 Sep 7]. Available from: http://whc.unesco.org/en/conventiontext/

[88] CITES. CITES - Convention on International Trade in Endangered Species of Wild Fauna and Flora [Internet]. 2021 [cited 2021 Sep 7]. Available from: https://cites.org/eng

[89] IPPC. IPPC - International Plant Protection Convention [Internet]. 2021 [cited 2021 Sep 7]. Available from: https://www.ippc.int/en/

[90] IUCN. PA Law Capacity Development | IUCN [Internet]. 2021 [cited 2021 Sep 7]. Available from: https://www.iucn.org/theme/environmental-law/resources/pa-law-capacity-development-0

[91] Morzaria-Luna HN, Cruz-Piñón G, Brusca RC, López-Ortiz AM, Moreno-Báez M, Reyes-Bonilla H, et al. Biodiversity hotspots are not congruent with conservation areas in the Gulf of California. Biodivers Conserv 2018 2714.

2018;27(14):3819-42. Doi: 10.1007/s10531-018-1631-x

[92] Deák B, Valkó O, Nagy DD, Török P, Torma A, Lőrinczi G, et al. Habitat islands outside nature reserves – Threatened biodiversity hotspots of grassland specialist plant and arthropod species. Biol Conserv. 2020;241:108254. Doi: 10.1016/J.BIOCON.2019.108254

[93] Su X, Han W, Liu G, Zhang Y, Lu H. Substantial gaps between the protection of biodiversity hotspots in alpine grasslands and the effectiveness of protected areas on the Qinghai-Tibetan Plateau, China. Agric Ecosyst Environ. 2019;278:15. Doi: 10.1016/J.AGEE.2019.03.013

[94] Bax V, Francesconi W. Conservation gaps and priorities in the Tropical Andes biodiversity hotspot: Implications for the expansion of protected areas. J Environ Manage. 2019;232:387-396. Doi: 10.1021/ie101194g

[95] Butchart SHM, Clarke M, Smith RJ, Sykes RE, Scharlemann JPW, Harfoot M, et al. Shortfalls and Solutions for Meeting National and Global Conservation Area Targets. Conserv Lett. 2015;8(5):329-337. Doi: 10.1111/conl.12158

[96] Oppel S, Beaven BM, Bolton M, Vickery J, Bodey TW. Eradication of Invasive Mammals on Islands Inhabited by Humans and Domestic Animals. Conserv Biol. 25(2):232-40. Doi: 0.1111/j.1523-1739.2010.01601.x

[97] Fagundes AI, Filipe A, Oliveira N, Andrade J. Control and erradication of invasive exotic plants, Report for task C5, Project LIFE+ Berlengas. Lisboa, Portugal; 2018.

[98] Harter DEV, Irl SDH, Seo B, Steinbauer MJ, Gillespie R, Triantis KA, et al. Impacts of global climate change on the floras of oceanic islands – Projections, implications and current knowledge. Perspect Plant Ecol Evol Syst. 2015;17(2):160-183. Doi: 10.1016/J.PPEES.2015.01.003

[99] Stone CP, Anderson SJ. Introduced animals in Hawaii's natural areas. In: Proceedings of the Thirteenth Vertebrate Pest Conference. Monterey, California: University of California-Davis; 1988. p. 28.

[100] Scott JJ, Kirkpatrick JB. Rabbits, landslips and vegetation change on the coastal slopes of subantarctic Macquarie Island, 1980-2007: Implications for management. Polar Biol. 2008 Mar 6;31(4):409-419. Doi: 10.1007/s00300-007-0367-y

[101] Eijzenga H. Vegetation change following rabbit eradication on Lehua Island, Hawaiian Islands. In: CR Veitch CR, Clout MN, Towns DR, editors. Proceedings of the International Conference on Island Invasives, Auckland, New Zealand, 8-12 February 2010. Island invasives: eradication and management. IUCN, Gland, Switzerland; 2011. p. 290-4.

[102] Canale DE, Dio V Di, Massa B, Mori E. First successful eradication of invasive Norway rats *Rattus norvegicus* from a small Mediterranean island (Isola delle Femmine, Italy). Folia Zool. 2019;68(1):29. Doi: 10.25225/fozo.060.2019

[103] Angel A, Wanless RM, Cooper J. Review of impacts of the introduced house mouse on islands in the Southern Ocean: are mice equivalent to rats? Biol Invasions. 2008;11:1743-54. Doi: 0.1007/s10530-008-9401-4

[104] Le Corre M, Danckwerts DK, Ringler D, Bastien M, Orlowski S, Morey Rubio C, et al. Seabird recovery and vegetation dynamics after Norway rat eradication at Tromelin Island, western Indian Ocean. Biol Conserv. 2015 May 1;185:85-94. Doi: 10.1016/j.biocon.2014.12.015

[105] Mouga T, Mendes S, Fagundes AI, Oliveira N, Cris P, Morais L. Recent Efforts to Recover *Armeria berlengensis*, an Endemic Species from Berlengas Archipelago, Portugal. Plants. 2021;10(498). Doi: 10.3390/plants10030498

[106] IUCN. Invasive species | IUCN [Internet]. 2021 [cited 2021 Sep 8]. Available from: https://www.iucn.org/theme/species/our-work/invasive-species

[107] Emerton L, Howard G. A Toolkit for the Economic Analysis of Invasive Species. Nairobi: Global Invasive Species Programme; 2008.

[108] Veitch CR, Clout MN, Martin AR, Russell JC, West CJ, editors. Island invasives: scaling up to meet the challenge international conference on island invasives 2017 Proceedings of the Occasional Paper of the IUCN Species Survival Commission N 62 °. Gland, Switzerland: IUCN; 2019. 734 p.

[109] Oh D-H, Kowalski KP, Quach QN, Wijesinghege C, Tanford P, Dassanayake M, et al. Novel genome characteristics contribute to the invasiveness of *Phragmites australis* (common reed). bioRxiv. 2021;2021.04.19.440155. Doi: 10.1101/2021.04.19.440155v1

[110] Simberloff D. History, impacts, and prospects for controlling a bevy of devastating invaders. Biol Invasions 2020 233. 2020;23(3):969-71. Doi: 10.1007/s10530-020-02395-2

[111] Heywood VH, Dulloo MR. *In-situ* conservation of wild plant species: A critical global review of good practices, IPGRI Technical Bulletin. Vol. n. 11. Rome, Italy: IPGRI; 2005

[112] IUCN/SSC. Guidelines on the Use of *Ex-situ* Management for Species Conservation. Version 2 [Internet]. Gland, Switzerland; 2014 [cited 2021 Sep 7]. Available from: www.iucn.org/about/work/programmes/species/publications/iucn_guidelines_and__policy__statements/

[113] Rao NK, Hanson J, Dulloo ME, Ghosh K, Nowell D, Larinde M. Manual of Seed Handling in Genebanks-Handbooks for Genebanks No. 8. Handbooks for Genebanks No. 8. Bioversity International, Rome, Italy. 2006. 147 p.

[114] Erwin Bergmeier. Seasonal pools in the vegetation of Gavdos (Greece)-*In situ* conservation required. Bocconea. 2001;13:511-516.

[115] Bramwell D. Conserving Biodiversity in the Canary Islands. Ann Missouri Bot Gard. 1990;77(1):28.

[116] Sommerville KD, Clarke B, Keppel G, Mcgill C, Newby Z-J, Wyse S V, et al. Saving rainforests in the South Pacific: challenges in *ex-situ* conservation. Aust J Bot. 2018;65(8):609-624. Doi: 10.1071/BT17096

[117] Werden LK, Sugii NC, Weisenberger L, Keir MJ, Koob G, Zahawi RA. *Ex-situ* conservation of threatened plant species in island biodiversity hotspots: A case study from Hawai'i. Biol Conserv. 2020;243:108435. Doi: 10.1016/j.biocon.2020.108435

[118] Russell JC, Kueffer C. Island Biodiversity in the Anthropocene. Annu Rev Environ Resour. 2019;5:48. Doi: 10.1146/annurev-environ-101718-

[119] Volis S. Conservation-oriented restoration – a two for one method to restore both threatened species and their habitats. Plant Divers. 2019;41(2):50-58. Doi: 10.1016/J.PLD.2019.01.002

[120] IUCN. IUCN Global Standard for Nature-based Solutions. A User-friendly Framework for the Verification, Design

and Scaling up Of NbS. Gland, Switzerland; 2020.

[121] Cohen-Shacham E, Walters G., Janzen C, (edt.). Nature-based Solutions to Address Global Societal Challenges, xiii + 97. [Internet]. Gland, Switzerland: IUCN; 2016 [cited 2021 Sep 5]. Available from: https://portals.iucn.org/library/sites/library/files/documents/2016-036.pdf

[122] Pedersen Zari M, Kiddle GL, Blaschke P, Gawler S, Loubser D. Utilising nature-based solutions to increase resilience in Pacific Ocean Cities. Ecosyst Serv. 2019 Aug 1;38:100968. Doi: 10.1016/J.ECOSER.2019.100968

[123] Duvat VKE, Magnan AK. Contrasting potential for nature-based solutions to enhance coastal protection services in atoll islands. In: Klöck C, Fink M, editors. Dealing wit h climate change on small islands: Towards effective and sustainable adaptation. Göttingen: Göttingen University Press; 2019. p. 45-75.

Chapter 4

Ecology of the Granular Spiny Frog *Quasipaa verrucospinosa* (Amphibia: Anura - Dicroglossidae) in Central Vietnam

Binh V. Ngo and Ya-Fu Lee

Abstract

We conducted a large-scale assessment at 35 primary forest sites and 42 secondary forest sites in Bach Ma National Park, central Vietnam, using the detection/non-detection data for each site over multiple visits, to quantify the site proportions that were occupied by granular spiny frogs (*Quasipaa verrucospinosa*). We additionally investigated the effect of site covariates (primary versus secondary forests) and sample covariates (temperature, humidity, and precipitation) to examine the environmental needs that may be incorporated for conserving rain forest amphibians in Vietnam. From the best model among all candidate models, We estimated a site occupancy probability of 0.632 that was higher than the naïve occupancy estimate of 0.403 and a 57% increase over the proportion of sites at which frogs were actually observed. The primary forest variable was an important determinant of site occupancy, whereas occupancy was not associated with the variable of secondary forest. In a combined AIC model weight, the detection model p (temperature, humidity, precipitation) included 90.9% of the total weight, providing clear evidence that environmental conditions were important sample covariates in modeling detection probabilities of granular spiny frogs. Our results substantiate the importance of incorporating occupancy and detection probabilities into studies of habitat relationships and suggest that the primary forest factor associated with environmental conditions influence the occupancy of granular spiny frogs.

Keywords: Anura, bootstrap, maximum likelihood, metapopulation, monitoring

1. Introduction

Studies on site occupancy of various species and their spatial patterns have been conducted in recent years to inform and develop amphibian conservation programs [1–4]. Data that were commonly adopted may include forest type, land ownership (e.g., state or private), soil class, and the proximity of a resource (e.g., streams and wetlands) to the sites occupied by a targeted species [4–6]. However, many studies of metapopulation dynamics that seek to understand for the factors that determine whether a species will exist at a location [7–9]. Large-scale monitoring programs for amphibian species [10–12] often relied on remotely sensed data (data on amphibians that has been gathered using biophysical variables derived from moderate resolution imaging spectroradiometers or from remote-sensing instruments on satellites) to depict spatial models in habitat occupancy [13–15]. Ignoring detectability may lead to biased estimations of site occupancy [16–18] and studies of habitat occupancy are often hampered by imperfect detectability for the species [1, 19–22].

Previous studies suggest that it is preferable to use a sampling method that involves multiple visits to sites (or patches) during the appropriate season in which a species can be detected [3] and the proportion of sites that are occupied by the species is assessed in the face of imperfect detection [21, 22]. In these cases, sampling sites may represent separate habitat patches in a dynamic context of metapopulations or sampling units (quadrats) regularly visited as part of a large-scale monitoring program [6] because the presence or absence of a species from a collection permits inference to the entire region of interest. Detection and nondetection models using multiple visits to each site permit assessment of detection probabilities and interesting parameters, including determining the ratio of sites occupied by the target species.

Human activities such as timber harvest, land conversion to monoculture crops or other developments, and manipulation of natural waterways have heavily eroded tropical primary rain forests and secondary forests in central Vietnam [23]. These actions have created forest fragments while severely impacting biodiversity and associated natural interaction and processes. Similar to other landscapes in this region, habitats in Bach Ma National Park have been manipulated and transformed, creating metapopulations of species, including the granular spiny frog (*Quasipaa verrucospinosa*), in the entire park. This species has been listed in the IUCN Red List as a near threatened species due to environmental degradation, loss of forest and stream habitats, global climate change, and overexploitation for consumption [24]. However, little is known about many aspects of the population ecology of *Q. verrucospinosa* in Vietnam and large-scale studies of occupancy models for this species are nonexistent. Information on site occupancy and microhabitat use in granular spiny frogs in primary and secondary forests of Bach Ma National Park is lacking even though terrestrial habitats have been identified as important to conservation programs [23, 25, 26].

In this study, We estimated site occupancy for *Q. verrucospinosa* in Bach Ma National Park, central Vietnam to (1) compare occupancy and detection probabilities for two specific habitat types (primary and secondary forests); (2) to obtain an overall estimate of site occupancy for the entire national park; and (3) to determine the number of microhabitat use individuals. I examined the effects of site covariates, including temperature, humidity, and precipitation, on the occupancy and detection of frogs, and tested the hypothesis that differences among habitat types result in different levels of detection. The presence of a primary forest canopy that better regulates forest temperature and soil moisture is crucial in determining anuran survival, reproduction, movement models, and species richness [4, 27–29], and amphibian species may be particularly sensitive to the effects of habitat fragmentation [30–32]. We predicted a greater abundance of granular spiny frogs in primary forests (undisturbed habitat) than in secondary forests (disturbed habitat).

2. Materials and methods

2.1 Study sites

The fieldwork took place in Bach Ma National Park (15°59’12″-16°16’09”N, 107°37’06″-107°54’14″E, approximately 37,487 ha in size), central Vietnam

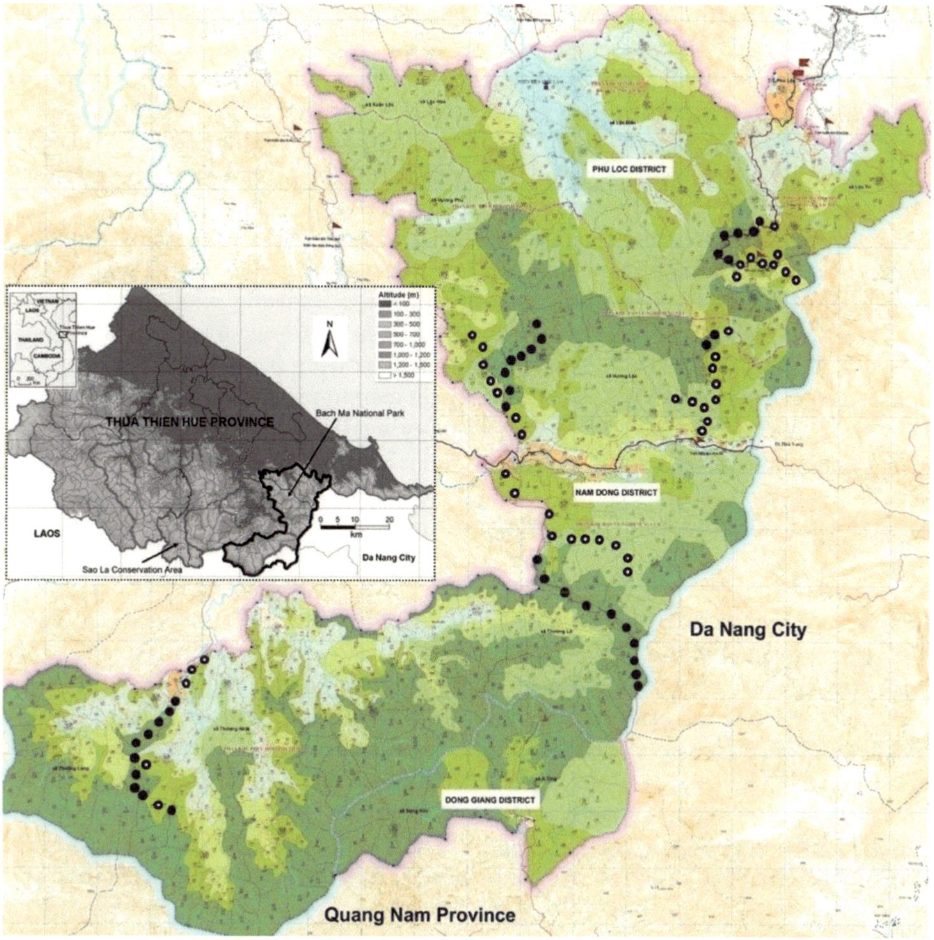

Figure 1.
Location of survey sites in Map of Bach Ma National Park, Thua Thien-Hue Province, Central Vietnam, showing 35 sites in primary forests (●) and 42 sites in secondary forests (○), where granular spiny frogs were monitored during the breeding season of 2013.

(**Figure 1**). The study area is dominated by montane rain forests at elevations of 700–1400 m a.s.l. and cloud forests from about 1400 m up to summits at 1712 m [23, 33]. Seasonal monsoons and a tropical climate characterize this study area, with annual temperatures averaging 22.6 ± 0.26°C (ranging from 16.9 ± 0.39°C in January to 26.6 ± 0.36°C in June) and an annual mean precipitation of 3492.1 ± 228.7 mm. Most of the rain is concentrated in the main rainy season from September to December (monthly mean: 629.2 ± 44.1 mm) and the little rainy season (monthly mean: 149.2 ± 14.5 mm) from May to August, whereas the period from January to April is relatively dry, with monthly mean rainfall of 94.6 ± 12.8 mm.

2.2 Field sampling

The study area comprised primary forests (ca. 32.2%; canopy is not fragmented), secondary and restored forests (54.0%; fragmented canopy), and administrative areas (13.8%; plantations; [23, 34]). From September to December 2013, we conducted seven surveys for *Q. verrucospinosa* during the peak breeding season [35], but only in primary and secondary forests (i.e., distributed regions of this species). We set 77 sampling plots of 20 × 50 m (1000 m^2 each site), 35 in primary forests and 42 in secondary forests (**Figure 1**). We only selected sample sites that contained water bodies, either a part of a stream or marsh where *Q. verrucospinosa* are commonly active, and each site was located about 300 m apart to ensure independence among sites. No addition, removal, or alteration of plots was made during the entire study period. We considered the primary and secondary forest variables as site covariates to describe habitat occupancy.

Each site was visited every two weeks, and each site was visited and sampled at the same time. In the night, a team of two people walked slowly with a roughly equal pace along the plot, and visually searched for frogs using spotlights from 19:00 to 02:00 hours for 50 m. We searched for *Q. verrucospinosa* in the water where they were visible and reachable, on land up to 10 m away from the stream or the marsh, and on tree trunks and vegetation. After locating frogs, we collected them by hand [36]. We also adopted the auditory survey method [37, 38] of using calls to detect granular spiny frogs and to count hidden individuals at each site.

2.3 Data analyses and model selection

To determine site occupancy, it is necessary to simply record whether an individual is detected "1" or not detected "0" during each survey at each site when visited. We estimated the effect of the secondary forest variable (with strong disturbance) on occupancy and detection probability. Using field observations on forest canopy gathered prior to this study, we determined the level of the secondary forest variable at each site, and a covariate of secondary forest was defined as 1 if the site showed evidence of the fragmentation of canopy and 0 otherwise. At each time a site was visited, I also recorded air temperature (temp), relative humidity (humi), and precipitation (rain). These variables were considered sample covariates to estimate detection and presence probabilities, respectively. When an individual frog was detected, we recorded the habitat in which it was found as aquatic, terrestrial, or arboreal. We used these variables to estimate microhabitat use in *Q. verrucospinosa*.

Each site has its own detection history that can be represented by a mathematical equation (Appendix 1). For example, supposing 30 sites were each sampled four times within a season and the target species (*Q. verrucospinosa*) was detected at site 1 during the first and last survey occasion (1001). The site was occupied (ψ), the probability of detecting the species during the jth survey was p_j, and the species was detected on the first and last surveys (p_1 and p_4) but not on the second and third surveys. We can write the probability of this detection history as follows: Pr $(H_i = 1001) = \psi p_1(1 - p_2)(1 - p_3)p_4$.

Sites 2 and 30 represent a case where the target species was never detected (detection history = 0000). These sites could either be unoccupied, which mathematically is $(1 - \psi)$, or they could be occupied but not detected. In this case, we can write the probability of this detection history as follows: $\psi(1 - p_1)(1 - p_2)(1 - p_3)(1 - p_4)$ or $\psi(1 - p_j)^4$. Thus, we can write the probability of detection history (0000) as follows: $\psi(1 - p_j)^4 + (1 - \psi)$.

If a site is not surveyed at the jth survey occasion (θ, the probability of miss detecting the species or missing observations), no information regarding the detection (or non-detection) of this species was collected from that site at that time. For example, the probability of detection histories (sites 3 and 4) could be expressed as: $Pr(H_3 = 10x0) = \psi p_1(1 - p_2) \times \theta \times (1 - p_4)$ and $Pr(H_4 = x0x1) = \psi \times \theta \times (1 - p_2) \times \theta \times p_4$.

Finally, the mathematical equation of all detection histories are combined into model likelihood as follows: $L(\psi, p/H_1, \ldots, H_{30}) = \Pi(i = 1, 2, 3, \ldots 30)Pr(H_i)$.

Maximum likelihood methods are incorporated in the program PRESENCE and this software was used to obtain estimates of occupancy and detectability for frogs at Bach Ma National Park.

We used the program PRESENCE and single-season occupancy models to assess occupancy and detection probabilities. This pattern assumes that sites or patches were closed to changes in occupancy between the first and last surveys of a given sampling season (i.e., no colonization or extinction events within the sampling season), and detection of the target species at a site is independent of detecting the species at other sites [3, 21]. We used the following parameters of interest for the present study: ψ is the probability of a site occupied by *Q. verrucospinosa* and p_j is the probability of detecting the species during the jth survey given that it is present.

We used two essential models for the present study. The first model that assumes that occupancy and detection probabilities with respect to *Q. verrucospinosa* are constant across sites and surveys [denoted $\psi(.)p(.)$]. The second model assumes constant occupancy among sites, but detection probabilities are allowed to vary among seven survey occasions [denoted $\psi(.)p(\text{survey})$]. Previous studies suggest that the detection probability ≥ 0.15 in the model $\psi(.)p(.)$ is needed for unbiased occupancy modeling [2, 39]. Our model $\psi(.)p(.)$ with a detection probability of 0.329 (SE = 0.035) is appropriate to pursue inference and to estimate the details of the parsimonious process of model selection. In the present study, detectability was either constant across all survey occasions and sites $p(.)$, or varied in three possible ways: among seven survey occasions $p(\text{survey})$, or across sites according to weather conditions $p(\text{temp, humi, rain})$, or both $p(\text{survey, temp, humi, rain})$. We also estimated the value of the coefficient for the secondary forest variable with respect to its influence on occupancy probability and our candidate set contains 16 models without considering interactions between factors (**Table 1**).

We used the Akaike Information Criteria for small sample size (AIC_c), the differences in the Akaike Information Criteria for a particular model when compared to the top-ranked model (ΔAIC_c), the AIC model weight (w), the number of parameters for each model (Np), and twice the negative log-likelihood value ($-2\,l$), to establish the process of model selection [40]. All models with AIC differences of <2.0 have a substantial level of empirical support and should be considered when making statistical inferences or reporting parameter estimates of the best models, and the AIC weights summed to 1.0 for all of the members of the model set [40].

We followed a two-step process to detect site occupancy patterns [4, 6, 41, 42]. We first examined occupancy patterns as a function of site covariates employing the best detection patterns (as indicated by AIC_c weight). Burnham and Anderson [40] recommend using AIC_c for model ranking to help account for small sample sizes. Then, we modeled detection probability as a function of the sample covariates while keeping occupancy constant [i.e., $\psi(.)p(\text{covariates})$]. Finally, we examined the null hypothesis, the model $\psi(.)p(.)$, and one alternative hypothesis, the model $\psi(.)p(\text{survey})$, using the χ^2-test by the linear interpolation [6, 43] prior to inference for the next models.

We employed the equation, $\Psi_{mle} = S_D/SP_{mle}$, to assess a constant detection probability. S_D is the number of sites where *Q. verrucospinosa* was detected at least once, S is the total number of sites, and P_{mle} is the estimated probability of detecting the species at least once during a survey given it is present. $P_{mle} = 1 - (1 - p_{mle})^7$, where p_{mle} is the maximum likelihood estimate of a constant detection probability in a single survey of an occupied site, and mle represents the maximum likelihood estimate of the respective model parameters. Thus, the probability of detecting at least once *Q. verrucospinosa* after k surveys (k = 7) of the site will be $p^* = 1 - (1 - p)^7$, where p assumes that *Q. verrucospinosa* is detected imperfectly and gives the probability of detecting *Q. verrucospinosa* in a single survey of an occupied site. This is one minus

Model	*Np*	AIC_c	ΔAIC_c	*w*	$-2\,l$	SF_1	±SE
$\psi(PF), p(sur, temp, humi, rain)$	11	322.33	0.00	0.242	300.33	—	—
$\psi(PF), p(temp, humi, rain)$	4	322.45	0.12	0.228	314.45	—	—
$\psi(PF), p(temp, humi, rain, SF)$	5	323.39	1.06	0.143	313.39	—	—
$\psi(SF), p(temp, humi, rain)$	5	324.05	1.72	0.102	314.05	−0.549	0.727
$\psi(SF), p(sur, temp, humi, rain)$	12	324.12	1.79	0.099	300.12	−0.401	0.781
$\psi(PF), p(sur, temp, humi, rain, SF)$	12	324.19	1.86	0.095	300.19	—	—
$\psi(SF), p(temp, humi, rain, SF)$	6	325.36	3.03	0.053	313.36	−0.221	1.141
$\psi(SF), p(sur, temp, humi, rain, SF)$	13	326.06	3.73	0.038	300.06	−0.341	0.910
$\psi(SF), p(SF)$	4	352.95	30.62	0.000	344.95	−0.219	1.562
$\psi(PF), p(SF)$	3	354.83	32.50	0.000	348.83	—	—
$\psi(PF), p(sur, SF)$	9	360.85	38.52	0.000	342.85	—	—
$\psi(SF), p(sur, SF)$	10	362.83	40.50	0.000	342.83	−0.225	1.555
$\psi(SF), p(.)$	3	371.27	48.94	0.000	365.27	−1.929	0.555
$\psi(SF), p(sur)$	9	381.26	58.93	0.000	363.26	−1.928	0.554
$\psi(.), p(.)$	2	383.14	60.81	0.000	379.14	—	—
$\psi(.), p(sur)$	8	393.13	70.80	0.000	377.13	—	—

Table 1.
The summary of AIC model selection from 77 sites in the primary and secondary forests of Bach ma National Park. ΔAIC_c is the difference in AIC value for a particular model when compared with the top-ranked model; w: *The AIC model weight;* Np *is the number of parameters;* $-2\,l$ *is twice the negative log-likelihood value;* SF_1*: The coefficient for the secondary forest variable with respect to its effect on occupancy probability; SE: Standard error; PF: Primary forest; SF: Secondary forest; temp: Temperature; humi: Humidity; rain: Rainfall; Sur: Survey.*

the probability of *Q. verrucospinosa* being undetected in all *k* surveys [6]. For the probability of occupancy given that a species is not detected at a site (i.e., that *Q. verrucospinosa* was present at a site given it was never detected), we used the following equation: $\Psi_{condl} = \psi(1 - p_j)^7]/[(1 - \psi) + \psi(1 - p_j)^7]$, where ψ is estimated occupancy probability, p_j is the detection probability estimates in the jth survey. We obtained the standard errors for Ψ_{mle} and Ψ_{condl} by application of the delta method, where the variance–covariance matrix for ψ and p given by the program PRESENCE.

To estimate the fit of the model to data, accounting for an over-dispersion in model selection, and to account for missing observations, we used a simple Pearson's Chi-square statistic to test whether there was sufficient evidence of poor model fit [5]. We calculated the observed χ^2 goodness-of-fit statistic and estimated an over-dispersion parameter ($\hat{c}$) from 10,000 bootstrap samples using the program PRESENCE for the global model (the most general model in the model set or the model with the most parameters) from the candidate models (i.e., AIC weight > 0.001). Over-dispersion is common in ecological models, and adjusting the model selection criteria is recommended [5, 6]. In the absence of a global model, the weighted-average (w_i) of $\hat{c}$ was used to portray a goodness-of-fit for all candidate models [40]. If the values of weighted $\hat{c}$ are greater than one, it suggests that there is more variation in the observed data than expected by the model (over-dispersed occupancy models), while values less than one suggest less variation. If the values of weighted $\hat{c}$ are equivalent to one, then the target model is an adequate description of the data [5, 6].

We analyzed the data using the program PRESENCE (USGS-Patuxent Wildlife Research Center, Maryland, USA). We used SPSS v.16.0 (SPSS Inc., Chicago,

Illinois, USA) for Windows 10 to analyze the data of microhabitat use, and set the significance level at $P \leq 0.05$ for all analyses. To test the number of individuals using microhabitats and among surveys, we used a one-way analysis of variance (ANOVA). We used the χ^2 tests to examine the significance level between the first model $\psi(.)p(.)$ and the second model $\psi(.)p$(survey) through the seven surveys. We tested the possible effects of climatic factors (air temperature, relative humidity, and precipitation) on the detection of individuals using multiple regression analyses. All data are presented as mean $\pm$ 1 SE (unless otherwise noted).

3. Results

Quasipaa verrucospinosa was detected at least once at 31 of the 77 sites, yielding an overall naïve occupancy estimate of 0.403, clearly indicating that detection probabilities are less than one. There conceivably can be a number of locations where granular spiny frogs were present but simply never detected during the seven survey occasions. Our detection-corrected occupancy estimates by site in the primary and secondary forests of the national park ranged 0.143–0.714 (average naïve occupancy = 0.351 $\pm$ 0.032). As a general approach, two essential models [$\psi(.)p(.)$ and $\psi(.)p$(survey)] need to consider before inferring next models. The first model assumes that occupancy and detection probabilities are constant across sites and surveys. The rate of sites occupied by *Q. verrucospinosa* from the constant model [$\psi(.)p(.)$] was 0.433 (SE = 0.061). The second model assumes constant occupancy among sites, but detection probabilities are allowed to vary among the seven surveys. The rate of sites occupied based on the second model of [$\psi(.)p$ (survey)] was 0.432 (SE = 0.061).

The estimated occupancy probability is very similar in both models, 0.433 and 0.432 from the first and secondary models, respectively. When estimating occupancy probabilities including only the two models [($\psi(.)p(.)$ and $\psi(.)p$(survey)], both models gave essentially the same results, and both are about 8% larger than the naïve occupancy estimate. The model-averaged estimate of occupancy probability between primary and secondary forest habitat categories was 0.433 (SE = 0.061). When examining the results in which parameter estimates only have the two models [$\psi(.)p(.)$ and $\psi(.)p$(survey)], a difference of 9.99 ΔAIC_c units between these two models [with the AIC weight value of 0.993 in the model $\psi(.)p(.)$] shows that the model $\psi(.)p(.)$ is the "best" model. However, the second model [$\psi(.)p$ (survey)] still has a reasonable relative level of support (the AIC model weight value of 0.007) and there is further evidence of this second model to pursue inference. We examined a likelihood proportion of the null hypothesis of detection probability being constant and the alternative hypothesis that detection probability differs among the seven survey occasions. The test statistic for this is 379.14–377.13 = 2.01 (**Table 1**), compared to the χ^2 distribution with 8–2 = 6 degrees of freedom, by the linear interpolation, resulting in a significant level of $P = 0.933$. Thus, there is insufficient evidence to reject the null hypothesis in this study.

Testing the global model (the model with the most parameters) from the candidate set (**Table 1**), the model ψ(secondary forest)p(survey, temperature, humidity, precipitation, secondary forest), does not show any evidence of over-dispersion (weighted $\hat{c}$ = 0.436), indicating insufficient evidence of the poor model fit using 10,000 bootstrap iterations. As a result, the adjustment has been made to the model selection procedure (AIC) and parameter assessments to estimate the details of this parsimonious process of model selection. Detectability varied among surveys and possibly among sites with previously disturbed and undisturbed histories (**Figure 2**).

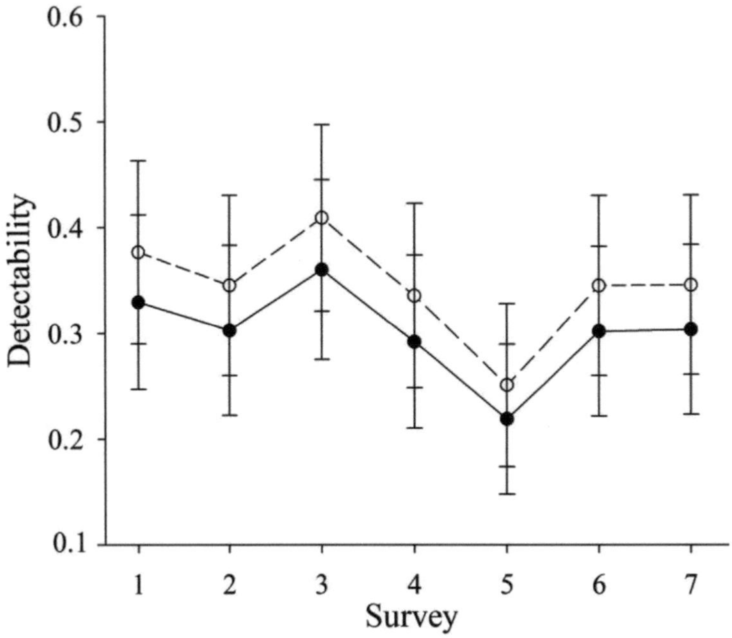

Figure 2.
Estimating the average pattern of detectability across surveys and among sites with different disturbance histories of granular spiny frogs. Undisturbed habitat (—○—) and disturbed habitat (—●—).

Our candidate set contained 16 models without considering interactions between factors due to limitation of software and complexity of models (**Table 1**). There was no single model that was demonstrably better than the others. As a general rule, the six top models are separated by less than 2.0 AIC units, which means that these models have substantial support and should be considered when reporting parameter estimates or making inferences (**Table 1**). The AIC model weight (w) was distributed across a number of models, indicating that a number of models may be reasonable for our collected data. In terms of model weights, the p(temperature, humidity, precipitation) models have 90.9% of the total, providing clear evidence that weather condition is an important factor in terms of accurately modeling detection probabilities. In terms of comparing hypotheses, the hypothesis that the detection probability varied among weather conditions, therefore, has much greater support than the hypothesis that it was constant. Many of the top-ranked models also contained the factor "survey" for detection probabilities, providing evidence that the survey occasions differed in their ability to find *Q. verrucospinosa* in the sites; a combined model weight for p(survey) models is 43.6% of the total. There was substantially less support for the hypothesis that the level of the secondary forest variable affected detection probabilities for *Q. verrucospinosa*, with a combined model weight of 23.8% (**Table 1**).

The primary forest variable ranked first among the set of models that accounted for differences in the survey, temperature, humidity, and precipitation to explain occupancy, and detection probabilities were approximately 2.5 times more likely than the next best model (evidence ratio [Akaike weight of top model/Aikaike weight of second best model] = 2.45). A model including temperature, humidity, and precipitation from primary forest sites ranked secondly among the set of models to explain the probability of occupancy and detectability were about 2.3 times more likely than the next competing model from secondary forest sites (evidence ratio [0.228/0.102] = 2.25). Estimating detection probabilities for each sampling covariate on each survey occasion is given in **Table 2**.

In terms of occupancy probability, based upon rankings and AIC model weights, the results are somewhat conclusive about the effect of secondary forest sites (29.2%) on the ψ(primary forest) model. The combined weight for the ψ(primary forest) models was 70.8%, and the ψ(secondary forest) models was 20.1% (**Table 1**). The coefficient value for the secondary forest variable with respect to its effect on occupancy probability, the eight AIC selection models showing the negative SF values (all values of $\hat{a}_2 < 0$), indicating certain evidence that the probability of occupancy is higher at the primary forest sites than at the secondary forest sites (**Table 1**). From the top-ranked model with $\Delta AIC_c < 2.0$ units, the model ψ(secondary forest)*p*(survey, temperature, humidity, precipitation) on the logit scale produces the following equation for estimating occupancy: Logit $(\psi_i) = 1 \times \hat{a}_1 + \hat{a}_2 \times SF_i = 1 \times 0.608 + (-0.401) \times SF_i$.

For a primary forest site (where the secondary forest variable = 0, according to the value of $SF_i = 0$), which gives the odds of occupancy of $e^{0.608} = 1.837$ (:1) and a probability of occupancy of 1.837/(1 + 1.837) = 0.647. The odds ratio for a secondary forest site being occupied (value $\hat{a}_2 = -0.401$) by *Q. verrucospinosa* is $e^{-0.401} = 0.669$. Thus, the odds of occupancy at a secondary forest site is 0.669 × 1.837 = 1.231 (:1) or a probability of occupancy of 1.231/(1 + 1.231) = 0.552. I also estimated a confidence interval for the influence of the secondary forest variable on site occupancy based upon the logit scale, an approximate two-sided 95% confidence interval is $-0.401 \pm 2 \times 0.781 = (-1.963, 1.161)$, giving an interval of $(e^{-1.963}, e^{1.161}) = (0.140, 3.193)$ for the odds ratio.

In terms of the overall estimate of site occupancy based upon the top-ranked model ψ(secondary forest)*p*(survey, temperature, humidity, precipitation), an average from the estimated occupancy probabilities for the primary forest sites (35 sites) and the secondary forest sites (42 sites), an overall estimate based on the influence of the secondary forest variable was {(35 × 0.647 + 42 × 0.552)/(35 + 42)} = 0.595, with an SE value of 0.114. This is approximately 48% larger than the naïve occupancy estimate (the fraction of sites where *Q. verrucospinosa* was detected) of 0.403. However, this is about 9% smaller than the occupancy estimate in the "best" model ψ(primary forest)*p*(survey, temperature, humidity, precipitation) of 0.632 (SE = 0.078). Clearly, accounting for detection probability has increased the estimated level of occupancy as expected (we discuss in detail below why the overall level of occupancy is larger than the naïve estimate). Based upon Bayes' Theorem, we also estimated the probability of a site being occupied, given the granular spiny frog *Q. verrucospinosa* was not detected there in any of the seven survey occasions (Ψ_{condl}), from the "best" model [ψ(.)*p*(.)], $\Psi_{condl} = 0.433 \times (1-0.329)^7]/[1-0.433 \times \{1-(1-0.329)^7\}] = 0.044$, with an estimated SE value of 0.021. The value of p^* in this model where the probability of detection is constant, the

Sampling covariates	Survey occasions						
	1	2	3	4	5	6	7
Temperature	0.283	0.259	0.306	0.241	0.176	0.257	0.254
Humidity	0.274	0.249	0.297	0.228	0.161	0.247	0.243
Precipitation	0.341	0.347	0.371	0.299	0.242	0.292	0.299

Table 2.
Estimating detection probabilities for each survey in Quasipaa verrucospinosa *from the AIC occupancy model selection for sampling-specific covariates.*

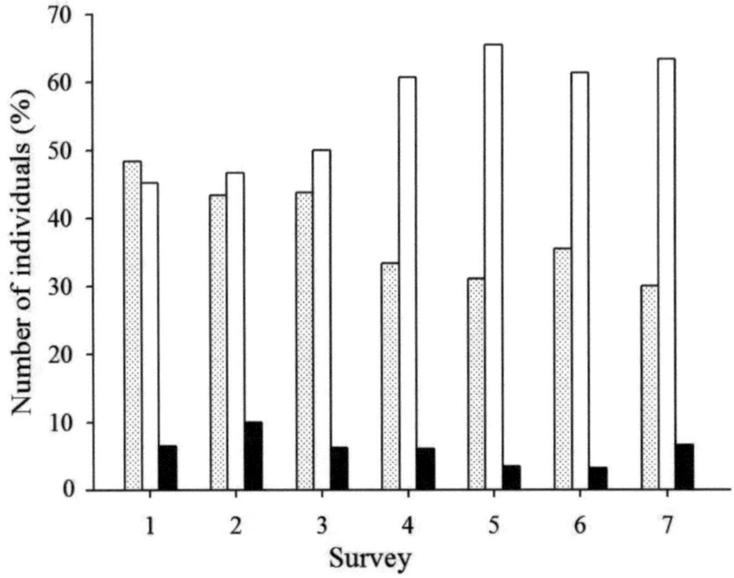

Figure 3.
Microhabitat use in granular spiny frogs from Bach Ma National Park, Central Vietnam. Aquatic (░), terrestrial (□), and arboreal (■) habitats.

probability of detecting *Q. verrucospinosa* at least once after *k* surveys of the site was $p^* = \{1 - (1{-}0.329)^7\} = 0.939$.

An overall estimate of microhabitat use in *Q. verrucospinosa* showed that the number of granular spiny frogs using the terrestrial habitat (121 individuals, 56.0%) was larger than the aquatic habitat (82 individuals, 38.0%) or the arboreal habitat (13 individuals, 6.0%). The number of individuals was significantly different among three habitat types ($F_{2,20} = 101.58$, $P < 0.001$; **Figure 3**). In total, we found 216 individuals during the seven surveys. The number of individuals was found among seven survey occasions were not significantly different ($F_{6,75} = 0.94$, $P = 0.472$). Multiple regression results for possible effects of air temperature, relative humidity, and precipitation on the detection of individuals were significant among surveys ($R^2 = 0.139$, $F_{3,251} = 27.92$, $P < 0.001$).

4. Discussion

Our results indicate that *Q. verrucospinosa* occupancy in the tropical forests of Bach Ma National Park is not associated with the secondary forest variable. Excluding the two last models, the six models include the secondary forest covariates (including both occupancy and detection probability) with the values of $\Delta AIC_c > 30.6$ units and all AIC model weights are equal zero. These findings were similar to those of previous studies that *Q. verrucospinosa* frogs were mainly found in primary forests in central Vietnam [24, 33–35, 44, 45]. In fact, we sampled a relatively broad gradient of forest types (42 sites were classified as secondary forest and 35 sites were classified as primary forest). However, *Q. verrucospinosa* frogs were only found at eight sites with eight respective individuals in a total of 42 sampling sites in secondary forests. Thus, we speculate that air temperature, relative humidity, and abundant precipitation during our sample season may have lessened forest type effects, because weather conditions and survey factors were important covariates for occupancy and detectability of *Q. verrucospinosa* in Bach Ma National Park tropical forests. The presence of a forest canopy that regulates air temperature and forest soil moisture appears more critical in determining survival of amphibians and movement models (e.g., [46, 47]). Previous studies show that some anuran species (especially juveniles) have indicated a preference for habitat types with forested canopies compared with fragmented forests or open-vegetation types [48, 49].

Wildlife occupancy relates only to site characteristics, whereas the probability of detecting a species during a single survey can vary with survey characteristics (e.g., temperature and precipitation) or site characteristics (e.g., habitat variables such as primary and secondary forests; [6]). An observed absence at a site occurs if either the species was truly absent, or the species was present at that site but not detected simply; while non-detection of a species does not mean that that species was truly absent unless the probability of detecting the species was 100%. That is the reason why previous occupancy studies of wildlife populations are often impeded by imperfect detectability [16, 21]. Thus, the rate of sites where a species of interest is detected will always be an underestimate with respect to the true occupancy level in the study region when detection is imperfect. Hence, inferences regarding the effects of site characteristics on habitat occupancy will be difficult or impossible to describe exactly [6, 16]. Our results from the best model ψ(primary forest)p(survey, temperature, humidity, rainfall) in the total candidate models were reliable, with the occupancy estimate of 0.632 (CI = 0.471–0.768) compared to the naïve occupancy estimate of 0.403. Although we did not consider colonization probabilities and local extinction factors, these two variables often influence parameter estimates in long-term monitoring programs of amphibians [21, 50, 51].

Moreover, our parameter estimates have satisfied normal assumptions of a model of single-species and single-season occupancy [6], including (1) the occupancy state of the sites does not change during the survey period, but can change between survey periods, (2) the detection of the target species in each survey occasion of a site is independent of detections within other survey occasions of the site, (3) occupancy probability is the same across sites or differences in habitat occupancy may be explained with site traits (covariates), (4) species detection probability at occupied sites is the same across all sites and surveys, or differences in detection probability can be explained with survey or site traits, and (5) the detection histories observed at each location are independent for a species of interest.

A brief examination of the estimated detection probabilities clearly indicates why the overall level of occupancy is estimated to be 49% larger than the naïve occupancy estimate, the estimation based simply on the number of sites where *Q. verrucospinosa* frogs were detected during the seven surveys. There is clearly a reasonable level of survey variation and substantial differences among sampling covariates, including temperature, relative humidity, and precipitation (see results). Furthermore, the detection of a species at each site is indeed indicative of the presence but non-detection of a species is not equivalent to the absence, unless the detecting probability of the species was one, and a species can go undetected at a site or some sites even when present. Therefore, non-detection sites of a frog represent a case where the target species (*Q. verrucospinosa*) was never detected. These sites could either be unoccupied, which mathematically is $(1 - \psi)$, or they could be occupied but we never detected the target species during k survey occasions, which mathematically is $\psi(1 - p_j)^k$. Both of these detectabilities have been included in maximum likelihood methods incorporated in the program PRESENCE to obtain estimates of occupancy and detectability. Although estimates in both essential models [($\psi(.)p(.)$ and $\psi(.)p$(survey)] are about 8% larger than the naïve occupancy estimate (suggesting that *Q. verrucospinosa* was never detected at one in every seven surveys), we believe that *Q. verrucospinosa* frogs can be more likely to occupy primary forest locations compared to secondary forest locations.

Parameter assessments and associated confidence intervals from pattern averaging indicated that the primary forest was an important determinant of *Q. verrucospinosa* occupancy in Bach Ma National Park. The appearance of both covariates (primary and secondary forests) in competing patterns is not a surprising result, but the weak relationship to habitat occupancy was unexpected. In terms of occupancy and detection probabilities, the negative values of the secondary forest variable indicated certain evidence with respect to its effect on occupancy and detectability of *Q. verrucospinosa*. The effects of forest and year factors on occupancy and detectability of tropical amphibians and habitat use emphasize the importance of conducting longer term researches for describing critical habitat relationships [52, 53]. Some populations of salamanders and frogs can fluctuate in the number of individuals (or even in number of species, genera, or orders) among breeding seasons [54–56], and with temperature and rainfall varying annually among forest categories, forest and year effects on occupancy are often common [4, 57].

Our results indicate that precipitation, temperature, and relative moisture were the most important sampling covariates for detection probabilities of *Q. verrucospinosa*. The importance of these environmental factors on amphibian breeding activity [58, 59], capture proportions [60], and calling of anuran species [61], and hence detection probability, has been well documented. In many cases, precipitation and temperature are expected to be good predictors of detectability for amphibian species. Although we did not analyze interactions between temperature, moisture, and precipitation on detectability in the present study, these variables often interact to affect amphibian timing [56, 62] and movement physiology [63]. A recent study indicated that detectability of anuran species is independently positively associated with temperature and precipitation, with temperature consistently having a greater effect [4]. The survey variable is also an important relative covariate for detection probabilities in this study, and detectability varied within the seven survey occasions and possibly among sites with different disturbance histories.

Missing observations are a common case in the model of the presence or absence of a species from a collection of sampling sites, and occur widely applied in wildlife and ecological studies [3, 4]. Missing observations may arise through a number of reasons, such as a vehicle or equipment breakdown or logistic difficulties in getting field personnel to all sites. Therefore, it may not be possible to survey all sites at all sampling occasions. These sampling inconsistencies can be accommodated using the proposed model likelihood. If a site is not surveyed at the jth survey occasion, no information regarding the detection (or non-detection) of this species has been collected from that site at that time. Our observed data (77 sites with the seven surveys were conducted), including 17 missing observations, the percentage of missing data for the probability of *Q. verrucospinosa* presence was 3.15%. According to MacKenzie et al. (2002) on average, the standard error of ψ increased about 5% with 10% missing observations, and about 11% with 20% missing observations. Thus, our occupancy estimates and the bootstrap standard error estimates in the present study were reliable and accounted well for the loss of information for *Q. verrucospinosa*.

A common method of estimating over-dispersion is to use the observed chi-squared goodness of fit statistic for a global model (the most complex model with the greatest number of parameters), which should be estimated for lack of fit first [6]. According to previous studies estimating the occupancy and detection probability, it should be demonstrated that a fitted model adequately describes the observed data [64, 65]. Substantial lack of fit in the model may lead to inaccurate inferences, either in terms of bias or in terms of precision (e.g., reported standard errors are too small; [5]). Our global model does not indicate any evidence of lack of fit using 10,000 bootstrap samples, with an estimated over-dispersion parameter $\hat{c}$ of 0.436. However, comparing our results with those assessing the fit of site-occupancy models given in MacKenzie and Bailey [2] and MacKenzie et al. (2006) showed that our models are suitable and that there is insufficient evidence of a poor model fit.

Amphibian species must choose terrestrial microhabitats that prevent loss of excessive water corresponding to each season and, thus, maintain hydration [66]. Moist environments are very important to the survival of juveniles [56, 67]. In this study, we detected frogs using all three habitat types during seven surveys in the main rainy season. In there, using terrestrial, aquatic, and arboreal habitats were 56%, 38%, and 6%, respectively. Although our sampling sites tended to remain moist throughout the sample period in the primary and secondary forests of Bach Ma National Park, there is evidence that in the summer (April to July), the rate of microhabitat use in the semi-aquatic model (about 60%) is larger than the terrestrial model (about 35%, B.V. Ngo, unpubl. data). This can explain why temperature, precipitation, and relative humidity were associated with detectability of frogs. We speculate that an overall moist forest environment (temperature and rainfall) coupled with species-specific behavioral adaptations (e.g., [56, 67]) allowed frogs to remain equally active across the range of precipitation events in our sampling period.

5. Conclusions

Based on the detection/non-detection data for each site over multiple visits for granular spiny frogs, from the best model among all candidate models, we estimated a site occupancy rate of 0.632 that was higher than the naïve occupancy estimate of 0.403 and a 57% increase over the rate of sites at which frogs were actually observed. The site variable of primary forest was an important determinant of site occupancy, whereas occupancy was not associated with the variable of secondary forest. The species detection model p(temperature, humidity, rainfall) included 90.9% of the total weight, providing clear evidence that environmental conditions were important sample covariates in modeling detection probabilities.

Acknowledgements

We thank the staff of A Luoi and A Pat Forest Ranger Stations, Bach Ma National Park, Border Stations 629 and 633, Phong Dien Nature Reserve, and Sao La Conservation Area, for support of our feldwork. We thank colleagues from Department of Life Sciences, National Cheng Kung University, Taiwan, for their support. The Ministry of Education and Training funded this study under grant number B2020-DHH-08. The authors also acknowledge the partial support of Hue University under the Core Research Program, grant number NCM.DHH.2018.10.

Appendix 1. Predictive details of site occupancy estimation of *Quasipaa verrucospinosa* living in Bach Ma National Park, Thua Thien-Hue Province, central Vietnam

Site or Patch	Survey Occasion			
	I	II	III	IV
Site 1	1	0	0	1
Site 2	0	0	0	0
Site 3	1	0	θ	0
Site 4	θ	0	θ	1
… …	—	—	—	—
Site 30	0	0	0	0

Author details

Binh V. Ngo[1]* and Ya-Fu Lee[2]

1 Department of Biology, University of Education, Hue University, Hue, Vietnam

2 Department of Life Sciences, National Cheng Kung University, Tainan, Taiwan

*Address all correspondence to: nvbinhsp@hueuni.edu.vn

References

[1] Bailey, L. L., J. E. Hines, J. D. Nichols, and D. I. MacKenzie. 2007. Sampling design trade-offs in occupancy studies with imperfect detection: Examples and software. Ecological Applications 17: 281-290.

[2] Bailey, L. L., T. R. Simons, and K. H. Pollock. 2004. Estimating site occupancy and species detection probability parameters for terrestrial salamanders. Ecol. Appl. 14:692-702.

[3] MacKenzie, D. I., J. D. Nichols, B. L. Gideon, S. Droege, J. A. Royle, and C. A. Langtimm. 2002. Estimating site occupancy rates when detection probabilities are less than one. Ecology 83:2248-2255.

[4] Roloff, G. J., T. E. Grazia, K. F. Millenbah, and A. J. Kroll. 2011. Factors associated with amphibian detection and occupancy in southern Michigan forests. Journal of Herpetology 45:15-22.

[5] MacKenzie, D. I. and L. L. Bailey. 2004. Assessing the fit of site-occupancy models. Journal of Agricultural, Biological, and Environmental Statistics 9:300-318.

[6] MacKenzie, D. I., J. D. Nichols, J. A. Royle, K. H. Pollock, L. L. Bailey, and J. E. Hines. 2006. Occupancy estimation and modeling: Inferring patterns and dynamics of species occurrence. Academic Press, New York, USA.

[7] Hanski, I. 1992. Inferences from ecological incidence functions. Am. Nat. 139:657-662.

[8] Scott, J. M., P. J. Heglund, M. L. Morrison, J. B. Haufler, M. G. Raphael, W. A. Wall, and F. B. Samson. 2002. Predicting Species Occurrence: Issues of Accuracy and Scale. Island Press, Washington, USA.

[9] Thompson, W. L., G. C. White, and C. Gowan. 1998. Monitoring Vertebrate Populations. Academic Press, San Diego, USA.

[10] Kaiser, K. 2008. Evaluation of a long-term amphibian monitoring protocol in central America. Journal of Herpetology 42:104-110.

[11] Seber, G. A. F. 1982. The Estimation of Animal Abundance and Related Parameters. Macmillan, New York, USA.

[12] Williams, B. K., J. D. Nichols, and M. J. Conroy. 2002. Analysis and Management of Animal Populations. Academic Press, San Diego, California, USA.

[13] Bisrat, S. A., M. A. White, K. H. Beard, and D. Richard Cutler. 2012. Predicting the distribution potential of an invasive frog using remotely sensed data in Hawaii. Diversity and Distributions 18:648-660.

[14] Carey, C., W. R. Heyer, J. Wilkinson, R. A. Alford, J. W. Arntzen, T. Halliday, L. Hungerford, K. R. Lips, E. M. Middleton, S. A. Orchard, and A. S. Rand. 2001. Amphibian declines and environmental change: Use of remote-sensing data to identify environmental correlates. Conservation Biology 15: 903-913.

[15] Shive, J. P., D. S. Pilliod, and C. R. Peterson. 2010. Hyperspectral analysis of Columbia spotted frog habitat. The Journal of Wildlife Management 74: 1387-1394.

[16] Gu, W. and R. K. Swihart. 2004. Absent or undetected? Effects of non-detection of species occurrence on wildlife-habitat models. Biological Conservation 116:195-203.

[17] Tyre, A. J., B. Tenhumberg, S. A. Field, D. Niejalke, K. Parris, and H. P. Possingham. 2003. Improving precision and reducing bias in biological surveys:

Estimating false-negative error rates. Ecological Applications 13:1790-1801.

[18] Weir, L. A., J. A. Royle, P. Nanjappa, and R. E. Jung. 2005. Modeling anuran detection and site occupancy on north American amphibian monitoring program (NAAMP) routes in Maryland. Journal of Herpetology 39:627-639.

[19] Dorazio, R. M., J. A. Royle, B. Söderström, and A. Glimskär. 2006. Estimating species richness and accumulation by modeling species occurrence and detectability. Ecology 87:842-854.

[20] MacKenzie, D. I., L. L. Bailey, and J. D. Nichols. 2004. Investigating species co-occurrence patterns when species are detected imperfectly. Journal of Animal Ecology 73:546-555.

[21] MacKenzie, D. I., J. D. Nichols, J. E. Hines, M. G. Knutson, and A. B. Franklin. 2003. Estimating site occupancy, colonization, and local extinction when a species is detected imperfectly. Ecology 84:2200-2207.

[22] Nichols, J. D., J. E. Hines, D. I. Mackenzie, M. E. Seamans, and R. J. Gutiérrez. 2007. Occupancy estimation and modeling with multiple states and state uncertainty. Ecology 88:1395-1400.

[23] Nguyen, V. C., H. Ngo, T. L. Le, and N. T. Phan. 2013. Thua Thien-Hue Monographs. Social Sciences Publishing House, Hanoi, Vietnam.

[24] IUCN 2021. The IUCN Red List of Threatened Species. Version 2021-1. h ttps://www.iucnredlist.org.

[25] Lehtinen, R. M. and S. M. Galatowitsch. 2001. Colonization of restored wetlands by amphibians in Minnesota. American Midland Naturalist 145:388-396.

[26] Weyrauch, S. L. and T. C. Grubb. 2004. Patch and landscape characteristics associated with the distribution of woodland amphibians in an agricultural fragmented landscape: An information-theoretic approach. Biological Conservation 115: 443-450.

[27] Gardner, T. A., M. A. Ribeiro-Junior, J. O. S. Barlow, T. C. S. AVila-Pires, M. S. Hoogmoed, and C. A. Peres. 2007. The value of primary, secondary, and plantation forests for a Neotropical herpetofauna. Conservation Biology 21: 775-787.

[28] Skelly, D. K., L. K. Freidenburg, and J. M. Kiesecker. 2002. Forest canopy and the performance of larval amphibians. Ecology 83:983-992.

[29] Vallan, D. 2002. Effects of anthropogenic environmental changes on amphibian diversity in the rain forests of eastern Madagascar. Journal of Tropical Ecology 18:725-742.

[30] Blaustein, A. R., D. B. Wake, and W. P. Sousa. 1994. Amphibian declines: Judging stability, persistence, and susceptibility of populations to local and global extinctions. Conservation Biology 8:60-71.

[31] Bradford, D. F., F. Tabatabai, and D. M. Graber. 1993. Isolation of remaining populations of the native frog, *Rana muscosa*, by introduced fishes in Sequoia and Kings Canyon National Parks, California. Conservation Biology 7: 882-888.

[32] Marsh, D. M. and P. B. Pearman. 1997. Effects of habitat fragmentation on the abundance of two species of leptodactylid frogs in an Andean montane forest. Conservation Biology 11:1323-1328.

[33] Ngo, B. V., Y.-F. Lee, and C. D. Ngo. 2014. Variation in dietary composition of granular spiny frogs (*Quasipaa verrucospinosa*) in central Vietnam. Herpetological Journal 24:245-253.

[34] Hoang, Q. X., T. N. Hoang, and C. D. Ngo. 2012. Amphibians and Reptiles in Bach Ma National Park. Agricultural Publishing House, Hanoi, Vietnam.

[35] Ngo, B. V., C. D. Ngo, and P.-C. L. Hou. 2013. Reproductive ecology of *Quasipaa verrucospinosa* (Bourret, 1937): Living in the tropical rain forests of central Vietnam. Journal of Herpetology 47:138-147.

[36] Ribeiro-Junior, M. A., T. A. Gardner, and T. C. S. Avila-Pires. 2008. Evaluating the effectiveness of herpetofaunal sampling techniques across a gradient of habitat change in a tropical forest landscape. Journal of Herpetology 42:733-749.

[37] Duellman, W. E. and L. Trueb. 1994. Biology of Amphibians. McGraw-Hill Book, New York.

[38] Lee, Y.-F., Y.-M. Kuo, Y.-H. Lin, W.-C. Chu, H.-H. Wang, and S.-H. Wu. 2006. Composition, diversity, and spatial relationships of anurans following wetland restoration in a managed tropical forest. Zoological Science 23:883-891.

[39] O'Connell, A. F., N. W. Talancy, L. L. Bailey, J. R. Sauer, R. Cook, and A. T. Gilbert. 2006. Estimating site occupancy and detection probability parameters for meso- and large mammals in a coastal ecosystem. Journal of Wildlife Management 70: 1625-1633.

[40] Burnham, K. P. and D. R. Anderson. 2002. Model Selection and Inference: A Practical Information-Theoretic Approach. Springer-Verlag, New York, USA.

[41] Kroll, A. J., K. Risenhoover, T. McBride, E. Beach, B. J. Kernohan, J. Light, and J. Bach. 2008. Factors influencing stream occupancy and detection probability parameters of stream-associated amphibians in commercial forests of Oregon and Washington, USA. Forest Ecology and Management 255:3726-3735.

[42] Olson, G. S., R. G. Anthony, E. D. Forsman, S. H. Ackers, P. J. Loschl, J. A. Reid, et al. 2005. Modeling of site occupancy dynamics for northern spotted owls, with emphasis on the effects of barred owls. J. Wildl. Manage. 69:918-932.

[43] Zar, J. H. 2010. Biostatistical Analysis. Prentice Hall, New Jersey, USA.

[44] Nguyen, S. V., C. T. Ho, and T. Q. Nguyen. 2005. Checklist of Amphibians and Reptiles. Hanoi Agricultural Publishing House, Hanoi, Vietnam.

[45] Nguyen, S. V., C. T. Ho, and T. Q. Nguyen. 2009. Herpetofauna of Vietnam. Edition Chimaira, Frankfurt am Main, Germany.

[46] Porej, D., M. Micacchion, and T. E. Hetherington. 2004. Core terrestrial habitat for conservation of local populations of salamanders and wood frogs in agricultural landscapes. Biological Conservation 120:399-409.

[47] Price, S. J., D. R. Marks, R. W. Howe, J. M. Hanowski, and G. J. Niemi. 2005. The importance of spatial scale for conservation and assessment of anuran populations in coastal wetlands of the western Great Lakes, USA. Landscape Ecology 20:441-454.

[48] Demaynadier, P. G. and M. L. Hunter. 1999. Forest canopy closure and juvenile emigration by pool-breeding amphibians in Maine. Journal of Wildlife Management 63:441-450.

[49] Rothermel, B. B. and R. D. Semlitsch. 2002. An experimental investigation of landscape resistance of forest versus old-field habitats to emigrating juvenile amphibians. Conservation Biology 16:1324-1332.

[50] Hanski, I. 1994. A practical model of metapopulation dynamics. J. Anim. Ecol. 63:151-162.

[51] Moilanen, A. 1999. Patch occupancy models of metapopulation dynamics: Efficient parameter estimation using implicit statistical inference. Ecology 80: 1031-1043.

[52] Skelly, D. K., E. E. Werner, and S. A. Cortwright. 1999. Long-term distributional dynamics of a Michigan amphibian assemblage. Ecology 80: 2326-2337.

[53] Skelly, D. K., K. L. Yurewicz, E. E. Werner, and R. A. Relyea. 2003. Estimating decline and distributional change in amphibians. Conservation Biology 17:744-751.

[54] Gibbs, J. 1998. Distribution of woodland amphibians along a forest fragmentation gradient. Landscape Ecology 13:263-268.

[55] Pechmann, J. H. K., D. E. Scott, R. D. Semlitsch, J. P. Caldwell, L. J. Vitt, and J. W. Gibbons. 1991. Declining amphibian populations: The problem of separating human impacts from natural fluctuations. Science 253:892-895.

[56] Wells, K. D. 2007. The Ecology and Behavior of Amphibians. University of Chicago Press, Chicago, USA.

[57] Semlitsch, R. D., D. E. Scott, J. H. L. Pechmann, and J. W. Gibbons. 1996. Structure and dynamics of an amphibian community: Evidence from a 16-year study of a natural pond. Pages 217-248 *in* M. L. Cody and J. A. Smallwood, editors. Long-Term Studies of Vertebrate Communities. Academic Press, San Diego, USA.

[58] Blaustein, A. R., L. K. Belden, D. H. Olson, D. M. Green, T. L. Root, and J. M. Kiesecker. 2001. Amphibian breeding and climate change. Conservation Biology 15:1804-1809.

[59] Reading, C. J. 2007. Linking global warming to amphibian declines through its effects on female body condition and survivorship. Oecologia 151:125-131.

[60] Demaynadier, P. G. and M. L. Hunter. 1998. Effects of silvicultural edges on the distribution and abundance of amphibians in Maine. Conservation Biology 12:340-352.

[61] Sullivan, B. K. 1992. Sexual selection and calling behavior in the American toad (*Bufo americanus*). Copeia 1992:1-7.

[62] Harding, J. H. 1997. Amphibians and Reptiles of the Great Lakes Region. University of Michigan Press, Ann Arbor, Michigan, USA.

[63] Preest, M. R. and F. H. Pough. 1989. Interaction of temperature and hydration on locomotion of toads. Functional Ecology 3:693-699.

[64] Lebreton, J.-D., K. P. Burnham, J. Clobert, and D. R. Anderson. 1992. Modeling survival and testing biological hypotheses using marked animals: A unified approach with case studies. Ecological Monographs 62:67-118.

[65] McCullagh, P. and J. A. Nelder. 1989. Generalized Linear Models. Chapman and Hall, New York.

[66] Seebacher, F. and R. A. Alford. 2002. Shelter microhabitats determine body temperature and dehydration rates of a terrestrial amphibian (*Bufo marinus*). J. Herpetol. 36:69-75.

[67] Rittenhouse, T. A. G., E. B. Harper, L. R. Rehard, and R. D. Semlitsch. 2008. The role of microhabitats in the dessication and survival of anurans in recently harvested oak-hickory forest. Copeia 2008:807-814.

Chapter 5

Management and Sustainability of Greek Wetlands

Paraskevi E. Mpeza

Abstract

Greece has today about 400 large and small wetlands. Some of them are international importance and some are considered as national importance. Several of them are composite and form wetland mosaics or complexes. The most common wetland types in Greece are: rivers, estuaries, deltas, lagoons, shallow lakes, shallow marine formations, marshes. Their total area is still quite large (210,000 ha) in spite of the heavy losses that occurred during the last two generations Threats of degradation are drainage, dam construction, irrigation networks, alteration in river morphology such as diversion flow, clearing of natural vegetation which alter hydrological regime and affect wetland function. Sustainable agriculture in the hydrological basins of important wetlands should be considered because these systems are threaten most.

Keywords: Wetlands Greece, agriculture, management, legislation

1. Introduction

Greece historic route, is depicted in wetland landscape. Since antiquity detailed description of the reedbeds of lake Copais is included in the botanical work 'Enquiry into plants' (IV:10–12) of Theophrastus from the 4th century BCE. Related flora and fauna are entailed in works on natural science of Theophrastus ('Enquiry into plants', 'On the causes of plants'), Aristotle ('History of animals', 'Parts of animals', "On the generation of animals"), Dioscorides ('On medical matters') [1]. First settlements were established along rivers and around lakes taking advantage of good grazing conditions and naturally irrigated land.

Contemporary history major social economic events, wars, economic crisis, have put a fingerprint in wetland landscape. For thousand year people ignored the diverse function that wetland perform and also put on them the label of malaria Farmers were interest in wetland trying to drain them for crops and to prevent flooding and to abstract irrigation water. Greece socioeconomic history is linked with wetland status After 1920, a rapid increase area loss was observed for example the Greek state implements a large scale land reclamation projects in the plain of Serres town. This coincides with the fact that period 1928 to 1936, was a great refugee problem that followed the Greco-Turkish War in 1922. Another example Lake Karla, Greece, was almost completely drained in 1962 both to protect surrounding farmlands from flooding and to increase agricultural area. The reclamation did not attribute the expected benefits [2]. Loss of wetland functions and values resulted in environmental, social, and economic problems [3]. An estimated loss 63% in original wetland area in Greece has been occurred over the twentieth century.

Large deltaic areas were shrinked with coastal erosion phenomena. These are attributed to hydrology modification due to dam construction [4].

A"first approximation" of National Wetlands Inventory was published in 1994, including a list of 400 wetlands, with inventory data on 271 of them by the Greek Biotope/Wetland Centre (E.K.V.Y) with the contribution of the former Greek Ministry for the Environment, Spatial Planning, and Public Works in 1994. Main focus of the original Inventory was the continental wetlands. Northern Greece (Anatoliki Makedonia, Thraki, Kentriki and Dytiki Makedonia and Thessalia) there are 118 wetlands, or 31% of the total number. The rest of the continental Greece (Ipeiros, Dytiki Ellada, Sterea Ellada, Peloponnisos and Attiki) contains 151 (40%), while 109 (29%) are in the islands and Crete (Ionia Nisia, Nisia Voreio Aigaio, Nisia Notio Aigaio and Kriti). As regards area or length, Northern Greece contains 48% (97479 ha) of the total wetland area and 56% (2389 km) of the total length of linear wetlands (e.g. rivers). The corresponding figures are, for the rest of continental Greece, 48% (97608 ha) and 37% (1588 km), and for the islands and Crete 4% (7530 ha) and 7% (294 km) [5].

However the morphology of Greece is unique with 6000 scattered islands and islets with a wealth of wetlands. Small wetlands which are spreading through the territory have received little attention and are not in the focus of conservation Nevertheless it is argued that play an important role in the maintenance of species biodiversity. Any loss reduces connectivity among species populations. Inadequate legal protection threatens its existence.

In an effort to fill this gap, in 2004 WWF Greece launched the "Conservation of Aegean Island Wetlands" project, striving to document the state of Greek island wetlands, highlight their importance and draw attention to whatever is needed for their preservation. The outcome of the project was the documentation and delineation of 824 natural and artificial wetlands (>0.1 ha) in 76 islands of Greece (100 on 8 Ionian Islands, 520 on 65 Aegean islands and 204 on Crete and 2 satellite Islands) have been documented and delineated. Of them, 602 are natural wetlands and 222 are artificial [6, 7]. Special issue needs to be mentioned is a priority freshwater habitat in Mediterranean the Mediterranean Temporary Pond (MTP) is a priority freshwater habitat type (3170*, NATURA 2000), that is mainly encountered in Mediterranean type arid and semi-arid climates. They are characterized by their ephemeral nature of their wet phase and the absence of any link with permanent aquatic body making them vulnerable to climate change. In Greece the MTP sites which are coincided in 18 Natura sites are concentrated in the southeastern part of the country [8].

Some of wetlands are international importance some are considered as European and national importance. Several of them are composite and form wetland mosaics or complexes. Ten Sites are designated as wetland of international importance as Ramsar sites with a surface area of 163,501 hectares varying size from 5,078 ha in lake Mikri Prespa to 33,687 ha in Messolonghi lagoons Greece the usual case is that the Natura 2000 site is much larger (average size 11,275 Ha) and engulfs the Ramsar site. A characteristic example of this relation between the two is the case of the lakes Kerkini and Mikri Prespa: the Ramsar boundaries are restricted to the water body whereas the relevant proposed Sites of Community Importance are large enough to include large part of the surrounding catchment area as well [6].

The most common wetland types in Greece are: rivers estuaries, deltas, lagoons, shallow lakes, shallow marine formations, marshes, springs, reservoirs. Their total area is still quite large (210,000 ha) in spite of the heavy losses that occurred during the last two generations [9]. Wetlands accordingly to their hydrology patterns are classified as precipitation dominated, groundwater fed and surface water dominated. With the exception of lake Mikri Prespa, which is not connected to a

Figure 1.
Ramsar sites in Greece [10].

major river basin, all the others have been proposed as representative examples of wetlands which play a substantial hydrological, biological or ecological role in the natural functioning of a major river basin or coastal system (**Figure 1**).

Nowadays threats of degradation are drainage, dam construction, irrigation networks, alteration in river morphology such as confinement of river beds, building embarkments, clearing of natural vegetation which alter hydrological regime and affect wetland function. Biodiversity governance is a continuous battle towards sustainable management in Greek wetlands despite the continuous threats.

2. Wetland protection legislation and administration

Environmental protection and legislation in Greece was very limited in the 1960s and 1970s. A basic national law for protected areas declaration is the Act on the Protection of the Environment of 1986 (No 1650) and its amendments by law 3937/2011 (ΦΕΚ 60/Α/31-3-2011). This complements previous legislation and introduces the designation further more classes of protected areas in Greek territory. These are characterized as Absolute Nature Reserve Area, Nature Reserve Area, National Park, Protected Significant Natural Formation and Protected Landscape and Ecodevelopment Area [11].

Local Authorities are called to play an important role in the implementation of the 1986 Act. The power to designate protected areas and to determine the boundaries of such areas may be transferred to the Prefectures (i.e. the representatives of the central government at local level) under article 21.5 of the 1986 Act. In addition, under article 27 of that same Act, any of the powers exercised by the Minister of the Environment under the Act may also be transferred to the Prefects.

Greece has moved one step forward, ratified international regulations for special commitment on the protection of the natural environment derived from international conventions. These conventions cover a) Wetlands of international importance according to the Ramsar Convention, b) World Heritage Sites (UNESCO), c) Biosphere Reserves (UNESCO, Man and Biosphere), d) Specially Protected Areas according to the Barcelona Convention, e) Biogenetic Reserves (Council of Europe) and f) Eurodiploma Sites (Council of Europe).

The turning point to Biodiversity birds is the adoption of Ramsar Convention. This entered into force on 21 December 1975. According to Convention "wetlands are areas of marsh, fen, peatland or water, whether natural or artificial, permanent or temporary, with water that is static or flowing, fresh, brackish or salt, including areas of marine water the depth of which at low tides does not exceed six meters". According to the same Convention, wetlands are also "riparian and coastal zones adjacent to the wetlands, and islands or bodies of water deeper than six meters at low tide lying within the wetlands". The Ramsar sites of Greece cover a total surface of 163,501 ha, and all of them also form part of Natura 2000 network.

Greece has currently 10 wetland complexes (consisting of 59 sites) as Wetlands of International Importance (Ramsar Convention Bureau, 1996) These are Kotychi lagoons, Amvrakikos, Mesolonghi laggons, Axios Loudias and Alakmonas delta, Lakes Volvi and Koronia, Artificail lake Kerkini, Lake Mikri Prespa, Nestos delta and adjoining lagoons, Lake Vistonis, Porto Lagos, Lake Ismaris and adjoining lagoons and Evros delta (Ramsar Sites Information Service). Moreover 100 sites were characterized as wetlands of national importance It However only ten wetlands of Greece are characterized as Ramsar sites International Importance and seven of them are included in Montreux Inventory with bad environmental status [9].

The trans-border Ramsar wetlands of Greece are the Evros delta and the lake Mikri Prespa (75% and 92% respectively belong to Greece). Three other sites, Axios delta, Nestos delta and lake Kerkini have trans-border watersheds. Eight out of the ten Greek Ramsar sites and their surrounding areas are a series of more than one wetlands. From the remaining two, lake Kerkini is a man-made wetland in an area that used to hold a series of marshes. Kotychi is an isolated wetland complex. Both of them are in strategic positions within migration routes for birds and thus are connected to other wetlands.

Designation of Special Protection Areas, according the Birds Directive 79/409/EC and pSCIs proposed Sites of Community Interest, according the Habitats Directive 92/43 are considered as keystones for Natura 2000 sites delineation in Greece. Sites Natura 2000 constitute environmental network of areas with high ecological, biodiversity or esthetic value. Greece includes at its National List 241 Sites of Community Importance (SCI) according to the EU Directive 92/43 and has declared 202 Special Protected Areas (SPA) according to EU Directive 79/409 [12, 13].

With the of Law 2742/99 there has been a shift, towards a more participatory approach in biodiversity governance. It was adopted mainly through the establishment of Management Agencies. Since 1999, 29 management agencies have been established in 94 of the 419 Greek Natura 2000 sites [14]. As a consequence, the majority of Greek Natura 2000 sites do not have a specific governance mechanism for their management. Management agencies should form the basis for periodic nationwide synthesis and reporting of information on the kinds, and outcomes of regulatory actions related to wetlands conservation. Stakeholders and local communities express their preference towards improving participation in decision making in conservation of Natura 2000 sites [15].

With the recently voted environmental law in Greece (Law 4685/2020) which is going to be the central Coordinator for all actions a new scientific, advisory and coordinating Organization, is established regarding protected areas in Greece. The new institution bears the name 'Organization for Natural Environment and Climate Change' and it is under the auspices of the Hellenic ministry of Environment and Energy Consolidation. In parallel this Organization is going to keep the 24 decentralized units, each one corresponding to a protected area [16]. It is argued consolidation of all management bodies to one central agency would improve administration of protected areas. In the other hand decentralized units must keep their dynamic and protect local wetlands with the active involvement of local communities.

People now want restore wetlands as in the case of Lake Karla, and take advantage of their benefits to the landscape [2, 17]. A National Strategy of Wetland Resources restoration, in 1990 was designed as a means for wetland conservation and restoration [18].

Wetlands cannot be understood hydrologically as a site in isolation but as an essential part of catchment system. Processes operating in the higher part of the basin or in the higher recharge zone to an aquifer controlled wetland feed the biotope. On the other coastal morphology and water quality is dependent of wetland function. Water Frame Directive as it has been harmonized in Greek legislation with the law 3199/2003 (ΦΕΚ280Α/09.12.2003) and the Presidential Decree 51/2007 (54/A/08.03.2007), puts the concept of integrated management at the geographic scale of River basins and helps understanding wetland function [19–23].

Wetlands occupy a dominant role in Environmental Education that is becoming increasingly important. They are considered particularly attractive areas for training in Environmental Education, because of their high scientific value and the fact that one can observe more and faster changes in images, sounds and events than in terrestrial ecosystems. It is a laboratory in the nature. Environmental Education Centers (EEC) is an institution with 30 year history in Greece consists of a network dispersed geographically in 53 distinct areas in country. Local wetlands ecosystems, biodiversity, protected areas conservation is on the core of provided environmental education. An outdoor component diversifies it from traditional education teachers, pupils and all the citizens. Despite difficulties arise from unsecure funding, unflexible school schedule, lack of specialized environmental knowledge centers seem to offer multiple benefits to the participants [23]. Operation of the Information Centre in ecosystem area, museum of natural and cultural history, points of observations in suitable locations, serve as environmental interpretation activities.

Contributors to the environmental education in the country are also NGOs (Non-Government Organizations), and Protected Area Management Agencies [24]. NGOs that are actively participate in the environmental protection are Hellenic Ornithological Society, the Greek branch of WWF, Medsos, Mom, Kallisto, Arktouros, Hellenic Society for the Protection of the Environment and the Cultural Heritage, EKBY, Archelon Medies (Mediterranean Education Initiative for Environment and Sustainability), and MIO-ECSDE (Mediterranean Information Office for Environment, Culture and Sustainable Development).

3. Aspects of Greek wetland management

Wetlands are normally not very easy to access so many people do not visit them. Photographers and film makers take snapshot of a beautiful scenery which thrives in watefowl, rare plants and reptiles. These powerful images attract the public but not educate them deeply in wetland science. In many case degradation could have been avoided if there was sufficient knowledge.

Although the high level of endemism, the biodiversity of Mediterranean wetlands is not well-known by the general public, with the possible exception of waterbirds, which gather in amazing numbers at some sites. The deltas of the Evros, Nestos, Acheloos, the double-delta area of the Arachthos and Louros, as well as the Lakes Kerkini, Shkodra and Prespa contain rich bird faunas often with more than 300 bird species per site. Delta Evros as a crossroad in waterfowl corridor is renowned as great value delta [25]. Illegal hunting is an activity related with wild bird fauna in delta regions as Amvrakikos gulf (Barelos, personal communication).

Wetlands constitute an ecological laboratory. A mosaic of phyto communities with varied chorological interest such as 'communities' saline and subsaline soils, freshwater soils periodically inundated and phytocommunities without particular habitat reference and riparian forests [26]. They possess one or more morphological or anatomic adaptions to enhance their ecological tolerance and adopt highly specialized life strategies in the wet dynamic environment. Aquatic bed plants, Nymphea alba, Potamogenon sp., Lemna sp., shrubs as Tamarix, Vitex agnus castus coastal dune such as Juniperus sp., Pancratium maritium, emergent plants as Phragmites austalis, Typha Latifoli, Salicornia and trees such as Populus Alba, Salix alba, Platanus orientalis Pinus pine asp. are the most conspicuous component of Greek wetland ecosystems (**Figure 2**). Some species are endemic and encountered in particular rare priority habitats with a great value for its conservation. As the base of the food chain and support a critical habitat for birds, fishes, macroinvertebrates of all the wetland communities [27, 28].

Wetland area have a great economic value in Greece. Wetland features and agriculture are intimately linked. Agriculture is by far the most important water consumer in Greece (89%).

The Evros, Pinios, Strymon plains as well as the lower Axios and Acheloos, including their main deltas, are fertile landscapes, intensively cultivated and densely populated. Water abstraction for irrigation purposes, construction of dams

Figure 2.
1. *Thickets of* Vitex agnus castus L.*in a natural habitat of community interest 2. Coastal dune with* Juniperus spp *(priority habitat). 3. Coastal dune Pancratium maritium 4.* Populus alba *5. Broadly distributed wetland grass, Phragmites australis 6. Salt tolerant* Tamarix alba. *(photos Mpeza).*

and ditches, fertile soils in the riparian valleys are the causes of an intense agriculture interest [29]. The Evros, Pinios drain the most intensively cultivated basins (53.4–40.6% of the basin). Plains of Serres (Strymon basin), Thessaloniki (lower parts of Axios and Aliakmon basins), Thessaly (Pinios basin) and Arta (lower Arachthos basin) have been designated as Nitrogen Vulnerable Zones [19].

About 60% of the total rice production and 2/3 of the total mussel production (>30 000 tons/year) of Greece occurs in the Axios Delta and estuary. The most important lakes for fisheries are production: 950 tons/year), Trichonis (500 tons/year), declining in recent years, Kerkini (150 tons/year) and Prespa (100 tons/year) [30]. It is worth noting that the former Karla Lake which was fed from Pinios river had an average annual production of 1000 tons. Even if the total quantity of fish caught in the most important lagoon system Amvrakikos has undergone a reduction because of a dramatic decrease in eel production, local fishermen catch great quantities of mullets (40.4%), sea bass (28.4%), eels (38.2%) sea bream catches have increased by 178.6% [31].

Spiritual and cultural values are arisen in wetland area. From antiquity people concentrate around water and stories of water civilization were written. Wetlands hosts archeological sites of major cultural, historic and scientific interests are, ancient theaters, mills, lighthouse (faros) monuments, geological formations, churches depicting history in the centuries from the Neolithic period to post-Byzantine period [32].

Small farmers, fishermen, aquaculturists are living around wetlands. Local communities develop traditional water management practices, use traditional tools and boats, fairy houses, old customs as the bird dance where children mimic the movements of birds was in the Lake Mikri Prespa, reminiscent of the past Greece [33].

A traditional shilt house (called "pelada" in Greek) is unique in the lagoon of Kotychi, Etoliko in Messsolonghi Central Greece, is made from lake plants Small boats with lack of keel, so that they are easily navigates and easily drawn in the land are all connected with civilization in the perimeter of the lagoons. These traditional boats called priaria in Western Greece and "plava" in Northern Greece [34].

Intangible values as seasonal changing color palette. Salicornia's reddish marsh fall color, outstanding Flamingos color in the winter landscape sounds of rustling leaves, bird song, water flow, are emerged from the unique landscape (**Figure 3**).

Wetland functions are seriously impeded from human activities in Greece. Hydrology regime is wetland signature. Are dependent on rainfall, runoff, and seasonal flooding for their water supplies. All Mediterranean wetlands suffer more often in prolonged periods of extreme, high temperatures which lead to diminution of water or total drought. Human induced climate change make signs to wetland function. Lowering of water table and extended shortage rainwater cause stress to vulnerable aquatic ecosystems. Signs of climate change make their appearance in local communities.

Kalodiki calcareous fen is an inland belonging to the western chain of Greek wetlands. It possesses a great ecological value that as it protects 8 species of the Nature Directives and 4 habitat types of the Habitats Directive and one priority habitat 7210* Calcareous fens with Cladium mariscus and species of the Caricion davallianae [35]. Soil moisture, water depth and to a lesser extent dissolved nitrogen determine their 18 vegetation types relative composition [35]. The wetland struggles for its survival as it might be dry much of the year, but that are maintained by repeated seasonal saturation or inundation, require protection even at times when they are completely dry if they are to retain their functions (**Figures 4** and **5**).

Ecological state of the whole ecosystem is strongly dependent on a small dam. Farmers often damage the dam in order to irrigate their fields. The dam is badly maintained and leaks. Phytocommunities respond impressive in water level

Figure 3.
Salicornia sp *colonized mud as sand in National Park of Amvrakikos gulf, NW. Greece. Habitat code 1310 (photo Mpeza).*

Figure 4.
Submerged plant Nymphea alba situated inland wetland of Kalodiki Natura 2000, code GR2120002 (Mpeza).

Height. When water levels exceed 4 m, the Phragmito Magnocaricetea communities disappear, while Potametea communities disappear when the level drops below the soil surface. However, disappearance of emergent plants and their communities due to excessively high water levels would influence bird species by the

Figure 5.
Dry Kalodiki fen at October 2019 (Mpeza).

absence of breeding places [36, 37]. Fen acts as carbon storage and take attention in mitigating climate change.

Coastal wetlands receive the burden of human activities like the flux of massive visitors or suffer from erosion phenomena, climate change, eutrophication and construction of small enterprises and roads or establishment of settlement area. Coastal systems land valuable habitats and contribute to biodiversity. Priority habitats such as type 2250" Salt dunes with *Juniperus spp*." it is encountered in Greek Natura sites such as Elafonisi, Falasairna, Gavdos, Acheronta Starits contributing to landscape integrity. The habitat increases coastal resilience retaining sand and halt erosion phenomena in parallel creating biodiversity areas. The habitat suffers from fragmentation with road construction, car parks and small business as result of intense touristic activity in the coast eradicate crucial habitas [38, 39].

Intense agriculture activity in the perimeter of wetland area is a non –point source for agrochemicals and pesticides inputs to receiving waters [40]. In the period 1995–1996, water samples from Louros estuary revealed a continuous presence of triazines, alachlor and metolachlor and sporadic peaks in May and June for other herbicides as, atrazine, simazine and degradation product desethyl-atrazine (DEA) The inputs of the five major herbicides, atrazine, simazine, alachlor, metolachlor and desethyl-atrazine (DEA) to the Louros River are mainly from tributaries and the agricultural area draining to the river estuary. Atrazine and its degradation product DEA are the most abundant herbicides discharged into Amvrakikos Gulf, followed by metolachlor, simazine and alachlor [41, 42].

Mussels were used in another study to assess possible pesticide pollution impacts in the Amvrakikos Gulf in the period 1992–1996. Around Louros and Arachthos rivers in the flood plain of Arta there is an intensive agriculture activity. Riverine flows are discharged for in the swallow lagoons of the deltaic formation. Scientists used mussels to detect general oxidative stress effects on the health status of mussels. They used special biomarkers of oxidative stress as decreased acetylcholinesterase activities that indicated exposure to organophosphate and carbamate pesticides. Responses of the antioxidant enzyme glutathione peroxidase suggested the presence of contaminants capable of reactive oxygen species generation that could be related to organochlorine pesticide contamination in the area [43, 44].

It has been considerable research on the ability of wetlands in agricultural settings to serve as sinks for fertilizers such as phosphate and nitrate and a limited number of studies show the potential for wetlands to adsorb agricultural pesticides [45].

Research confirms the continuous presence of pesticides in river waters in catchment with intense agriculture activity. As regarding pesticides, the most polluted rivers are the Axios and Aliakmon. S-triazines, amide herbicides and organophosphorous insecticides are the most frequently detected, while organochlorine pesticides as legacy pesticides (banned in Greece in 1972) occur at very low concentrations [30, 45].

Wetlands as a land between terrestrial and open water ecosystems have proven to play a key role in trapping plastic liter, including large items Rivers are an important pathway for plastic litter transport. High flow takes away large items towards the coast while during low flow, plastic waste is stranded on riverbanks During high flow, thinner plastic bags are many times trapped from the overhanging vegetation at the bank of the rivers (**Figure 6**). Furthermore, plastic items on the river route are obstructed at dams [17, 46].

Rural communities, farmers, cattlemen, fishermen throw their wastes into rivers and coastal lagoons. Illegal hunting takes place in areas rich in waterfowl species and an example is Amvrakikos [33]. Agriculture nets, plastic films, pesticide empty bottles, ropes are abandoned in the field from farmers. These items are degraded under sun break into smaller pieces and are carried away from wind and surface run

Figure 6.
Illegal dumping site in river banks in an intensively cultivated area in Natura 2000 site B. "Christmas tree" transport solid waste with river flow C. Old Delta of Kalamas river, NW. Greece D. dumping site in lagoon n delta Kalamas (Ieronimaki).

off into wetland Causes are arisen from inadequate solid waste management scheme and lack of environmental awareness. While a lot of research is conducted in coastal litter, little is known about the accumulation of the plastic litter in the transition zone of wetland. Moreover, little is known about the mechanisms that control the transport of microplastics and their accumulation in on wild life [47].

Many rivers receive untreated effluents from rural communities that are not connected with WWTPs (Water Waste Treatment Plants) and this causes stress to all receiving water bodies. For example, the Aoos basin, most of the catchment remains in a wild, almost untouched state with restricted agriculture, forestry, cattle breeding and some aquaculture. The river receives untreated effluents from five urban settlements (Konitsa, Permet, Argirokastro, Tepelen, Mamalje, Selenica), small-scale industrial areas and by-products of petroleum extraction in the lower section [30].

Nevertheless, wetland function can be beneficial for agriculture also. Wetland ecosystems are characterized with outstanding biodiversity, longer and more complex food chains which may reveal biotic interactions useful for designing pest management strategies. Birds that live in these ecosystems, nearby cultivations eat insects which cause harm to crops. Special mention is made for soil borne pests as they are serious pests eating seeds and seedlings grown in winter. This also diminishes toxic insecticides usage in field and adverse impacts to the ecosystems.

Pollinating insects find water in humid wetland soils especially in arid regions. Crops such cotton, sunflower is adapted to insect pollination, although insects are nuisance for farmers in the area. A unique crop which occurs exclusively around lake Prespa "Phaseolus multifolio" takes advantage of the pollinator's abundance in the neighborhood wetland [48].

Plants of wetland such as riparian woodland (*Platanus orientalis*, *Alnus glutinosa*, *Populus sp.*, *Ulmus minor*, *Fraxinus sp.*, *Salix sp*). Reedbeds (*Phragmites sp*). halophytic, semihalophytic and shrub–scrub vegetation (A*rthrocnemum glaucum*, *Juncus sp.*, *Salicornia sp.*, *Tamarix sp)*. Marshes and wet meadows (L*olium perenne*, *Menta pulegium*, *Plantago major*, *Carex sp.*) are used for nutrient removal and organic contaminants degradation [2].

Alternative new ideas about purification potential of wetlands have been recognizing. Halophytes in conjunction with associated entophytic and rhizosphere bacteria are involved in organic contaminants biodegradation in contaminated water and soil. Species of genus Tamarix and its associated bacteria have been shown to contribute to degradation of bisphenol-A a widespread xenobiotic and endocrine disruptor [49, 50].

Depositional features of deltas such as Lagoons, sand bars, thin land forms towards the sea, barrier islands, are well formed in the geological past. Physical or human interventions alter river's geomorphology, destroy these geoforms. Construction of two high dams inland in river Nestos and diversion of the flow to the east resulted to drying river channels and erosion of coastal landforms. This affects crop's yield in the vicinity of the river and fish catch in lagoons [51].

A positive example of human activities in the conservation of wetlands are found in the wet meadow issue a fringe wetland biotope in the lake Mikri Prespa in the core of the National Park of Prespa. Wet meadow has been a valuable biotope in the ecosystem. Since the mids 1970's a dramatic reduction of wet meadow area and a concomitant expansion of reed beds was observed in the littoral zone. Tall helophytes such as P. australis, T. angustifolia, Schoenoplectus lacustris and related taxa tend to expand on drier substrates excluding less competitive species.

The drivers for this ecological change was water level fluctuations and dynamics of vegetation management. Habitants abandoned their traditional occupations such as livestock grazing, stop use of reeds as an animal feed and as construction material for ceilings, fishing moved to deeper waters from littoral zone and they diverted farming exclusively in beans.

An innovated restoration programme the LIFE Nature project titled "Conservation of Priority Bird Species at Lake Mikri Prespa, Greece was planned and implemented with encouraging results. A combination of adjustment water level and vegetation management with cutting and grazing by water buffalo herds gave rehabilitation to wet meadow biotope. Wet meadows are key habitats for spawning fish, amphibians, and feeding ground for endangered water birds, while they also hold socio-economic importance related to fish populations (carp) and grazing [52].

Wetland restoration and conservation is complicate ecological-socioeconomic projects in Greece. Restoring wetland value for the society goes together with wetland functions which are physical, chemical, and biological processes that are performed in the area and the interconnections in the whole catchment processes [2, 53].

4. Conclusions

Wetlands are ultimately linked from the ancient time with history culture and local economy. Despite the wetlands value there is a long road to the sustainable management and their harmonized relations with human activities. Although they suffer from degradation in the past, they still support endemic and threatened species and encompass priority habitats with a great value of conservation. Main economic activities as agriculture, aquaculture, fishing, livestock are taking place in estuaries in lakes and ponds in deltaic formations, which constitute wetland areas.

A wealth of small wetlands is scattered in the numerous remote islands in the Greek landscape which provide ecosystem services which cannot be ignored. With the law Law 4685/2020 a new institution 'Organization for Natural Environment and Climate Change' which is under the auspices of the Hellenic ministry of Environment and Energy Consolidation is going to be the central Coordinator for all actions in protected areas in Greece.

COVID-pandemic may influence the attitudes and the ideas about development and nature conservation in all over the world. In the new era Mediterranean Wetlands face ecological challenges with diminution and pollution which are intensified in the view of climate crisis. Physical issues as increasing average and maximum temperatures declining total precipitation increasing frequency and intensity of storms and sea level rise under the umbrella of climate crisis and the risks they pose for the survival of Greek wetlands must be foreseen.

New priorities and reevaluation of the National Strategy for the conservation of wetlands must be put a barrier in the Interface between science and policy science and local communities need to be strengthen.

In this direction implementation of agro-environmental (sustainable) management measures in rural areas neighboring wetlands as wise use of agrochemicals, choice of crops requiring fewer inputs. Assimilation of new knowledge about wetlands with systematic monitoring of water quality data, census of bird population trends mapping every plant role in wetland phytocommunity, harmonization with the use artificial intelligence for data processing and drone technology for large scale monitoring of monitoring of trends in bird population, characterize genetic material from biological populations. Novel education wetland projects need to be delineated which embedding wetland curriculum in school education and engage citizen science projects with activities and conservation action in the field.

Author details

Paraskevi E. Mpeza
Department of Agriculture, University of Ioannina, Arta, Greece

*Address all correspondence to: bmpeza@uoi.gr

References

[1] Pim de Klerk & Hans Joosten How ancient cultures perceived mires and wetlands (3000 BCE – 500 CE): an introduction IMCG Bulletin 2019-04 (May-July) pages 4-15.

[2] Zalidis G., Takavakoglou V., Panoras A., Bilas G., Katsavouni S., (2005). Re-Establishing a Sustainable Wetland at Former Lake Karla, Greece, Using Ramsar Restoration Guidelines Environmental Management Vol. 34, No. 6, pp. 875-886.

[3] Zalidis, G., Gerakis A.,(1999). Evaluating sustainability of watershed resources management through wetland functional analysis. Environmental Management 24:193-207.

[4] Krina A., Xystrakis F., Karantininis K., Koutsis N., (2020). Monitoring and projectingLand Use/ Land Cover Changes of Eleven Large Delta Areas in Greece from 1945 on Wards. Remote Sens. 12, 1241; doi:10.3390/rs12081241

[5] Greek Biotope /Wetland Center, the Goulandris Natural History Museum EKBY Web Site / Natura 2000 network.

[6] Panagiota Maragou and Daphne Mantziou WWF-Greece, (2000). Assessment of Greek Ramsar Wetlands, WWF, Greece.

[7] Paragamian, K., Giannakakis, T., Georgiadis, N., Catsadorakis, G., Poursanidis, D. ,Kardamak A., Noidou, M., Vrettou, F., Tziritis, E. and Pediaditi, E., WWF Greece (2017). A5 Island wetlands of Greece. From ignorance to international recognition. 12th Society of Wetland Scientists Europe Chapter Meeting.

[8] Dimitriou E., Karaouzas I., Skoulikidis N., Zacharias I., (2006). Assessing the environmental status of Mediterranean temporary ponds in Greece. Ann. Limnol. Int. J. Lim. 42 (1), 33-41.

[9] Drosiadis V., (2016). Protection and management of Greek Wetland areas. School of Spatial Planning and Development. Faculty of Engineering A.U.Th

[10] Ramsar Sites Information Service http://www.wetlands.org Accessed on 10 May 2021.

[11] Law No. 1650 On the protection for the environment.

[12] EEC, 25/04/1979. Council Directive 79/409/EEC of 2 April 1979 on the conservation of wild birds. Official Journal, L 103, P. 0001eP. 0025.

[13] EEC, 22/07/1992. Council Directive 92/43/EEC of 21 May 1992 on the conservation of natural habitats and of wild fauna and flora. Official Journal, L 206, P. 0007eP. 005.

[14] Natura 2000. Council Directive 92/43/EEC on the conservation of natural habitats and of wild fauna and flora and Directive 2009/147/EC of the European Parliament and of the Council of 30 November 2009 on the conservation of wild birds PRIORITISED ACTION FRAMEWORK (PAF) FOR NATURA 2000 FOR THE EU MULTIANNUAL FINANCING PERIOD 2014-2020 GREECE December, 2014 [http://www.ypeka.gr/ LinkClick.aspx? fileticket=sgwYR 3v9BWU%3D&tabid=539& language=el-GR

[15] Apostolopoulou E, Drakou EG, Pediaditi K., (2012). Participation in the management of Greek Natura 2000 sites: evidence from a cross-level analysis. J Environ Manage 113:308- 318.

[16] Tzatzaki V.M. (2020). Recent Developments in Environmental Law in

Greece: A Commentary. International Journal of Environmental Protection and Policy. Vol. 8, No. 3, 2020, pp. 66-69. doi: 10.11648/j.ijepp.20200803.13.

[17] Andrea, V.; Tampakis, S.; Tsantopoulos, G.; Manolas, E (2014). Environmental problems in protected areas. Management of Environmental. Quality. International Journal, 25, 723-737.

[18] Greece, National Strategy for Wetland Resources. Recovered from national_wetland_policies_-_greece.pdf (ramsar.org).

[19] EEC. Council Directive 91/676/EEC of 12 December concerning the protection of waters against pollution caused by nitrates from agricultural sources: 1991b.

[20] EEC. Council Directive 91/676/EEC of 24 September 1996 Integrated pollution prevention and control: 1996.

[21] Council Directive 91/676 EC concerning the protection of waters against pollution caused by nitrates from agricultural sources.

[22] European Communities, (2007). LIFE and Europe's wetlands Restoring a vital ecosystem

[23] Yanniris, C. (2015). 20+ Years of Environmental Education Centers in Greece: Teachers' Perceptions and Future Challenges. Applied Environmental Education & Communication,14(3), 149 - 166. presented at the 2015/07/03/. Retrieved from http://dx.doi.org/10.1080/15330 15X.2015.1067578.

[24] Valavanidis A., Vlachogianni T., (2015). Ecological Network for Nature Protection and Protected Areas in Greece. Problems of Management, Environmental Conservation and Biodiversity.

[25] Handrinos, G., Athanasiadis, A., and Rigas, Y. 2005. Evros Delta princess of Thrace's nature. ANEE A.E., Oikos LTD, KOAN Press.Sarika, M., Dimopoulos, P., Yannitsaros, A. (2005). Contribution to the knowledge of the wetland flora and vegetation of Amvrakikos Gulf, W Greece Willdenowia, 35(1), 69-85.

[26] Sarika, M., Dimopoulos, P., Yannitsaros, A. (2005). Contribution to the knowledge of the wetland flora and vegetation of Amvrakikos Gulf, W Greece Willdenowia, 35(1), 69-85.

[27] Cronk J.K., Fennessy M.S., (2009). Wetland Plants. Encyclopedia of Inland Waters, 590-598. https://doi.org/10.1016/B978-012370626-3.00060-0

[28] Mediterranean Wetlands Observatory, 2012. Biodiversity – Status and trends of species in Mediterranean wetlands. Thematic collection, issue # 1. Tour du Valat, France. 52 pages. ISB N: 2-910368-58-0.

[29] Stefanidis K., Christopoulou A., Poulos S., Dassenakis E., and Dimitrioiu E., (2020). Nitrogen and Phosphorous. Loads in Greek Rivers: Implications for Management in Compliance with the Water Frame Directive. Water 2020, 12, 1531; doi:10.3390/w12061

[30] Skoulikidis N., Economou A. Gritzalis K., Chapter 11in Rivers of Europe First Edition Elsevier, 2009.

[31] Coastal lagoon management in Greece (Fao.org)

[32] Papayannis T. and Pritchard, D. E. (2011), Culture and Wetlands in the Mediterranean: an Evolving Story, Athens, Med-INA.

[33] Gerakis A., & Kalburtjib K., (1998). Agricultural activities affecting the functions and values of Ramsar wetland

sites of Greece. Agriculture, Ecosystems and Environment 70 119±128.

[34] Siampani M., (2019). Towards a Model for the Development of the Mesolonghi-Aitoliko Sea Lagoon in Greece. Journal of Cultural Heritage, Volume: 4 issue: 2,page(s): 208-228

[35] Site facts sheet for Elos Kalodiki, EUNIS.

[36] Dimopoulos P., Sýkora K., Gilissen C., Wiecherink D., & Georgiadis T. (2005). Vegetation ecology of Kalodiki Fen (NW Greece). Biologia, Bratislava, 60/1: 69—82

[37] Kagalou I., Papastergiadou E., Beza P., Giannouris E., (2006). Assessment of the trophic state of Kalodiki wetland, Western Greece. Fresenius. Environmental Bulletin Volume 15 (2), 136-140.

[38] Kazakis G., Ghosn D., Remondou E., (2010). Visitors Impact Assessment on Coastal Dunes with Juniperus spin Crete and the South Aegean Greece eUNICOAST, LIFE07NAT/GR/000296.

[39] Picchi S., (2008). Management of Natura 2000 habitats. *Coastal dunes with Juniperus spp. European Commission.

[40] Georgiadis, Th. (ed.), 1994. Ecological impacts from the agricultural activities and management in three wetlands of Western Greece (Kalamas and Acheron Delta, Kalodiki Fen). University of Patras, Ministry of National Economy (ENVIREG), Patras, 463 pp. (In Greek)

[41] Albanis T.A., Dimitra G. Hela D.G., (1998.). Pesticide Concentrations in Louros River and Their Fluxes into the Marine Environment. International Journal of Environmental Analytical Chemistry,70:1-4,105-120,DOI: 10.1080/03067319808032608.

[42] Albanis, T. A., Danis, T. G., & Hela, G. H. (1995). Transportation of pesticides in estuaries of Louros and Arachthos rivers (Amvrakikos Gulf, N.Wite fct sheet for Elos Kalodikiite. Greece). The Science of the Total Environment, 171, 85-93. doi:10.1016/0048-9697(95)04667-5

[43] Tsangaris, C., Cotou E., Papathanassiou, E. Nikolaidou A., (2010). Assessment of contaminant impacts in a semi-enclosed estuary (Amvrakikos Gulf, NW Greece): Bioenergetics and biochemical biomarkers in mussels. Environment Monitoring Assessment 161:259-269 DOI 10.1007/s10661-008-0743-2.

[44] Valavanidis A., (2018). Environmental Pollution of Marine and Costal Areas in Greece. Review on marine pollution, monitoring and quality of seawater. WEBsite : www.chem-tox-ecotox.org, January 2018

[45] Barelos, D. Determination of Levels of Pesticide Residues in Surface Waters and Sediments in the Wetlands of Arachthos river-Amvrakikos gulf and the lagoons of Rodia-Tsoukalio–Logarou. Master Thesis, University of Ioannina, Ioannina, Greece, September 2006.

[46] Ieronimaki, M. Tracking of Solid Waste on the Amvrakikos' Coast Line. Assessment of the Transfer of Pollution through the Arachthos River. Master Thesis, University of Thessaly faculty of Health Sciences Veterinary Department, Volos, Greece, 2017.

[47] Gil-Delgado, J.A.; Guijjarro, D.; Cosalvez, R.U.; Lorez-Iborra, G.M.; Ponz, A.; Velasco, A. (2017) Presence of plastic particles in water bird faeces collected in Spanish lakes. Environmental Pollution 2017, 220, 7]–736.

[48] Gerakis, P. 1992. Conservation and management of Greek wetlands:

Proceedings of a Greek wetlands workshop, Thessaloniki, Greece, 17-21 April 1989. IUCN, Gland, Switzerland, 493 pp.

[49] Christofilopoulos, S., (2017), SBR unit with halophytes for bisphenol A treatment. Postgraduate Theses in Community School of Chemical and Environmental Enginnering. Technical University of Crete, Chania.

[50] Datta et al., (2013), Phytoremediation protocols: An overview, Plant-Based Remediation Processes (pp.1-18), Published from Springer-Verlag Berlin Heidelberg

[51] Xeidakis G., Georgoulas A., Kotsovinos N., Delmani P., Varaggouli E., (2010). Environmental Degradation of the Coastal Zone of the West part of Nestos River delta, N. Greece. Bulletin of the Geological Society of Greece, Volume 43, 1074-1084.

[52] Kazoglou Y., Koutseri I., Malakou M., (2004). Conservation Management of Wet Meadows At The Greek Part Of Lake Mikri Prespa. Balwois, Ohrid,25-29 May 2004.

[53] Vougioukaloua M., Akriotisa T., Dimalexis A., (2011) Waterfowl habitat restoration in a seasonal island wetland. Ecological engineering: from concepts to applications. Procedia Environmental Sciences 9 191-195.

Section 3

Marine Protected Areas

Chapter 6

The Role of Marine-Protected Areas as A Life Support for Fishery Communities: Indonesian Perspective

Djumanto, Muhammad Erdi Lazuardi, Imam Musthofa Zainudin and Sukma Ridarwati

Abstract

This chapter examines relevant literature on marine-protected areas (MPAs) development and their benefit to support fishery communities in Indonesia. The MPAs concepts experienced since Indonesia's kingdoms eras, continuing the Dutch Colonial period, the next post-independence, and the period from 2000 to the present. One of the functions of MPAs is as a source of livelihood for fishery communities. The size of MPAs in the year 2000 was around 2.6 million hectares (ha) and significantly increased in 2021 up to 23.3 million ha. The size of MPAs is growing along with forming the Ministry of Marine Affairs and Fisheries. The Indonesian government targets MPAs of 32.5 million ha or 10% of the territorial sea of the archipelago. The involvement of stakeholders in the utilization and management of MPAs ensures the area's sustainability and environmental safety. It improves the welfare of fishermen through the availability of fish resources. Therefore, in the purpose of protecting ecological assets, even though the literature sources were limited, our finding suggests that many MPAs involving local communities and traditional management can become the foundation of the fishery community's livelihood.

Keywords: economic, fisheries, livelihood, marine-protected areas

1. Introduction

Indonesia is located between Australia and Asia as well as between the Indian Ocean and the Pacific Ocean, at 6°08′ North (N)–11°05′ South (S) and 94°45′ East (E)–141°01′ E. As a geographical position, where located between two major continents, it causes the climate in Indonesia influenced by the environment of mainland Asia and Australia. Indonesia is situated between two vast oceans, causing sea breezes from the Pacific and Indian Oceans, carrying a lot of water vapor and causing high rainfall. The direction of the wind changes every half year, along with the Sun's position concerning the equator so that it has two seasons. When the Sun's position is south of the equator, most areas in Indonesia experience the rainy season. But, when the Sun's position is north of the equator, most areas in Indonesia experience a dry season. Indonesian waters are located at the equator so that it has a tropical climate that receives enough sunlight throughout the year. Indonesia is one of 17 mega diversity countries with 4813 fish, 1592 birds, 781 reptiles, 270 amphibians, and 515 mammal species. Besides that, there are 590 hard coral reefs, 202 mangroves, and 15 seagrasses. In addition, aquatic biota consists of 2500 species of mollusks, 2000 crustaceans, and six species of turtles [1].

Indonesia is an archipelagic country with many large and small islands totaling 17,504 islands, of which 7870 islands have been named, and 9634 have not been named. A total of five large islands are owned by Indonesia, including Sumatra (473,481 km^2), Kalimantan (743,330 km^2), Java (128,297 km^2), Sulawesi (180,681 km^2), and West Papua (102,946 km^2). The proportion of the total area of these large islands reaches about 85.8% of the total land area. The coastline is 99,093 km long, and the land area reaches 1,910,931 km^2, so the ratio of the length of the coast to the land area is 51 m/km^2. In addition, the territorial sea reaches 5,800,000 km^2 consisting of 800,000 km^2 territorial sea, 2,300,000 km^2 archipelago seas, and 2,700,000 km^2 Exclusive Economic Zone [2]. Indonesia also has various coastal water habitats, such as estuaries, lagoons, coastal rivers, bays, tides, coastal dunes, mangroves, coastal swamps, wetlands, seagrass beds, raised soils, and coral reefs.

The main ecosystems of coastal areas, including mangroves, seagrasses, and coral reefs, have an essential role in supporting coastal communities' lives, especially for fishermen. These ecosystems can provide food, industrial raw materials, medicines, and ecosystem services. Coastal ecosystems have many functions, such as ecologically, biologically, economically, and sociologically [3]. The ecological process of coastal ecosystems is feeding ground, spawning ground, and nursery grounds of various aquatic organisms. The biological function of coastal ecosystems is to recycle organic matter into nutrients that can be utilized by aquatic biotics, and provide oxygen through photosynthesis and carbohydrates as the primary food source for herbivores. The economic function is food and medicines, industrial raw materials, tourist areas, and environmental services. Sociological function is a place to perform ceremonies or activities related to belief, religion, or worship.

Excellent and healthy coastal ecosystem conditions can provide abundant food, industrial raw materials, and environmental services for the community [4]. Stakeholders can utilize fishery resources in healthy coastal ecosystems to the maximum recommended and sustainable limits. The increasing demand for fulfilling the population's needs causes coastal ecosystems utilization to increase so that the quality of coastal ecosystems decreases. Therefore, it is necessary to establish marine-protected areas (MPAs) to maintain the quality of the coastal ecosystem to remain good. In addition, the establishment of marine-protected areas is expected to provide opportunities for aquatic biotic to grow and reproduce without any disturbance from fishing activities. Aquatic biota breeds in conservation areas are expected to be a source of germplasm and seedlings suppliers in buffer zones and other utilization areas.

Several provinces or regencies in the archipelago have created conservation areas that are passed down from generation to generation and obeyed by the community. Residents in several areas have established regulations for the controlled use of fishery resources. Rules are made in writing or unwritten, which are strictly adhered to by the community. Laws are made by the district representing stakeholders and apply to the general public. People who violate these regulations will be subject to social sanctions and pay fines, determined based on these regulations. Several provinces have developed models for the use and management of fisheries resources that have been in effect since the royal era, for example, "panglima laot" in Province Aceh, "lubuk larangan" in Sumatra, "kelong" in Batam, "mane'e" in North Sulawesi, "sasi" in Maluku and Papua, and "awig-awig" in Lombok [5].

There are many environmentally friendly fisheries resource management models developing among the people in each region. Panglima Laot, which has existed since the twelfth century, is a traditional institution that connects fishermen with the government in Aceh Province. Panglima Laot is an institution that makes provisions for customary sea law that applies to fishermen throughout Aceh Province. The Panglima Laot is tasked with regulating procedures for catching marine fish, resolving fishing disputes, and other tasks related to the sustainability of fishery resources in Aceh Province [6]. Panglima Laot regulates environmentally friendly fishing procedures, prohibits damage to the marine environment, and stipulates abstinence from the sea on certain days. Abstinence from the sea has implications for the sustainability of fishing and the sustainability of fishery resources. The Panglima Laot stipulates a ban on going to see every August 17, Friday, Eid al-Fitr/Adha, December 26, and on the day of khanduri/sea alms.

Another form of local wisdom is the existence of the prohibition pit in West Sumatra. Lubuk Larangan is a part of the river-protected area from fishing activities and is a place for fish to spawn. The prohibition on fishing using all types of fishing gear applies for a certain period to allow fish to grow and reproduce. Communities living along the Subayang River have agreed to maintain and preserve the river area. The agreement is stated in the customary law that applies to the indigenous community of Rantau Kampar Kiri. Lubuk prohibition is opened once a year in the dry season before religious holidays [7]. The public may catch fish using environmentally friendly fishing gear when the pit is open.

The Ministry of Marine Affairs and Fisheries/Kementerian Kelautan dan Perikanan (MMAF/KKP) has determined that the area of marine conservation areas in Indonesia in 2030 is about 10% of the total area of Indonesian waters, which is around 32,500,000 ha. The increase in marine conservation areas was initially prolonged, then experienced a very rapid rise. The size of marine conservation areas in 1945 was less than one million ha, and 50 years later, the area became 2.6 million ha consisting of 24 conservation areas. In 2005, the area of marine conservation areas had increased to 5.5 million ha; then in 2015, the size of marine conservation areas increased drastically to 17.3 million ha consisting of 154 conservation areas. Furthermore, in 2020, marine conservation areas have grown to 23.14 million ha consisting of 196 regions [8]. Indonesian water conservation areas in early 2021 have reached 23.34 million ha or 7.18% of Indonesian waters [2, 9]. The development of marine-protected areas in Indonesia over the last two decades can be seen in **Figure 1**. Studies on the development of marine-protected areas in Indonesia are still scarce. Therefore, this chapter aims to examine relevant literature related to the development of marine-protected areas during the royal period, the period from 1600 to 1945 or the Dutch colonial period, the period from 1945 to 2000, and then the period from 2000 to the present. This chapter is expected to be a reference for stakeholders to manage marine-protected areas.

1.1 The roadmap of fishery resources conservation

The development of natural resource conservation in Indonesia can be grouped into four models or periods according to the conditions developed at that time, each of which has a different character. The four periods of conservation development, namely:

1. The era of the kingdom,

2. The era of Dutch occupation,

3. The era of independence, and

4. The age of Reformation.

During the kingdom, local people have made efforts to preserve natural resources by giving a haunted label to forests or lakes considered haunted and protected. People label the location as a haunted place so as not to get disturbed. The title haunted means that certain areas, such as lakes that are a source of drinking water for the community, have guards, namely ghosts, who can harm anyone who does terrible things to the lake. People believe that a sacred area becomes haunted, causing only handlers or trusted people to guard the area against daring to enter it. The haunted label causes residents not to dare to kill fish, animals, and cut down trees in the area. Someone who dares to violate the prohibition will get harm or disturbance by spirits, demits, or demons who guard the haunted area. An area is made a haunted area because the area has water sources, protected areas, ancient relics, places of pilgrimage, and places of worship. Areas that humans rarely visit cause no disturbance or deforestation so that forest vegetation grows densely, which can store water reserves for residents [10].

Conservation of natural resources that existed during the kingdom began before the sixteenth century. Today, two models are still adhered to by the community: the sacredness of an object and an agreement to preserve common natural resources. The sacredness of certain things does not use written rules but is based on mystical stories circulating in society, oral traditions, or advice from elders. There are mystical stories in certain places where natural resources are a source of livelihood for many people or communities, for example, in Telaga Renjeng, Pandansari Village, Paguyangan Subdistrict, Pekalongan District. In the lake, there is a catfish (*Clarias batrachus*) which is very sacred. The lake is a source of raw water for the surrounding community. The existing aquatic biota becomes a source of germplasm so that people protect it by maintaining its sacredness. The myth that circulates is that the fish is a lake-dwelling creature so that anyone who disturbs the fish or catches it, then he/she and his/her family will experience disaster. The rumor was that if someone fell ill after catching fish in the lake and then became healthy after returning it, so the mystical story was valid. In almost every area, mystical stories are built

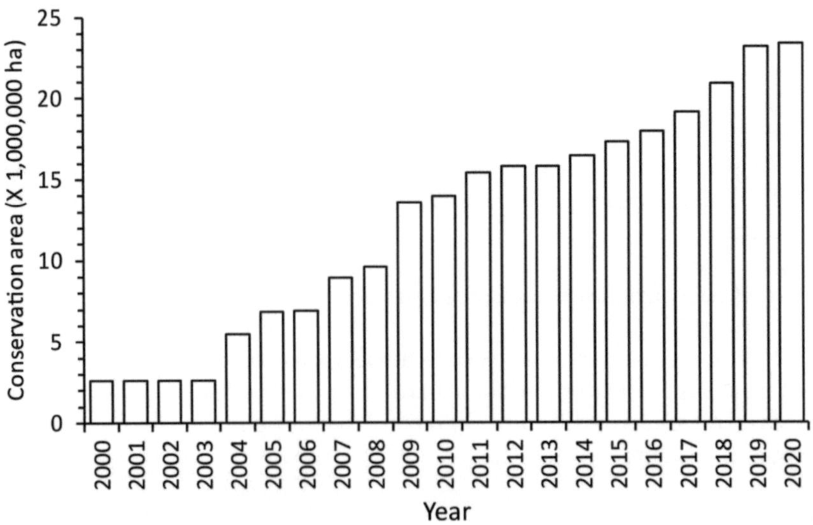

Figure 1.
Development of Indonesian marine and fishery conservation areas in the period 2000–2020.

by the ancestors to protect the community's natural resources so that they remain sustainable for posterity [11].

The next model for conserving natural resources is an agreement between the community and its leaders to regulate the best way to use them. The community entrusts the leadership in managing as well as possible the utilization and preservation of aquatic resources. The leader consists of a group of people who are considered capable and represent various groups. Leaders determine how to use natural resources, harvest amount, harvest time, location, and sanctions for those who do not comply with the rules. Regulations on natural resources have become models of local wisdom, for example, "Panglima Laot" in Aceh and several models of local wisdom in several provinces in the archipelago. The community more embraces conservation by conserving natural resources in specific locations, so this local wisdom model is more successful than other models [6, 7].

During the Dutch occupation, the colonial government cooperated with the indigenous population to exploit the forest. Timber from the forest was used to build office buildings and other infrastructure for Dutch needs, which reached its peak in the early nineteenth century. Uncontrolled forest use causes forest areas to be deforested, which occurs in several locations on the island of Java. For example, the forests on Mount Merbabu, Sumbing, and Sundoro are almost bare. Deforestation causes landslides and flash floods during the rainy season, causing heavy casualties and losses. On the other hand, the water supply is significantly reduced during the dry season, and droughts occur in various places. Deforestation causes farmers to suffer from water and food shortages during the dry season [12].

In the mid-nineteenth century, there were hunting activities that were carried out by the Dutch and and native. Hunting activities were carried out to meet food needs, trade, protect crops, and as a hobby. Uncontrolled hunting of wild animals causes the population to decline very sharply. Wild animals that have high selling value are threatened with extinction in their natural habitat.

Deforestation and wild hunting of animals prompted the Dutch colonial government to find solutions to save animals and plants who are critical populations. In 1909, the colonial government carried out conservation by issuing Staatsblad No. 497 and 594 to protect animals and plants in the Cibodas nature reserve. The law protects and prohibits hunting all animals, except for animals deemed dangerous and disturbing, such as monkeys and orangutans.

Staatsblad No. 497 and 594 exclude orangutans so that populations of orangutans, Javan rhinoceros, and rare birds are threatened with extinction. Furthermore, the colonial government issued Staatsblad No. 134 and 266 of 1931, which prohibits the export of protected animals and their derivatives, and Staatsblad No. 17 of 1932 concerning the establishment of Nature Reserves and Wildlife Sanctuaries. Complete protection of animals and plants is applied to the nature reserve area, while limited use can be carried out in the wildlife reserve area. According to Staatsblad, several wildlife sanctuaries were inaugurated, such as the Baluran conservation forest on the island of Java. Staatsblad remained in effect and was adopted after Indonesia's independence [13].

At the beginning of independence, Indonesia inherited conservation regulations from the Netherlands, valid for approximately 35 years. Conservation of natural resources at that time was in charge of the forestry sector. In 1971, the Ministry of Forestry established the Directorate of Nature Protection and Conservation to manage nature protection activities. In 1973, the Indonesian government ratified the Convention on International Trade in Endangered Species of Wild Flora and Fauna (CITES), then the government ratified CITES with Law no. 43 of 1978. Since independence until the reform era, the government has inaugurated 10 new National Parks, but no single water conservation law has been issued. In 1990, the government issued Law No. 5/1990 concerning the conservation of natural resources and their ecosystems. With the promulgation of this law, 35 years since independence, the Indonesian government issued regulations on natural resource conservation that focus on terrestrial flora and fauna [14].

During the Reformation period, which began in 1998, the government formed a new ministry to manage marine resources through Presidential Decree No. 355/M of 1999 dated October 26, 1999, in the 1999–2004 Cabinet. The government has refined the name of the Department of Marine Exploration (DEL) several times. The name of the marine exploration department was changed to the Department of Marine Affairs and Fisheries (DKP) based on Presidential Decree no. 165 of 2000, dated November 23, 2000. Coastal and marine resources have received greater attention since the establishment of the Department of Marine and Fisheries Exploration in 1999.

After the Ministry of Marine Affairs and Fisheries was formed, attention to the conservation of marine and fishery resources began to emerge. From 2000 to 2020, the government has issued regulations related to marine and fishery resource conservation. The government has made five laws, two government regulations, and regulations of the minister of marine and fisheries as many as 11 numbers. In addition, the government makes regulations in the form of Decrees of the Minister of Marine Affairs and Fisheries with six numbers, Director-General Regulations with two numbers, and Director General Decrees with two numbers. In the last two decades, the Ministry of Marine Affairs and Fisheries has issued more than 28 types of regulations at the ministerial/Director-General level, directly or indirectly related to the conservation of marine and fishery resources [14, 15]. Some of the regulations related to the conservation of fish and aquatic resources are as follows:

1. Law of The Republic of Indonesia No. 31 of 2004 concerning fisheries, which contains provisions on preserving fishery resources in articles 11 to 14.

2. Law of The Republic of Indonesia No. 27 of 2007 regarding the management of coastal areas and small islands. This Law contains the conservation of coastal areas and small islands in articles 28 to 35.

3. Law of The Republic of Indonesia No. 45 of 2009 concerning Amendments to Law No. 31 of 2004 regarding fishery. In this Law, the provisions on conservation contained in the Law Number 31 have not changed.

4. Law of The Republic of Indonesia No. 1 of 2014 concerning Amendments to the Law Number 27 of 2007 concerning management of coastal areas and small islands. In this regulation, the conservation provisions contained in Article 30 have been refined.

5. Law of The Republic of Indonesia No. 32 of 2014 concerning Marine Affairs. The conservation provisions are contained in Article 11 paragraph (1), which reads, "The Unitary State of the Republic of Indonesia has the right to preserve and manage biological wealth on the high seas."

6. Government Regulation No. 60 of 2007 concerning conservation of fish resources. This government regulation contains the conservation of fish resources, which includes 54 articles of conservation provisions.

7. Regulation of the Minister of Marine Affairs and Fisheries No. 13/PERMEN-KP/2014 concerning Marine-Protected Area Network.

8. Regulation of the Minister of Marine Affairs and Fisheries No. 21/PERMEN-KP/2015 concerning partnership for management of aquatic conservation areas.

9. Decree of the Director General of Coastal and Small Islands Marine Affairs No. Kep. 44/KP3K/2012 concerning technical guidelines for evaluation of the effectiveness of management of marine, coastal, and small islands conservation areas.

10. Regulation of the Director General of Coastal and Small Islands Marines No. 02/PER-DJKP3K/2013 concerning technical guidelines for boundaries of coastal conservation areas and small islands (KKP3K).

The government issued various regulations as the basis for determining marine-protected areas (MPAs). The establishment of marine-protected areas to protect and preserve marine and fishery resources, encourage the economy through natural water tourism programs, and social responsibility for the community's welfare. The existence of marine-protected areas is expected to maintain optimal fish stock populations. The Indonesian government targets the area of conservation areas in 2030 to reach 32.5 million ha. In 2021, the area of marine conservation areas in Indonesia will reach 23.9 million ha with a total area of 201 units. The area of marine-protected areas that are operated and used sustainably reaches 9.9 million ha [16].

1.2 Strategy for the management of fishery resource conservation areas

Indonesia's coastal areas have various types of ecosystems that can be managed and developed into productive areas. These various types of ecosystems, such as mangroves, coral reefs, seagrass beds, river estuaries, sandy beaches, can be developed for capture fisheries, aquaculture, tourism, or other purposes that generate community income. In addition, these ecosystems provide various types of resources that can be used directly and indirectly and environmental services. Various types of ecological services are provided by coastal ecosystems; for example, it can protect the coast from natural disturbances from the sea, provide habitat for various types of organisms on land and water, and provide a fresh environment, unique environment, and other environmental services. Coral reef ecosystems, mangroves, and seagrasses are interconnected and vital ecosystems for aquatic biota. The loss or destruction of one of these ecosystems can result in the disruption of other ecosystems. The subsequent impact disrupts the life cycle of marine life. The following describes the ecosystems of coral reefs, mangroves, seagrass beds, and the distribution of fish communities.

The polyp is an individual coral animal shaped like a tube and has a mouth at the top surrounded by tentacles. Coral animals can produce lime or $CaCO_3$, which becomes a reef. Some polyps form colonies that number in the thousands. The type of coral animal affects the shape of the reef and the direction of growth and color. The types of coral animals in Indonesian waters reach more than 590 species, and as many as 195 species are endemic. Coral animals have a symbiotic relationship with zooxanthellae or algae that are capable of photosynthesis. Therefore, coral reefs can live in warm, shallow, and clear waters in the tropics for the photosynthesis of symbiotic algae [17].

Indonesian waters are part of the world's coral triangle, including six countries, namely Indonesia, Malaysia, Philippines, Papua New Guinea, Timor Leste, and Solomon Islands (**Figure 2**). Coral reef ecosystems in the world's coral triangle are

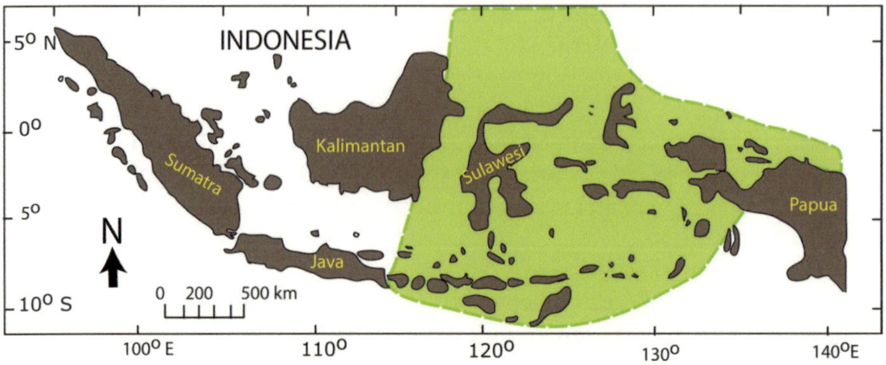

Figure 2.
Map of Indonesia's coral triangle area (green shadow), which is part of the world's coral triangle, has more than 500 coral species and more than 3000 fish species.

the most diverse, with more than 500 species of coral, more than 3000 species of fish, 202 species of mangrove, and 15 species of seagrass. Indonesia's coral reefs reach more than 5 million ha, accounting for 18% of the world's total coral reefs and 65% of the coral reefs in the coral triangle [18]. The coral triangle area becomes a spawning ground, a breeding ground, and a feeding ground for the most economically important fish, such as tuna and skipjack. The position of the coral triangle area is very strategic. It is the center of marine biodiversity, so it is crucial to maintain an excellent ecosystem to produce fish for the welfare of humankind.

Healthy coral reefs are essential so that their functions run well, namely as a habitat for biotic organisms, feeding ground, especially economically important fish, nursery ground, and other uses. Coral reef health is determined based on the percentage of live coral cover. Based on the proportion of coral cover, the category of coral cover is differentiated into four categories, namely excellent (76–100%), good (51–75%), moderate (26–50%), and poor (0–25%). The condition of coral reefs in the excellent category is 6.39%, the good category is 23.4%, the moderate category is 35.06%, and the lousy category is 35.15% [19].

Mangroves are plant communities that tolerate changes in salinity from fresh to very salty and can excrete excess salt due to substrate absorption. Mangroves are found in coastal areas, estuaries, and river estuaries that are connected to the sea. Mangroves can grow in coastal areas that are affected by tides, and the bottom of the water is muddy sandy mud or gravel sand. Mangrove plant communities form a distinctive ecosystem in tidal zones in coastal areas, live in large numbers, have aerial roots or taproots, and bear fruit. The mangrove forest ecosystem benefits directly and indirectly for the surrounding community. Mangrove forest ecosystem can provide many valuable services, such as supporting various ecosystem services, including soil formation, primary production, nutrient cycling. Mangroves as providers of environmental benefits, such as mangroves as a habitat for juvenile fish for consumption or ornamental fish, are spawning grounds and nursery grounds. During photosynthesis, mangroves assimilate CO_2 from the atmosphere to produce carbohydrates as the basis of the food chain. Photosynthetic activity can reduce greenhouse gas emissions so that it can inhibit global warming.

Indonesia's mangrove forest area reaches 3,310,000 ha, which is spread unevenly on the west coast of Sumatra Island, some parts on the north coast of Java Island, along the coast of Kalimantan Island, Sulawesi Island Coast, the southern coast of Papua, and several other small islands. The most extensive mangroves are located on the island of Papua, reaching an area of 1,497,724 ha (45.2%). Then followed by Sumatra Island covering an area of 666,439 ha (20.1%), Kalimantan Island covering an area of 735,887 ha (22.2%), Maluku Island 221,560 ha (6.7%), Sulawesi 118,891 ha (3.6%), Java Island 35,991 ha (1.1%), Bali and Nusa Tenggara 33,508 ha (1.0%). Mangrove forest area has decreased over time caused by natural and anthropogenic factors. Natural factors causing mangrove damage include natural disasters and abrasion, while anthropogenic factors include overexploitation, such as conversion, reclamation, pollution, and waste disposal from urban areas [20].

Seagrasses are flowering plants, monocots rooted in rhizomes, leaves, flowers, and fruits. Seagrass can grow in a shallow marine environment, have high salinity, be permanently submerged, and get enough light. Seagrass requires a substrate for root attachment in the sandy bottom, muddy sand, soft mud, and coral. Seagrasses consist of 2 families, 12 genera, and 48 species [21]. The type of substrate, water depth, and tidal conditions affect the type of seagrass that can grow. The same species that grow in different habitat conditions will have different growth patterns.

Seagrass bed ecosystems are essential ecologically and economically. Seagrass serves as a spawning ground habitat, nursery ground, feeding ground for various aquatic organisms, especially fish, crustaceans, and shellfish, which are economically significant. Seagrasses can be used as food for herbivores, such as rabbitfish and dugongs. In addition, seagrasses can produce carbohydrates through photosynthesis, which is the basis of the food chain. Seagrasses also play an essential role in supporting the life of coral reef and mangrove organisms through interconnection.

Seagrass cover levels were categorized into three parts based on the percentage of seagrass cover, namely high (60–100%), medium (30–59.9%), and low (0–29.9%). Seagrass ecosystems that have a high percentage of cover indicate a healthy ecosystem. The current health level of seagrass based on the cover portion is around 41.79%, indicating a moderate level of health. The health level of seagrass is influenced by various factors, significantly decreasing environmental quality. The decrease in seagrass areas is caused by natural factors and the impact of human activities on the coastal environment. Natural factors affect seagrass areas, such as waves, strong currents, and storms. Meanwhile, human activities that affect the decrease in seagrass areas are beach reclamation, dredging, and sand mining [21].

Indonesia has various species of freshwater fish that inhabit many types of flowing and flooded ecosystems. Past geological events influence the kinds of fish that inhabit each island. The types of fish that inhabit each island are very different, separated by the Weber, Wallace, and Lydekker lines. The distribution of terrestrial flora and fauna, including fish, is separated by an imaginary line called the Wallace line between western and central Indonesia. In contrast, the diversity of fish species between east and central Indonesia is divided by the Weber line (**Figure 3**). Lydekker's imaginary line separates the variety of flora and fauna between eastern Indonesia and flora and fauna of the Australian type. The Wallace line runs between the islands of Kalimantan and Sulawesi and between Bali and Lombok. The Weber line stretches from the north in the Maluku Islands to the south through the Sahul shelf toward the east side of East Nusa Tenggara. The Lydekker line is an imaginary line that follows the contour of the depth between 180 and 200 m at the edge of the Sahul Shelf [22].

Indonesia has marine waters with potential fish resources reaching 12.5 million tons/year [23]. Excessive use of fish resources causes damage to ecosystems in coastal areas. In the era around 1990, extreme fishing used fishing gear that was not environmentally friendly, which caused fish stocks and catches of fishermen to drop drastically. Furthermore, in 1990, the Republic of Indonesia issued Law Number 5 of 1990 concerning the conservation of biological natural resources and their ecosystems. Law Number 5 of 1990 is an effort to prevent damage to the aquatic environment. It aims to regulate the protection of life support systems, and preserve the diversity of plant and animal species and their ecosystems, and the sustainable use of natural resources.

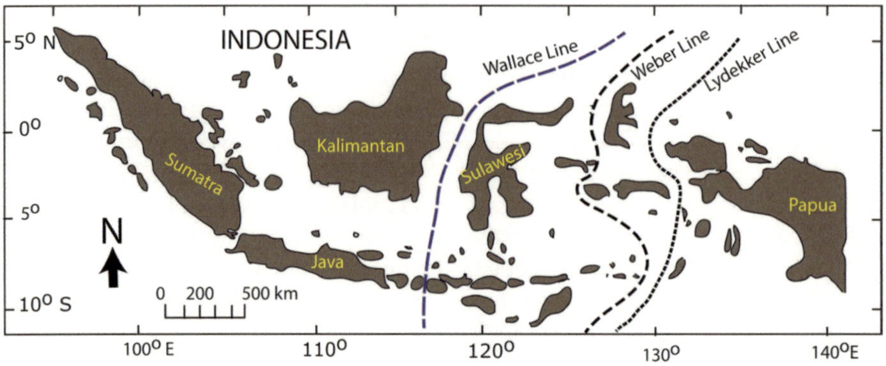

Figure 3.
Weber, Wallace, and Lydekker imaginary lines that separate the flora and fauna that inhabit the islands in western, central, and eastern Indonesia.

As previously explained, Indonesian waters have abundant and diverse fishery resources. Fishery resources can be used to improve the community's welfare and standard of living, provide employment opportunities, and meet domestic protein needs and export needs. However, the exploitation of fish resources must pay attention to sustainability so that the utilization rate of fish biomass must be balanced with the growth rate. Management of fishery resources is an effort to control fisheries business so that it is sustainable. In the last 50 years (1970–2020), the population of Indonesia has doubled, causing fish for consumption and other needs. In the same period, the utilization rate of fish resources has tripled. Overexploitation and unsustainable fishing gear cause most types of fish to be overfished. Capture fisheries production has reached saturation point and tends to decline or fluctuate since the 2000s. In turn, conservation needs to be done to save fishery and marine resources. The Indonesian government has developed a strategy to conserve fishery resources so that marine-protected areas can provide benefits and welfare for the wider community. The established techniques are: (a) strengthening of human resources and fisheries and marine institutions integrated between sectors; (b) maximum and sustainable management of marine and fishery resources; (c) increase the productivity of fishery resources based on research and science; and (d) expanding partnership network, including domestic and international markets as a source of funding.

1.3 Utilization of conservation areas

Marine and freshwater protected areas are protected water areas that are managed with a zoning system as an implementation of sustainable management of fish resources and the environment. The establishment or development of marine-protected areas is one of the efforts to increase aquatic biodiversity and control the sustainable use of fishery resources. Following the provisions of the Government Regulation of the Republic of Indonesia Number 60 of 2007 concerning conservation of fish resources, the determination of marine-protected areas is carried out with the aim of the following: (a) protecting and conserving fish resources and important ecosystem types in the waters to ensure the sustainability of their ecological functions; (b) realizing the use of fish resources and their ecosystems as well as sustainable environmental services; (c) preserving local wisdom in the management of fish resources in and around marine conservation areas; and (d) improve the welfare of the community around the water conservation area.

Protection or conservation areas are intended to protect fishery resources to maximize and sustain their utilization for the community. Therefore, several zones within the conservation area have been defined: (a) core zone, (b) sustainable fishery area, (c) utilization zone, and (d) other zones as needed. Within the core zone of a conservation area, research and education activities can only be carried out on the principle of not disturbing living things in the area. In the sustainable fisheries zone, capture fisheries activities can be carried out that prioritize the protection of habitat conditions for fish resources and the breeding cycle of fish species that prioritize local wisdom. Fish farming is also allowed in this area, considering the carrying capacity and environmental conditions of fish resources for the selection of fish species to be cultivated, feed management, technology, and business scale. In utilizing marine-protected areas, the following activities are carried out (a) fishing; (b) fish farming; (c) marine nature tourism; or (d) research and education [24].

Tourism activities in the utilization zone of marine conservation can be in mangrove ecotourism, marine tourism, or other forms of tourism. In the last few decades, there has been a very significant decrease in the area of mangroves due to anthropogenic activities. Mangroves have many functions for coastal communities, but the total size of mangroves is decreasing. The rate of decline in the area of mangroves reaches around 54,000 ha/year. Stakeholders can make efforts to maintain or increase mangrove areas, including reforestation, restoration, and utilization for ecotourism.

Mangrove ecotourism has been carried out by many people outside the marine conservation area. The Province of the Special Region of Yogyakarta does not have a mangrove area; however, the people of Pasir Mendit Village, Kulonprogo District, independently plant mangrove trees on the side of the Congot Lagoon to make the Congot Beach shady with mangrove plants (**Figure 4A**). The people of the Pasir Mendit Village have turned a sandy beach that was initially arid into an area that is overgrown with lush green mangroves (**Figure 4B** and **C**).

The coastal area of Congot Village eventually became a mangrove tourism spot, which was very crowded with tourists. The Congot mangrove tourism area is exciting to visit because its condition was originally an arid dune. Its condition has now drastically changed to a lush expanse of mangroves. Successful community efforts to plant mangroves on sandy land have encouraged other local communities, such as Baros Village, Bantul District, to carry out similar activities [25].

In marine-protected areas or conservation of fishery resources, tourism activities can be carried out in the utilization zone, while in other zones for transportation, security, and other activities. Managers of marine-protected areas or communities can carry out sustainable tourism activities in other use zones. In the early 2000s, when the economy improved, many people engaged in mass tourism activities, involving large numbers of tourists or groups of people. Mass tourism can have a negative impact in the form of degradation of the natural environment and positively impact the economy of the community visited by tourists. The negative impact caused by mass tourism, for example, is the presence of visitor waste scattered in various places. Furthermore, conservation area managers develop environmentally friendly and sustainable tourism activities that do not harm marine resource conservation areas.

Tourism development in marine-protected areas can provide many positive benefits. The positive benefits of tourism activities include increasing funds for maintaining marine-protected areas and opening new jobs for the surrounding community. Another benefit is to become a place of education and research for students. Marine-protected area managers can obtain development funds by selling visit tickets, area entry permits, and other services. On the other hand, the

Figure 4.
The map of the Congot coast of the special region of Yogyakarta, as a location for planting people's mangroves of Pasir Mendit Village, has finally become a beautiful tourist area (A). Mangroves planted by the community in early 2000 became a shelter for mangrove tourism areas (B). The community independently planted and cared for mangroves (C).

community earns income by selling services, such as tour guides, renting boats, renting diving facilities, selling food, selling souvenirs, and other services. There are excellent benefits for the community, area managers, and other stakeholders, so they must carry out tourism activities in an integrated, controlled, and sustainable manner. There are 183 tourist sites in marine-protected areas (**Figure 5**) whose core business is mangrove tourism [26].

The management of marine-protected areas has built many mangrove tourism areas, and the most widely built is on the island of Java, with a total of 67 locations. The development of mangrove tourism areas on the coast of Java is very profitable because the development capital will soon return. The creation of a mangrove tourism area along the coast of the island of Java brings many benefits to stakeholders. The main benefits for the manager of marine conservation areas are that the mangrove ecosystem is getting better, the population density of mangrove plants is increasing, the mangrove trees are safe or not disturbed by the community, and the managed funds are raised. Communities involved in mangrove tourism get income that can meet their household needs [27].

2. Utilization of conservation areas

MPAs in Indonesia have some beneficial purposes such as fisheries and marine tourism opportunities. It is covered in the use zone, where various activities include

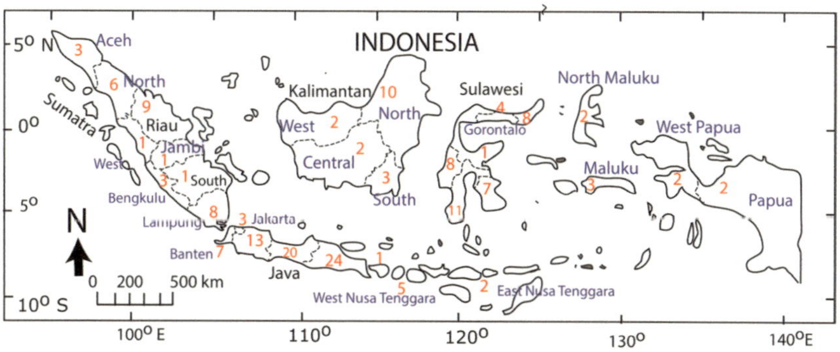

Figure 5.
The number of mangrove tourism areas developed by managers of marine conservation areas in each province (red numbers).

sustainable fisheries and responsible marine tourism. However, the data of fisheries and marine tourism activities within MPAs in Indonesia are limited.

Although the data are limited, unlike FMA, fisheries in specific MPAs are relatively still sustainable. The Ministry of Marine Affairs and Fisheries (MMAF) Republic of Indonesia maintains 35 MPAs priorities as pilot examples for MPAs management effectiveness [28]. The data show that the utilization of potential sustainable fisheries reached 42%. Therefore, 58% of sustainable fisheries commodities can be utilized with eco-friendly fishing gear.

For example, sustainable potency of pelagic fisheries in MPA Laut Sawu (3,355,352.82 ha), a national MPA in East Nusa Tenggara Province, managed under a national government named BKKPN Kupang (Balai Kawasan Konservasi Perairan Nasional/National MPA Office of Kupang) with 3.3 million ha (**Figure 6**). Fisheries stock reaches 156,000 tons/year with utilization of 65,332 tons/year (42%), while demersal fisheries stock reach 84,000 tons/year with utilization 17,779 tons/year

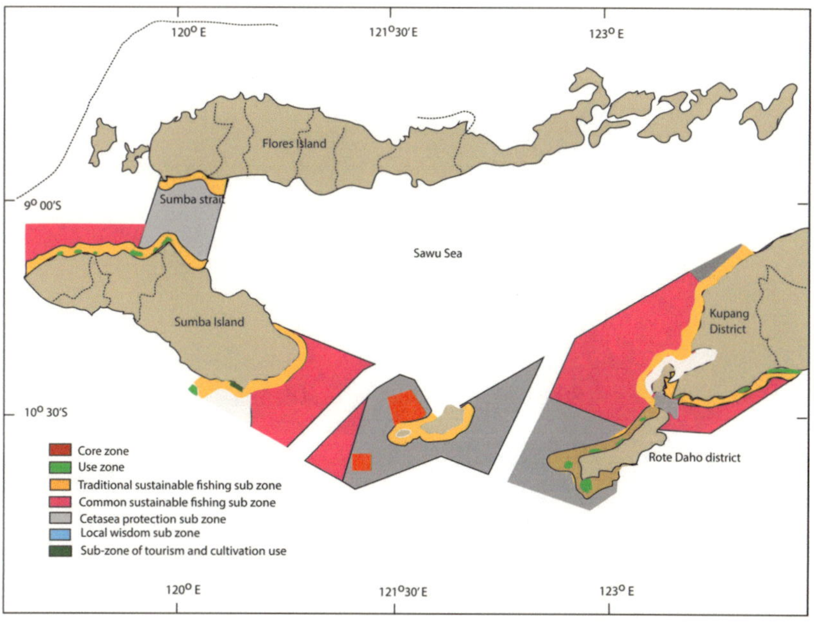

Figure 6.
The map of Laut Sawu national MPA, East Nusa Tenggara Province (Kepmen-KP No. 5/2014).

(21%) [29]. Thus, there is a potential to increase catch per unit effort by strengthening small fisheries groups.

Another example is the Alor District of East Nusa Tenggara Province, where nearly 75% of waters are MPA, named Selat Pantar MPA (276,693.38 ha) (**Figure 7**). It was established in 2015, but the management authority was established in 2019. Fisheries potency in Alor District reaches 45,715 tons/year, where only 18% have been utilized [29].

Although fisheries utilization in many MPAS has lower than sustainable potency, there are some threats in sustainable fisheries, such as destructive fishing by using explosive and poison gears. Most MPAs are managed under the Marine and Fisheries Agency in Provincial Government (22 out of 34 provinces) with no specific management unit entity [30]. As a result, many MPAs were established with a lack of management, so that destructive fishing still exists, for example, in Selat Pantar MPA. In 2014–2017, the percentage of rubble corals increased from 21% to 28%. In contrast, the portion of hard-coral cover was relatively stable, at 37%, but decreased to 27% in 2021. However, fish abundance and biomass showed growth (fish abundance increased from 746 individuals/hectares (ha) in 2014 to 1755 individuals/ha in 2021; fish biomass risen from 381 kg/ha in 2014 to 600 kg/ha in 2021) [31]. An example of a fisherman's profile is presented in **Figure 8**.

Fisheries management can also have a positive impact in a smaller area within MPAs that are managed by the local communities with local wisdom, commonly in Indonesia called "sasi." Sasi is known as traditional regulation for an open and closed season of fisheries utilization.

Anambas Islands MPA (1,262,686 ha) was established in 2014 and managed under a national government named Loka Kawasan Konservasi Perairan Nasional (LKKPN/National MPA Office) Riau Province (**Figure 9**). Apart from formal MPA, community-based conservation management such as Territorial Use Rights in Fisheries (TURF) practices exists in Mesabang Island. It is an example of how

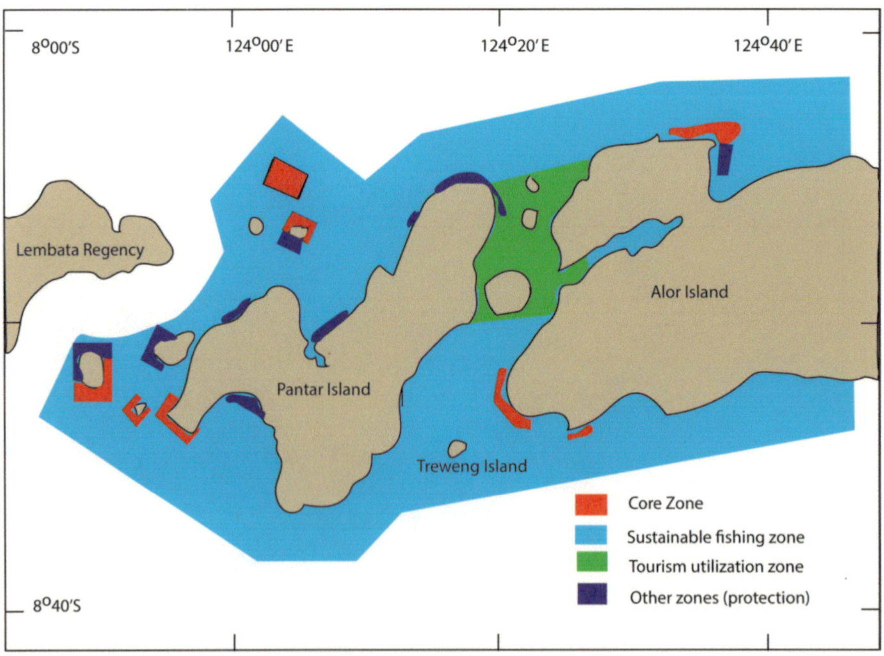

Figure 7.
Map of the Selat Pantar MPA of East Nusa Tenggara Province with its zoning based on the decree of the minister of marine and fisheries No. 35/2015.

Figure 8.
Fisherman's boats in the Selat Pantar MPA, Alor District, East Nusa Tenggara Province catching fish using environmentally friendly fishing gear.

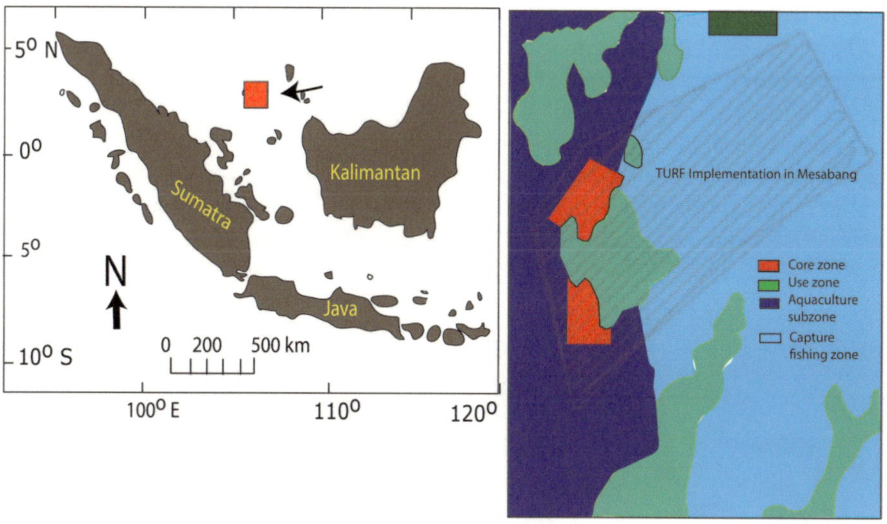

Figure 9.
Map of TURF practice in Mesabang Island within Anambas MPA Islands, Riau Islands province.

legal MPA can be integrated with community-based conservation. The purpose is to increase fisheries sustainability and produce higher-quality catches. Several regulations applied, such as the minimum size of reef fish (body girt should be up to 10 cm) and lift net only operated 6 days a month [32]. Sea surveillance is supported by villagers. As a result, from monitoring 2015–2016, live coral cover increased by 4.5%, mangrove forests cover by 5.8%, and seagrass beds cover by 7% [32].

Selat Pantar MPA was established in 2015, and the management authority under the provincial government is named KCD (Kantor Cabang Dinas/Representative Office of MAF Province Government). Community-based conservation management also exists within Selat Pantar MPA. For example, Baranusa, a traditional kingdom acknowledged by the local government, lived in five Pantar Barat Island villages.

Formal recognition of the marine tenure rights of the Baranusa Kingdom is acknowledged by a local regulation [33]. It is stated that the Alor Regency Government recognizes this traditional tenure and is committed to allocating funds to finance and strengthen traditional institutions.

In response, as part of the MPA design, the tenure scheme has been integrated into the zonation and management of SAP Selat Pantar. Their resources are to be covering reef fishes, coral reefs, clams, snails, and sea cucumbers. The communities manage their marine resources through the "Mulung" system (open-close system). Hading Mulung and Hoba Mulung are a combined system to open or close the fishery to harvesting. Hading Mulung is the closed season, while Hoba Mulung is the open season. Baranusa customary law also supports local MPA regulations such as restricting gear use to traditional fishing gears. During the closed season, fishers usually fish outside the closed areas (**Figure 10**) or focus on seaweed farming [34].

The positive impact based on a community perception study found that the implementation of Hading Mulung and Hoba Mulung increases the fisherman's income and catch (23% strongly agree; 73% agree). In addition, an ecological survey conducted in the Baranusa shows that high-value invertebrate species density increased from 231 individuals/ha in 2015 to 277 individuals/ha in 2017 within the Mulung area (**Figure 11**). Over the same period, outside the Mulung, the invertebrate density decreased from 520 to 100 individuals/ha. The key fisheries species (grouper, snapper, sweetlips) increased from 329 to 507 individuals/ha in Mulung area and from 245 to 460 individuals/ha outside Mulung areas [34].

Therefore, local initiative or local wisdom as community-based conservation management is a good example of how MPA benefits local people. It is also showing that community involvement in MPA management is needed.

In terms of marine tourism opportunity, in East Nusa Tenggara Province (**Figure 11**)—where has 4.8 million ha of land and 20 million ha of waters—the number of tourist visitors varies from 600 thousand in 2017, increased 800 thousand in 2018, then decreased in 2019, and become 570 thousand visitors per year.

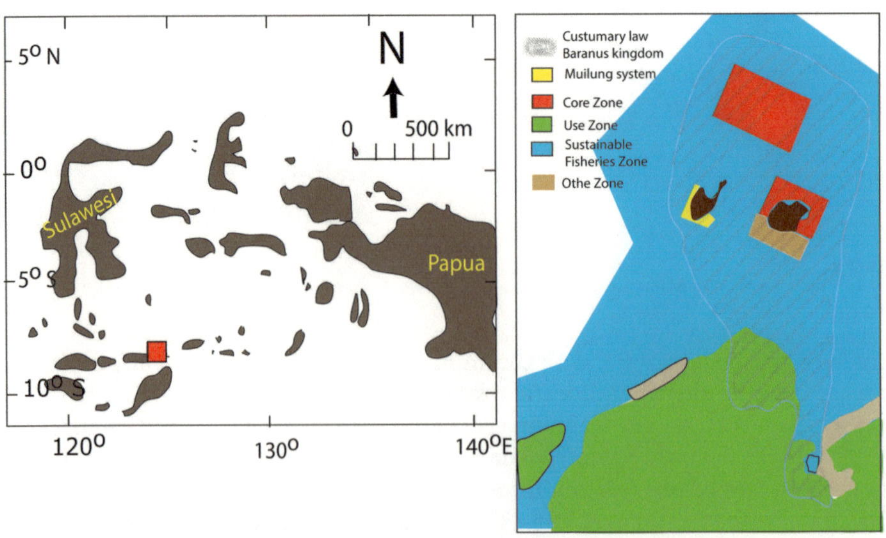

Figure 10.
Map of Baranusa kingdom within Selat Pantar MPA in Alor District, East Nusa Tenggara Province.

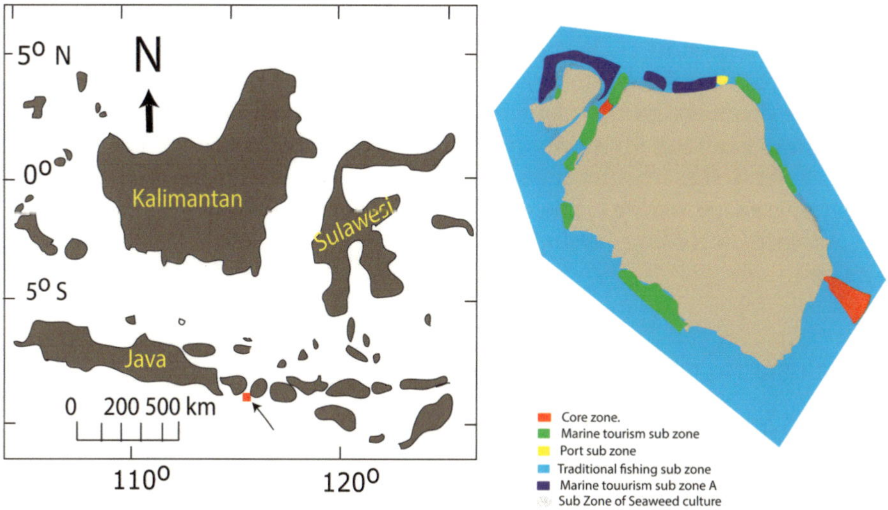

Figure 11.
Map of Nusa Penida MPA, Bali Province with its zoning based on the decree of the minister of marine and fisheries No. 90/2018.

However, it is only 9% of visitors in Bali 2019, whereas the land of Bali has only 12% compared to East Nusa Tenggara Province. Therefore, in many provinces where MPAs exist, marine tourism may have good potential to be developed. Surely, MPAs must be equipped with responsible marine tourism principles such as ecotourism, reducing plastic use, supporting local social, economic, and cultural, and avoiding any negative impacts from its activities in terms of environmental and social effects.

Nusa Penida MPA (20,057—Aquatic Tourism Park) was established in 2014 (renewed in 2018) and is located in Bali Province. The ecosystems and marine life targets for conservation in Nusa Penida include coral reefs, mangrove forests, seagrass beds, manta rays, sunfish, sea turtles, sharks, whales, and dolphins. These natural assets have become attractive destinations for tourists. Marine tourism activities have developed quite fast, with the number of domestic and international visitors to the islands growing almost every year, from nearly 100,000 in 2011 to nearly 300,000 tourists in 2017 [35]. However, as many other places have been facing the COVID-19 pandemic situation, the number of tourists also decreased since mid of 2020.

Moreover, facilities to support the growing tourism industry have expanded, with the number of hotels and accommodations on the islands more than doubling between 2015 and 2017 resembled mass rather eco-tourism. This condition has raised challenges in the management of Nusa Penida MPA, as some of the facilities and recreational activities that support tourism threaten the natural resources, the MPA is intended to conserve [35]. Increasing damage of the seabed was caused by pontoons—moored structures with shower and toilet facilities that accommodate dozens or even hundreds of visitors at the same time, increasing waste and lack of environmental awareness. In addition, the growing number of speedboats and divers at popular sites such as Crystal Bay and Manta Point posed a threat to coral reefs, manta rays, and sunfish. On the other hand, marine patrols to enforce MPA regulations were limited. One of the problems is the change of authority from Klungkung District to the Bali Provincial Government (as regulated in Act No. 23/2014). Therefore, many challenges appeared to manage Nusa Penida MPA even though a regulation such as the marine tourism code of conduct exists for the MPA [35].

However, management authority and Nusa Penida stakeholders are still handed in hand to address those challenges. A carrying capacity study has been conducted and hoped that all stakeholders agreed and regulation can be implemented soon. Sustainable financing also developed from tourist entrance fees to ensure management authority can run MPA activities/programs such as marine patrols, reef health monitoring as well as resource use monitoring. Moreover, enough resources and capacity to manage marine tourism activities are needed for the MPA management authority, including development plans for sustainable marine tourism in MPAs [35].

3. Conclusion and recommendation

Communities in several archipelagic areas have created marine-protected areas, and conservation of inland fishery resources has been passed down from generation to generation since the fifteenth century. Protected areas function as an effort to preserve natural resources. The area of marine-protected areas in Indonesia has overgrown since 2000. The government targets a marine-protected area of 32.5 million ha, and by 2021, it has reached 71.81%. Community involvement in managing marine-protected areas can preserve ecosystems and improve the welfare of the people involved. MPAs give benefits to local people through fisheries and marine tourism livelihood, although the current data or information are limited. In some cases, MPAs also can be integrated with local initiatives or local wisdom as community-based conservation management (named “sasi”) within MPAs. MPAs may still have threats such as destructive fishing activities due to a lack of management capacity within MPA management authorities/units.

It is recommended to involve a broader range of stakeholders in managing marine-protected areas, especially those living within the region. Wider community involvement is expected so that the community gets more significant benefits, the community can protect and save marine-protected areas, and marine-protected areas grow better and can generate income to finance activities.

Acknowledgements

We thank the Ministry of Marine Affairs and Fisheries, the Provincial Governments of Yogyakarta, Riau islands, East Nusa Tenggara, and Bali, the District Government of Kulonprogo, Anambas, Alor, and Klungkung. We also thank Mr. Warso and Sancoko for providing information on the strategy of planting and managing mangroves in sandy coastal areas, from arid conditions to dense green, and has become an attractive place for tourists.

Conflict of interest

The authors declare no conflict of interest.

Author details

Djumanto[1*], Muhammad Erdi Lazuardi[2], Imam Musthofa Zainudin[2] and Sukma Ridarwati[3]

1 Faculty of Agriculture, Department of Fisheries, Gadjah Mada University, Yogyakarta, Indonesia

2 Marine and Fisheries Program, WWF-Indonesia, Denpasar, Indonesia

3 Biology Laboratory, State High School 1, Yogyakarta, Indonesia

*Address all correspondence to: lely4192@yahoo.com

References

[1] BPS-Statistics Indonesia. Statistics of Marine and Coastal Resources 2016. Jakarta, Indonesia: BPS-Statistics Indonesia; 2016. p. 259

[2] Jepson P, Whittaker RJ. Histories of protected areas: Internationalization of conservationist values and their adoption in the Netherlands Indies (Indonesia). Environment and History. 2002;**8**(2):129-172

[3] Cribb R. Conservation in colonial Indonesia. International Journal of Postcolonial Studies. 2007;**9**(1):49-61. DOI: 10.1080/13698010601173817

[4] Boomgaard P. Oriental nature, its friends and its enemies: Conservation of nature in late-colonial Indonesia, 1889-1949. Environment and History. 1999;**5**(3):257-292. DOI: 10.3197/096734099779568245

[5] Persada NPR, Mangunjaya FM, Tobing ISL. Sasi as a culture of natural resource conservation in the Maluku Islands. Jurnal Ilmu dan Budaya. 2018;**41**(59):6869-6900

[6] Puspita M. Local wisdom in coastal and sea resources management laot customary law and the panglima laot institution in Nanggroe Aceh Darussalam. Journal of Cultural Studies. 2017;**3**(2):1-15. DOI: 10.14710/sabda.3.2.%p

[7] Yuliaty C, Priyatna FN. Lubuk Larangan: Local community knowledge dynamics of Inland fisheries resources management in Lima Puluh Kota District. Jurnal Sosial Ekonomi Kelautan dan Perikanan. 2014;**9**(1):115-125. DOI: 10.15578/jsekp.v9i1.1189

[8] Handayani CN, Andradi-Brown DA, Iqbal M, Estradivari, Rusandi A, Hakim A, et al. Status and Trends in Indonesian Protected Area Coverage of Marine Ecosystems. In: Ministry of Marine Affairs and Fisheries, editor. Management of Marine Protected Areas in Indonesia: Status and Challenges. Jakarta, Indonesia: Ministry of Marine Affairs and Fisheries and WWF Indonesia Foundation; 2020. pp. 61-86. DOI: 10.6084/m9.figshare.13341476

[9] Ministry of Environment and Forestry. Statistics of the Directorate General of Natural Resources and Ecosystem Conservation. 2019. p. 117

[10] Dunggio I, Gunawan H. An overview on the history of national park management policy in Indonesia. Jurnal Analisis Kebijakan Kehutanan. 2009;**6**(1):43-56

[11] Prameswari D, Sudarmono. Structure and composition of vegetation in Telaga Ranjeng nature reserve and their implication for conservation. Jurnal Konservasi Hutan Dan Konservasi Alam. 2011;**8**(2):189-196

[12] Kamilia I, Nawiyanto. Deforestation and rise of environmental movement at slope of mount Lamongan, Klakah 1999-2013. Publika Budaya. 2015;**1**(3):72-85

[13] Law of the Republic of Indonesia Number 5 Year 1990 Concerning Conservation of Biological Resources and Its Ecosystems. Jakarta, Indonesia: The Government of Indonesia.

[14] Adi WP. Implementation of CITES (Convention on International Trade in Endangered Species of Wild Fauna and Flora) in handling illegal slow loris trade in Indonesia. Journal of International Relations. 2017;**3**(4):21-31

[15] PSHK B et al. Legal and Institutional Frameworks in the Fisheries Sector in Indonesia. Yayasan Baruna Nusantara. Jakarta, Indonesia; 2019. p. 184

[16] Wiadnya DGR. Stages Towards Planning and Managing Marine Conservation Areas (KKP) in Indonesia. Focus Group Discussion dated July 20, 2011, at DKP Pasuruan. Malang, Indonesia: FPIK-UB Malang; 2011. p. 12

[17] Tuhumena JR, Kusen JD, Paruntu CP. Structure of coral community and association of biota in coral reef area in Minanga Village, Malalayang Ii District And Mokupa Village. Tombariri District. 2017;**1**(3): 6-12. DOI: 10.35800/jplt.1.3.2013.2842

[18] Edrus IN, Utama RS, et al. Development of reef fish community structure in coral waters Waktobi Islands National Park. Jurnal Penelitian Perikanan Indonesia. 2021;**27**(1):43-55

[19] Ardiansyah EF, Hartoni, Litasari L. Condition of hard coral cover and soft coral reef on Pramuka Island, Thousand Islands Administrative District, DKI Jakarta. Jurnal Maspari. 2013;**5**(2):111-118

[20] Khairunnisa C, Thamrin E, Prayogo H. Species diversity of mangrove vegetation in Dusun Besar Village Pulau Maya District, Kayong Utara District. Jurnal Hutan Lestari (2020). 2020;**8**(2):325-336

[21] Ansal MH, Priosambodo D, Litaay M, Salam MA. The structure of the seagrass padang community in the waters of the Waisai Islands, Raja Ampat District, West Papua. Jurnal Ilmu Alam dan Lingkungan. 2015;**8**(15):29-37

[22] Beron P. The Arachnogeography and the "lines" (of Wallace, Lydekker, Weber). Historia Naturalis Bulgarica. 2015;**22**:5-30

[23] Decree of the Minister of Marine Affairs and Fisheries of the Republic of Indonesia Number 50/Kepmen-Kp/2017. Concerning Estimation of Potential, Allowed Catch, and Level of Use of Fish Resources in the Fisheries Management Area of the Republic of Indonesia. Jakarta, Indonesia: The Minister of Marine Affairs and Fisheries of the Republic of Indonesia.

[24] Government Regulation of the Republic of Indonesia No 60 Year 2007 Concerning Conservation of Fish Resources. Jakarta, Indonesia: The Government of the Republic of Indonesia

[25] Djumanto. Conservation Management of Planted Mangroves through Evaluating Ecosystem Services in Baros Village Bantul Regency Mangrove Ecosystem Restoration, Sahadev Sharma. Indonesia: IntechOpen; 2020. DOI: 10.5772/intechopen.93780. Available from: https://www.intechopen.com/online-first/73463

[26] Muchlisi. Potential Development of Mangrove Ecotourism in Tanjung Batu Village, Derawan Island District, Berau District. Jurnal Manusia & Lingkungan. 2017;**24**(1):23-30. DOI: 10.22146/jml.22939

[27] Fatima SN, Anggoro S, Sulardiono B. Strategy for developing integrated mangrove conservation areas in Bedono Village, Demak. Jurnal Manajemen Sumberdaya Perairan. 2018;**7**(2):215-221. DOI: 10.14710/marj.v7i2.22544

[28] Veron JEN, DeVantier LM, Turak E, Green AL, Kininmonth S, Stafford-Smith M, Peterson N. The Coral Triangle. 2014. Coral Reefs: An Ecosystem in Transition (pp.47-55). Switzerland: Springer Nature; DOI: 10.1007/978-94-007-0114-4_5

[29] Ministry of Marine Affairs and Fisheries Database. Protected Area Data. 2021. Available from: http://kkji.kp3k.kkp.go.id/index.php/basisdata-kawasan-konservasi/details/1/75 [Retrieved: 6 August 2021]

[30] Lazuardi ME, Razak TB, Jack-Kadioglu T, Iqbal M, Rusandi A, Hakim A, et al. Formal marine protected area governance structure. In: Kementerian Kelautan dan Perikanan, editor. Management of Marine Protected Areas in Indonesia: Status and Challenges. Kementerian Kelautan and Perikanan and Yayasan WWF Indonesia: Jakarta, Indonesia; 2020. pp. 3-20. DOI: 10.6084/m9.figshare.13341476

[31] Nurbandika N, Iqbal M, Sukandar H, Leatemia DE, Fauzi N. Coral reef health monitoring for evaluation of management impacts in the Pantar Strait and Surrounding Seas of Marine Nature Reserve (SAP). In: Alor Regency 2021. Jakarta, Indonesia: WWF Indonesia Foundation; 2021

[32] Muawanah U. Case Study 7.C Territorial Use Rights in Fisheries (TURF) Implementation in TWP Anambas Islands, Riau Islands in Muawanah, U., Habibi, A., Blue, ME, Yusuf, M., Andradi-Brown , DA, Krueck, NC, Sjahruddin, FF, Iqbal, M., Rusandi, A., Hakim, A., Sapari, A., & Estradivari. (2020). Fishers and Marine Protected Areas. In: Ministry of Marine Affairs and Fisheries, editor. Management of Marine Protected Areas in Indonesia: Status and Challenges. Jakarta, Indonesia: Ministry of Marine Affairs and Fisheries and WWF Indonesia Foundation; 2020. pp. 173-192. DOI: 10.6084/m9.figshare.13341476

[33] Alor District Regulation. Regional Regulation of Alor Regency Number 4 of 2018 concerning Recognition and Empowerment of Customary Institutions. Indonesia: East Nusa Tenggara; 2018

[34] Lazuardi ME, Muawanah U. Case Study 7.B Marine Tenure Rights of the Baranusa Kingdom in Alor Regency, Nusa Tenggara Timur in Muawanah, U., Habibi, A., Lazuardi, M. E., Yusuf, M., Andradi-Brown, D. A., Krueck, N. C., Sjahruddin, F. F., Iqbal, M., Rusandi, A., Hakim, A., Sapari, A., & Estradivari. (2020). Fisheris and Marine Protected Areas. In: Kementerian Kelautan dan Perikanan, editor. Management of Marine Protected Areas in Indonesia: Status and Challenges. Jakarta, Indonesia: Kementerian Kelautan dan Perikanan and Yayasan WWF Indonesia; 2020. pp. 173-192. DOI: 10.6084/m9.figshare.13341476

[35] Welly M, Sanjaya W, Currier K. Case Study 8.B Development of Marine Tourism in TWP Nusa Penida, Bali *in* Amkieltiela, Yusri, S., Veverka, L., Papilaya RL, Lazuardi, ME, Andradi-Brown, DA, Aminuddin, I., Rusandi, A., Hakim, A., Sarmintohadi, Putra, HB, Sapari, A., & Estradivari. (2020). Tourism and Marine Protected Areas. In: Ministry of Marine Affairs and Fisheries, editor. Management of Marine Protected Areas in Indonesia: Status and Challenges. Jakarta, Indonesia: Ministry of Marine Affairs and Fisheries and WWF Indonesia Foundation; 2020. pp. 195-216. DOI: 10.6084/m9.figshare.13341476

Chapter 7

Measuring Marine Protected Areas' Conservation Effort: A Different Look at Three Deeply-Rooted Illusions

Jean-Eudes Beuret and Anne Cadoret

Abstract

As a major tool for policies to protect biodiversity, the current idea of Marine Protected Areas is based on a triptych (a status, a perimeter, and regulations) that is intended to ensure their effectiveness, with the conservation effort assessed by adding up the classified surface areas. Based on an international comparative analysis using 13 differentiated case studies, we take another look at three founding illusions according to which (a) the MPA status corresponds to protection (b) on the level of the classified perimeter, and (c) founded upon regulations laid down to be respected. Our analysis shows that the status is an activatable capital, whose activation may encounter various obstacles that we have listed; that we should distinguish between two levels and types of protection, active and passive, rather than stick to the classified perimeter; that the lack of specific regulations means nothing with regard to the lack of protection; and that MPAs with a legal arsenal at their disposal use these rules first and foremost as a medium for dialogue with stakeholders, with various aims. This analysis leads us to specify what MPAs actually are, and to suggest new means and indicators to assess the conservation efforts made.

Keywords: Biodiversity, Case Study, Conservation Policies, Governance, Marine Protected Areas, Political Ecology

1. Introduction

Biodiversity's alarming decline across the planet affects marine species as much as it affects species on land. According to the WWF Living Planet Report [1], marine species declined by 39% between 1970 and 2010,[1] especially in tropical zones and the Southern Ocean where sea turtles, large migrating birds and sharks. In light of this, in 2010, the 168 member states of the Conference of the Parties to the Convention on Biological Diversity pledged to attain a ratio of 10% of marine and coastal zones designated as protected areas by 2020.[2] And indeed, while 0.7% of oceans were officially protected in 2000, this had risen to 7,6% by January

[1] These figures are based on measurements covering 3,132 populations of 910 species of mammals, birds, reptiles and fish since 1970.

[2] According to the "Aichi Targets" that are among the 17 Sustainable Development Objectives of the United Nations Agenda by 2030.

2020 [2]: the surface area covered by Marine Protected Areas (MPA) is constantly increasing. By creating MPAs in its territorial and deep-sea waters, France is leading the way, announcing in 2017 that more than 22% of French waters are covered by at least one MPA. Then, in May 2019, it announced that it wished to increase the proportion of its territory classified as marine and terrestrial protected areas to 30% by 2022 (compared with 20% today), a third of which are to be protected as "fully natural". But what is this "surface area one-upmanship" worth, when Féral [3] observes that the increase in MPA surface areas comes at the expense of their normativity? And is an area that has been granted protection actually protected?

The effectiveness of MPAs is considered highly variable [1, 4]. In Ecuador [5], Colombia [6], Italy [7], Brazil [8], the Philippines [9] and elsewhere in the world, based on a review of research conducted on this subject [10], many studies question the effectiveness of the implementation of MPAs and the reality of their effects on conservation.[3] Aichi Target 11 refers to "*effectively and equitably managed, ecologically representative and well-connected systems of protected areas and other effective area-based conservation measures*": in addition to quantitative targets are often-overlooked qualitative criteria. The pursued targets are themselves called into question: are the efforts to achieve these internationally set targets being made to tackle the loss of biodiversity or for economic reasons [11–13], security issues, or to assert territorial control [3, 14, 15], etc.? Case studies offer more nuanced responses, and generally, protected areas do indeed protect biodiversity "*but not exclusively, and not everywhere, insofar as countries' strategies, management approaches and local practices may sometimes limit the scope of this objective*" [16].

Examining the effectiveness of MPAs – and above all, the conditions of their effectiveness – is crucial now that they are proliferating. While ecologists' studies make the connection between MPAs' effectiveness and regulatory [17], the social sciences are focusing on the question of MPAs' legitimacy, understood as "*the ability of a political action, in this case an MPA, to be perceived as right and just by the various people who are involved, interested and/or affected by it*" [18]. Many studies show the determining nature of this legitimacy, in Mayotte [13], Malaysia [19] and Canada [18]. While some studies point out the existence of specific regulations [17], others insist on the conditions of respecting these regulations: in particular, they highlight the inclusion of local and fishing communities as elements that allow for both a better acceptance of MPAs and forms of social control of its uses [10, 20, 21]. This brings us back to challenging the way in which the ocean conservation effort via MPAs is assessed because effective protection only exists under certain conditions, and because classified marine areas are not necessarily protected.

More generally, discrepancies can be observed between the way in which MPAs are considered on the basis of a triptych (perimeter, status and regulations), and the way in which these three elements are experienced. What is an MPA? And in addition to its indications, how are the status, perimeter and regulations that constitute an MPA used in reality? To answer these questions, an international comparative analysis was conducted based on 13 case studies, in 11 countries. It brings us back successively to three founding illusions of MPAs described thus: (a). A status offers protection, the areas under this status being the key indicator used by decision-makers to highlight their conservation efforts: we will return to the long processes of institutionalisation and construction of the social acceptance of MPAs, which are often classified as such but which offer few protection guarantees

[3] Questions relayed in articles with evocative titles in the professional press ("Are MPAs really protected", Le Marin, 2019), general press ("Classified but not sufficiently protected marine areas", Le Monde, 22nd October 2019) and satirical press ("Only the percentage was sufficiently protected", Le Canard Enchaîné, 23rd October 2019).

until these processes have been consolidated; (b). MPAs offer perimeter-wide protection: returning to this idea, we will suggest a different way of assessing what is effectively under protection; (c) Protection relies on regulations established with the aim of being respected: understanding the mechanisms of the use of regulations in MPAs reveals a more complex reality in which the regulation is primarily a medium for dialogue with actors whose contribution is crucial for the conservation effort, and this leads us back to the idea both that the non-respect of the regulation is a flaw and to the idea according to which the conservation effort is proportional to the degree of MPAs' regulatory. We will therefore examine both the basis of MPAs and the way to assess how they contribute to the conservation effort. Having described our topic, the analytical approach and the selection criteria for the MPAs studied, we will then address in turn each of the three illusions that mask nuanced realities for which this analysis offers avenues to explore regarding the qualification and improved effectiveness of MPAs.

2. How are marine protected areas conceived?

More recently than on land, the creation of MPAs began in the 1960s and accelerated in the mid-1970s. Today, MPAs have become one of the key tools in ocean management, used throughout the world to protect species and habitats, maintain the functioning of ecosystems and ensure a sustainable use of marine resources [18].

An MPA is first and foremost a legal status intended to provide it with visibility, prerogatives and stable integration in the local institutional context. However, within a country, or from one country to another, statuses are extremely diverse. The IUCN typology [22] shows a gradient between "full protection" status (nature without humankind), the legacy of a Western school of thought that sets humankind against wild nature [23], and zones for "the sustainable use of natural ecosystems", open for multiple uses, such as the Multi-Use Marine and Coastal Protected Areas in Chile. However, other points of differentiation exist, including either directly via conservation, or via fishery management (the Marine Fisheries Management Area in Cambodia, Extractive Reserves in Brazil). Behind the diversity of statuses is hidden the idea that a status can be considered as protection, an idea present in narratives in which classified surface areas are added up and considered protected. However, according to a global assessment of MPAs in 1995, only 29% of them achieved their objectives [24]. Jameson et al. [25] highlight two causes: their location, with MPAs subject to too many uncontrollable external influences (atmospheric, land- or ocean-based), and their management, limited by weak institutional and community-based capacities as well as inappropriate size with regard to the issues at stake. Furthermore, many MPAs throughout the world are qualified as "paper parks", meaning that they are legally designated but do little for conservation [18]. If this status does not necessarily provide protection, what does it provide, and what role does it play (or not) in a dynamic to devise measures?

An MPA is then a perimeter and regulations. MPAs were initially thought of as a "setting aside" of maritime areas, with the IUCN having first defined them as "*any area of intertidal or sub-tidal terrain, together with its overlying water and associated flora, fauna, historical and cultural features, which has been reserved by law or other effective means to protect part or all of the enclosed environment*" [26]. The law, by means of regulations combined with zoning, is the primary "effective means" of action envisaged. The definition of MPAs evolved in 2008 when the IUCN assimilated them with Protected Areas (PAs) defined as "*a clearly defined geographical space, recognised, dedicated and managed, through legal or other effective means, to achieve the long term conservation of nature with associated ecosystem services and cultural values"* [27]. Although it is no longer a question of "setting aside", the use of legal regulations remains central: it is a constitutive element of MPAs along with the perimeter. With regard to perimeters, although it may be relatively straightforward on land, this geometrical vision of space poses various challenges at sea. Among the specificities linked to the marine character of an MPA identified by Day et al. [22] are the fact that boundaries are difficult to establish, that the protected elements are not always visible, that the scale of marine connectivity between ecosystems and habitats is vast, and that monitoring activities is made more complex by the fact that there are many more access points to a specific area than there are on land. Consequently, what is the reality of a perimeter and how should it be considered? An MPA is defined by its perimeter, and managed with the aim of achieving a higher level of protection than the areas surrounding it [28], but how, in reality, do managers come to terms with these limitations?

3. Material and methods

3.1 Case studies

Moving on from the way in which MPAs are conceived to how they are experienced involves going to observe them, which is what we did for 13 case studies. The case study is an empirical research approach that consists of investigating a trend, an event, a group or a set of individuals selected non-randomly in order to obtain a precise description and an interpretation that exceeds its terms of reference [29]. The aim is to identify patterns with a view to generating theories [30–32]. What both gives this its validity and makes it interesting is studying a trend (here the implementation of an MPA) without dissociating it from its context [29, 33]; attempting to understand how this trend functions through immersion in its constitutive elements [34]; combining several sources of data, encouraging the analysis of different facets of the same trend making it possible to corroborate or extend the analyses by causing new questions to emerge [31, 35]; and making it possible to identify unexpected trends [29]. Here we make a comparative analysis [33] of cases marked by the embedding of units of analysis (the territory, the MPA and its integration in the area, dialogue and confrontation concerning the project or that interact with it, and categories of actors). According to Musca [32], an embedded design makes it possible to share out the risks of closure of data access between several sub-units, which is a considerable advantage for our study. This approach is adapted for studying complex processes in public spaces open to multiple social interactions, as is the case for MPA creation processes.

According to Yin [33], the scientific rigour of the case study is based not on the use of a single type of dataset but on the combined exploitation of quantitative and qualitative data from various sources such as field studies, archive analysis, interviews, observations and so on. This allows certain data to be validated by triangulation [35]. Several sources were therefore mobilised for each case study, first analysing very diverse documents (management plans, maps, national conservation framework policy documents, scientific literature, and press articles describing conflicts), conducted prior to the field study based on records of in-situ observations, semi-directive interviews and sometimes participation in key consultation moments (in 4 cases). In total, 201 interviews were conducted, but their number varied depending on the case study from 6 to 28 (**Table 1**): while problems gaining access to stakeholders sometimes restricted the scope of the study, the number was above all determined in each case by factors such as the decisive or otherwise role of

DOI: http://dx.doi.org/10.5772/intechopen.98933

Name	Country	Size (Km^2)	Classified	Priorities/Conservation	Number of interviews
National Marine Park of Zakynthos(NMPZ)	Greece	134	1999	*Caretta caretta* (loggerhead) sea turtle	6
Port Cros National Park (PNPC): core of park + membership area	France	46/230	1963	Marine fauna, Posidonia meadow, rocky or sandy beds, forest, avifauna	23
Multi-Use Marine and Coastal Protected Area of Isla Grande de Atacama (MCPA-MU-IGA)	Chile	12443	2004	Marine biodiversity, wetlands, fossils, geology	13
Tyre Coast Nature Reserve (TCNR)	Lebanon	116,8	1998	Birds, turtles, underwater archaeological sites	11
Gulf of Mannar Biosphere Reserve/Marine National Park(GMMNP)	India	10500/560	1986	Marine biodiversity: coral, sea grass beds, mangrove	9
Koh Rong Archipelago Marine Fisheries Management Area (MFMA)	Cambodia	405	2016	Coral, fish, mangrove, sea grass beds	11
Taean Haean National Park (THNP)	South Korea	840	1978	Beaches, sand dunes, landscape, 17 endangered species	8
Hauraki GulfMarine Park (HGMP)	New Zealand	13900	2000	Marine biodiversity, fishery resources	19
Moorea Marine Space Management Plan (PGEM)	French Polynesia	50	2002	Coral ecosystems	28
Multi-Use Marine and Coastal Protected Rapa Nui/Motu Motiro Hiva Marine Park	Chile, Rapa Nui	150000/720000	2018/ 2010	Coral reefs, fishery resources	14
Rosario and San Bernardo Corals National Natural Park (RSBCNNP)	Colombia	1200	1974	Coral reefs, marine biodiversity	28
Barra Do Una Sustainable Development Reserve / Litoral Centro Marine Park/Jureia Itatins patchwork of conserv. Units	Brazil	14,8 / 4230 / 844	2013/2008/2006	Atlantic Forest, mangrove, marine biodiversity	11
Saloum Delta Biosphere Reserve/National Park	Senegal	1800/760	1981/1976	Mangrove, marine avifauna	20

Table 1.
Case studies.

national authorities, the heterogeneity of stakeholders, and points of view within each category of actors, the aim being to cover as well as possible the diversity of stakeholders' positions. Although almost a third of the people encountered were MPA managers (officials or elected representatives), our sample also included community and organisation representatives (22%), economic actors (20%), scientists (10%), sector administration officials (9%), and local authority officials or elected representatives (7%).

3.2 Selecting case studies

The case studies were selected based on the extent to which they displayed certain shared characteristics so that they remained comparable, and also to maximise the diversity of the situations observed. In our selection process, we follow Dougherty [30] according to whom the rule of selection should be the contrast: this encourages the revelation of recurrences in the way in which MPAs are implemented, and, here, how their components (status, perimeter and regulations) are seen. The diversity of the case studies should make it possible to see if we find similar processes and mechanisms despite the fact that the stakeholders, issues at stake and contexts (geographical, cultural and socio-economic) radically differ. Yin [31] talks of replications: according to him, in a comparative analysis approach, each case should be selected either to predict similar results (literal replication), or to predict contrasting results for predictable reasons (theoretical replication) [31].

The comparability of case studies depends on shared characteristics. The MPAs we selected were: (a). Subject to high or low but never inexistent anthropic pressure, open to various uses and characterised by objectives to use ecosystems sustainably; (b). Both coastal and marine, given that coastal areas are often strategic spaces in marine species' cycles and that managing land-sea interactions is crucial for MPAs to be effective; (c). Of significant size, with the smallest perimeter measuring 50 km2; (d). In existence for more than five years, in order to be able to observe a trajectory of institutionalisation.[4] Then, to maximise situations, several geographical zones were selected (Europe, Asia, Oceania, Latin America and Africa), while at the same time choosing cases marked by the very variable levels of economic development and social capital of the human groups concerned. We also attempted to maximise the cultural diversity whilst still making sure the areas were comparable. The choice of three sites from the Polynesian arc thus aimed to compare the way in which MPAs contend with similar cultural references but in different ways depending on the countries. With the MPAs covering a variety of statuses, sizes and thematic approaches, these criteria were also taken into consideration: thus, our sample contained MPAs whose size ranged from 50 km2 to 700,000 km2, highly diverse statuses (marine parks, national parks, reserves, patchworks of conservation units, sustainable development reserves, fisheries management areas, etc.), the priorities of which were either directly to protect biodiversity, or to defend and manage a type of fishing in view of protecting both fishery resources and ecosystems. The case studies are presented in the **Table 1**.

The data collected were used to construct a chronological analysis and compiled into case study sheets to support a cross-sectional thematic analysis.

[4] The case of Cambodia is an exception, but although the status was not obtained until 2016, the process began in 2011.

4. Results: What is an MPA? Three founding illusions

4.1 First illusion: classification offers protection

4.1.1 MPA status: an activatable capital

In seven of the case studies, the MPAs are granted management authority and specific regulations that define the "interior" and "exterior" of the MPA. However, in three of the cases studied, the statuses exist and the areas are considered to be Marine Protected Areas but they do not enjoy any specific protection. These MPAs have neither management authority, nor means of action or management plans for various reasons: (a). The studies conducted after classification and the management plan were never completed; (b). Local actors rejected the top-down creation of an exogenous MPA, and it cannot exist without their contributions; (c). The MPA received funding and existence, but funding per project was interrupted and protection measures no longer exist. Lastly, in three other cases, the protection measures and specific management regulations are subject to the mobilisation and agreement of third party bodies over which the MPAs have no hierarchical authority. This may be intended, as in the case of the Hauraki Gulf Marine Park (HGMP), New Zealand: with the exception of five small marine reserves that cover only 0.3% of its perimeter, the policy principle of the HGMP is specifically to try and influence sectorial or territorial administrative acts via a forum, in the idea that what they decide has more impact for the environment than any conservation policy carried out without them. However, it is not always intentional, as is the case for the Saloum Delta in Senegal, where the biosphere reserve has no authority over initiatives (outside its central zone, which is a national park) and depends on satellite initiatives over which it has little control. These third-party bodies may mobilise their support, as in Senegal where they give *de facto* content to the biosphere reserve, or not: in this case, the status offers no protection.

This illustrates the fact that a status does not offer protection *a priori*: it is activatable capital, which can be likened to a specific territorial asset [36, 37]. The status transforms a resource (i.e. a latent potential) into an asset, whose value for a conservation measure depends on the way it is – or is not – mobilised. It is a specific asset in that it is attached to this area and enables it to set itself apart. However, its creation offers no guarantee of its activation.

4.1.2 Thwarted institutionalisation processes: three key factors

Its activation (or otherwise) is the result of the MPA's institutionalisation process, which we will endeavour to reconstitute, and which often faces stumbling blocks of various kinds. Three major obstacles emerge from the case studies.

Funding by project may lead to activation that is either uncompleted or temporary. For example, it is uncompleted in Chile in the Multi-Use Marine and Coastal Protected Area of Isla Grande de Atacama where, after a funding phase of several years, neither the management plan nor the governance body nor the funding mechanism were stabilised. In Lebanon, the Tyre Coast Nature Reserve has management authority but so little funding that its actions depend on projects that are periodically activated. It is also temporary in India, where the Gulf of Mannar Biosphere Reserve was active for six years before its funding was slashed to a tenth, leaving only enough to maintain an office located outside the perimeter: here, the periods of activation and eclipse are of longer duration. The same is true for Senegal where the IUCN was the *de facto* manager of the Saloum Delta Biosphere Reserve before withdrawing because of lack of funding: but in this instance, the conservation initiative was taken on by a diverse group of admittedly uncoordinated actors, some of whom joined in view of the stakes highlighted by the status, which thus contributes to revealing a specific environmental value.

Social non-acceptance is a second factor that hampers the activation of the specific territorial asset that the MPA status represents. In Brazil, the trajectory of the Jureia Itatins patchwork of marine and coastal conservation units provides a distressing illustration. In 1986, a full protection area was established thanks to the action of environmentalists who thus managed to put an end to property development and nuclear projects along the coast. However, its promoters had "forgotten" the long-established presence of a Caiçara population, literally attached to its territory and traditional lifestyle and activities, especially fishing. A conflict resolution emerged in the early 2000s thanks to a "patchwork of conservation units" that allows for the coordinated and contiguous existence of "full protection" areas and Sustainable Development Reserves (RDS) where the Caiçara population would be assisted to adapt its activities to environmental issues. However, while the fully protected reserve is unacceptable for the Caiçaras, the sustainable development reserve is unacceptable for naturalists, for whom the state of Sao Paolo acted as spokesperson by filing a lawsuit against this measure. The patchwork was created in 2006, cancelled in 2009 while the participative drawing up of a plan was underway, re-established several months later, suspended in 2013 by a new lawsuit and reinstated in 2014: to date, none of the patchwork's conservation units has a management plan and the institutionalisation process has stalled.

In other cases, no process of this kind has seen the light of day, as in the example of the Motu Motiro Hiva Marine Park in Chile, an MPA granted a status and recognised as such. Rejected by the Rapa Nui community, which was not consulted despite its considering this marine area as its own, it is theoretically managed by the national fisheries department but has neither initiatives nor a management plan. Nevertheless, a "rebound effect" was observed: although this Marine Park remains inactive, the resulting conflict marked the "Rapa Nui's social conservation boom" according to a community leader: the community appropriated the issue and took responsibility for the creation of an MPA. Admittedly, it is divided as to the form that this should take, but in 2018, obtained the creation of a Rapa Nui Multi-Use Marine and Coastal Protected Area.

Inter-institutional conflicts are the third major deciding factor in non-activation or de-activation and the relegation of a conservation initiative even if it has already been launched. In New Zealand, the Hauraki Gulf Marine Park is struggling to find its place in the institutional landscape. Intended to influence sectorial or territorial policies implemented on the level of the Gulf, it faces challenges from public administrations that deny it any legitimacy. This is the case for the Department of Fisheries, which defends exclusive prerogatives, refusing to consider its actions on a territorial scale while it manages fishery stocks, with a representative stating that the HGMP is "a small town that wants to influence a province". The forum is also beset by divisions between representatives of the Tangata Whenua (Maori people) who want equal representation and elected officials who do not. In the background are conflicts between the Ministry of the Environment and the Fisheries Department,[5] in addition to conflicts about democratic, representative, participative or customary legitimacy. This opposition caused the action to be suspended, but it bounced back to some extent elsewhere, beyond the field of the HGMP: a marine spatial planning mission launched by the HGMP and later suspended was taken up in its own right by an authority offering equal participation to the Tangata

[5] Scott (2016) describes potential – and sometimes actual – conflicts.

Whenua. As for the situational analysis produced by the HGMP, this led to the setting up of a funding offer for environmental measures from the North Foundation. Here, again, the MPA status has not necessarily offered protection of the area, and the specific territorial asset it represents has not been activated: however, extensions exist that can be qualified as "rebound effects".

In Cambodia, the institutionalisation process that was already well underway has been undermined by inter-institutional opposition. The Marine Fisheries Management Area (MFMA) created in 2016, run by the fisheries administration and an international NGO, risks being diluted in a Marine Park created in 2018 in the same zone by the Ministry of the Environment, with different objectives and without any consultation. And when the MFMA examined the idea of creating a tourist tax to fund the inspection of activities, the Ministry of Tourism pipped them to the post and created the tax for its own benefit. Everything that had been gradually constructed with local stakeholders has been undermined.

An MPA status does not, then, guarantee protection of the area involved because the institutionalisation processes of the MPAs are often disrupted. Three factors that hamper or suspend the activation of the potential offered by the MPA status in terms of conservation initiatives have been identified. Nevertheless, rebound effects sometimes occur that allow a certain activation, elsewhere, in a different way and/or via actors who are not MPAs, as illustrated by the cases studied in Rapa Nui, Senegal and New Zealand.

4.2 Illusion 2: the classified surface area is equal to the managed surface area

Measuring protected marine areas in each country takes into account the entirety of perimeters classified as MPAs in both political discourse and scientific literature [38]. However, even when the institutionalisation process is completed, a large proportion of the perimeter is not subject to any specific measures. Our analysis reveals the need to distinguish two perimeters rather than a single one, one for active protection and one for passive protection, the meaning of which we shall clarify using several cases.

4.2.1 The actively protected perimeter: various configurations

An actively protected perimeter is one where, on one hand, the management authority is proactive, and on the other, specific measures exist to tackle issues and pressures, exceeding the mere acquisition of knowledge. This perimeter is often only a fraction of the area classified as an MPA, with varied, linear or "patchwork" geographical configurations.

Patches may exist within the perimeter: for example, Moorea's Marine Space Management Plan is based on 8 no-take zones in addition to two regulated fishing zones and zones with various vocations (mooring, species' feeding grounds, cetaceans' resting grounds, etc.), but a large proportion of the perimeter remains under national law with no specific protection. In the Hauraki Gulf, these patches are very small no-take zones: five marine reserves with scientific, recreational and educational objectives as well as a temporary no-take area managed by the Maoris under fishery laws. These areas are the result of non-coordinated initiatives: the rest (over 99.5% of the perimeter) currently enjoys no actions or specific regulations linked to the MPA status. The actively protected perimeter may only be a single patch, as is the case for the case study in India where the central zone of the Biosphere Reserve is a National Park, monitored by rangers, and the peripheral zone is no longer the target of specific initiatives. This patch is sometimes limited to "whatever is left" after being whittled away despite the MPA. In Lebanon, for example, the Tyre Coast

Nature Reserve (TCNR) concerns both land and sea but its marine zone has neither regulations nor zoning nor specific initiatives and the TCNR includes a tourist zone besieged by pop-up restaurants each summer, a Palestinian refugee camp that is considered an entrenched camp and divides the North and South zones while its access road divides it from East to West, a border at risk of the construction of a motorway, zones used by farmers since the war and then a nature zone. The Reserve can only manage a very limited fraction of its perimeter, where it attempts to protect turtle nesting areas in particular.

Another configuration, this time linear, is the narrow, coastal strip such as the Taeanhaean National Marine Park in South Korea. Despite the MPA classification of a perimeter of around 326 km2, 89% of which is marine, no zoning or specific regulations are connected to the park and it does not work in coordination with the fishing industry. An oil spill led it to monitor seawater quality, but its area of intervention is essentially a narrow ribbon along the coast: it offers services to visitors (car park, campsite, tours) and carries out interventions for developing coastal paths, dune restoration, coastal reforestation, urban monitoring, monitoring of the oil spill recovery and environmental education. This is due to the country's history. Its industrialisation, led by an extremely interventionist regime, has given rise to environmental conflicts about living conditions around industrial complexes: in this context, National Parks have been created as areas for relaxation and recreation for urban dwellers, hence the focus on highly organised and landscaped access to heritage that is both natural and cultural, with Nature seen as inspired and inspiring. The park is a place to recharge ones batteries and contemplate a scene from nature (to a far greater extent than to protect biodiversity), which is why its interventions are focused on accessible zones and "viewpoints", in this case, the coastal linear strip.

4.2.2 The passively protected perimeter

The actively protected perimeter therefore makes up only a fraction of the classified perimeter. Elsewhere, the MPA management authority is neither proactive nor even active, but the rest of the classified perimeter nevertheless receives passive protection that depends on the mechanisms that we will describe with the help of several cases. For example, in Tyre, the perimeter's classification as a coastal nature reserve, combined with Tyre's UNESCO world heritage listing, has repeatedly been used to oppose "artificialisation" projects: the status has thus been used by the International Association for the Safeguard of Tyre to contest the route of the South Lebanon motorway in 2002, 2005 and 2010 and to postpone this threat. It is passive protection: actors are seizing on the status to oppose a threat to the environment affecting the perimeter without the intervention of the MPA.

In reality, passively protected perimeters exceed the borders of the perimeter covered by the status, although it is impossible to assess the area precisely, as we observed in India, Colombia and Korea. In the vicinity of the Taeanhaean National Marine Park, the Korean Government wished to create a 520 MW tidal-powered factory, but clashed with the KFEM (the largest federation of environmentalist organisations) and fishers before accepting the zone's classification as a Marine Park. The proximity of the Taeanhaean National Marine Park was used in the arguments of opponents who claimed that the project would have consequences within its perimeter. In Colombia, the existence of the Rosario and San Bernardo Corals National Natural Park was used in the same way by naturalist movements to oppose the digging of a branch of the Canal del Dique which crosses the country transporting polluting products; building the branch would have threatened the Varadero Reef that lies on the northern border of the National Natural Park. This is also passive protection that benefitted the asset without the intervention of the MPA. And lastly, there is the iconic case of India, iconic both because of the sheer scope of the project and the history of oppositions, the flames of which were fanned by the existence of the MPA. The brainchild of the colonial period, the Sethusamudram Project aims to dig an offshore canal to open up a shipping route to trade goods along the South East coast of India although today, the passage between India and Sri Lanka is impassable. After numerous disputes, in 1999, the State announced that it would complete the project in three years, then in 2005 announced the inauguration of the work, and approved the budget: but for the moment, the work that had started has come to a standstill because of opposition from various sources. These sources are religious (the route would affect Ram Setu, the causeway between India and Sri Lanka said to be built by the god Rama and his army of monkeys and a squirrel according to Hindu belief), socioeconomic (the state of Tamil Nadu reported the fears of fishers, and the Central Marine Fisheries Research Institute was concerned about the effects on fisheries resources), and environmental. It is interesting to note that the presence of a National Park and the Gulf of Mannar Biosphere Reserve (which the route is planned to pass through or close by) are systematically underscored in the arguments put forward. Despite the fact that to date, the Biosphere Reserve is almost inactive outside the central zone made up of the National Marine Park, its status is mobilised by various actors to protect the area. This represents passive protection.

Above all, the quantitative assessment of each country's conservation effort should take into account the surface areas under active protection, which are quantifiable, and secondly take into account the areas under passive protection. Without being strictly measurable, they can, at the least, be assimilated into the classified perimeter and nearby marine area.

4.3 Illusion 3: the regulatory illusion: are the rules meant to be respected?

4.3.1 When the level of protection is associated with the level of regulation

MPAs' effectiveness is often considered based on the legislative arsenal available to them. For this reason, Horta e Costa et al. [39] rank MPAs depending on the level of their regulatory protection, and qualify as unprotected any zones declared as MPAs when no legislative difference exists within and outside these zones for activities that may have an impact. The level of protection is associated with the level of regulation, making it an indicator that can "unambiguously distinguish the impacts of uses" ([39], p. 192). However, among our case studies, the most regulated MPAs are sometimes those where a culture of illegal harvesting develops, either from lack of monitoring or from the lack of alternatives for very vulnerable populations. More generally, simple observations in situ reveal that, whatever the MPA's level of institutionalisation, many regulations are not respected: does this render them meaningless? A more detailed analysis shows that managers distinguish between regulations that they insist are respected and regulations that are in place but not pursued in the event of violations, and instead used as a medium to inform, raise awareness or negotiate with those responsible for certain pressures. The reality is, then, more complex that a simple dichotomy between MPAs that are effective because based on regulations and MPAs that are not.

An initial approximation lies in the idea that an MPA without specific regulatory means is unprotected. Firstly, we observe that, among our case studies, if an MPA is a more regulated area than elsewhere, it is either because it is endowed with specific rules, or because it activates rules that are in force there as elsewhere, but elsewhere are neither known nor enforced. For example, in Moorea, the manager of the

Marine Space Management Plan strives to ensure that, outside the no-take zones, regulations are applied regarding the mesh size of nets, which remain ignored elsewhere but which are applicable on a national level. The same is true in Tyre, Lebanon: the city hall, which chairs the Reserve's management committee, carries out inspections to ensure that the prohibition of fishing with nets less than 500 m from the shore within the Reserve is respected – a regulation that applies everywhere, in theory. Generic regulations are revealed and activated independently of the existence of regulations specific to the MPA. Our case studies then reveal numerous mechanisms of collective self-discipline based on tacit conventions or self-regulation, which exist independently of any legal regulation: co-constructed with or without the MPA and appropriated by stakeholders, they are often far more effective than the law but invisible to the eyes of those who examine only legal regulations, such as Horta e Costa [39] and Zupan et al. [17]. For example, in India's Gulf of Mannar Biosphere Reserve, seaweed harvesters have established their own rules (it is forbidden to damage the coral reefs, to make fires in the mangrove, a 12-day period per month with no harvesting has been established, etc.), respect for which is based on social pressure and, in some villages, community-based surveillance. Trawl fishing has also drawn up rotation regulations in an attempt to limit over-fishing, and these regulations are respected without the need for a law. A second approximation involves the idea that the impacts of certain uses are reduced when covered by regulations: this assumes that, on the one hand, these rules are respected, and on the other, that they were established to be respected. However, comparative analysis reveals recurrences in the multiple functions attributed to the regulations, often used for dialogue purposes rather than to compel people to obey them.

4.3.2 To raise awareness, impose sanctions, establish dialogue and negotiate ... how regulations are really used

The MPAs studied in Greece, Colombia and France are consolidated and dispose of a specific legal framework and monitoring means at sea. However, all these case studies converge in two observations: everywhere, sanctions are rare, and the regulation is first and foremost a medium for dialogue.

In France, in one of the central zones of the Port Cros National Park, it has emerged from a specific study [40] that while 9,800 infractions were observed from 2010 to 2018 (offshore in 45% of cases), 97.3% of these infractions led only to verbal warnings and sanctions were applied in only 2.1% of cases: the regulation is primarily a medium for information and raising awareness, and it is mentioned in each annual activity report from 2010 to 2016 that the National Park's policy is to "favour information and awareness raising over sanctions".[6]

In Greece, in the case of the National Marine Park of Zakynthos (NMPZ), an observation survey enabled us to confirm the non-respect of mooring regulations, turtle observation distances and offshore speed limits, despite the fact that the NMPZ ensures that other regulations are respected, particularly those concerning the no-go zone. The massification of "3S" (Sea, Sand and Sun) tourism requiring recreational activities in the very same location as conservation issues has made it necessary to draw up regulations to share the area, and the NMPZ is managing to ensure these are respected: however, faced with the boom in demand, operators are showing an inventiveness that quickly renders obsolete the rules defined at any given moment. Henceforth, the manager chooses regulations on which he concentrates his inspections, while using the entire set of rules, respected or otherwise,

[6] PNPC Activity Reports (2010, p.15; 2011, p.17; 2012 p.22; 2015, p.16; 2016, p.14).

as a negotiating tool to prevent the most serious infringements. Some regulations become part of a transactional game. When the NMPZ renounced enforcing a rule to prohibit mooring in a zone because of lack of means but also of alternatives to offer (it hoped to build a mooring platform but did not have the funds to employ the staff needed to run it), it was in order to focus on other offences and ensure the enforcement of regulations prohibiting access to crucial nesting areas. Certain transactions are more-or-less explicit, such as with professional fishing, where for ten years, a status quo was respected, and described as such by the protagonists: the respect of regulations concerning zoning in return for the lack of new restrictions. Dialogue recently began to change this status quo: the regulations are used as a medium for dialogue.

In Colombia, the Rosario and San Bernardo Corals National Natural Park (RSBCNNP) is also besieged by tourism described by its manager as "overwhelming", with 1,300,000 visitors a year. Here, again, observation reveals that the PNN tacitly selects regulations that it enforces, renouncing others that are used as a negotiating tool. Admittedly, this is due to a lack of inspection means for the classified perimeter (despite 55 officials and the involvement of the national navy) but also, as in Greece, due to the marketing of a very attractive natural landscape that stimulates the emergence of new uses, creating uncontrollable situations.[7] Lastly, from the point of view of the inhabitants, an excess of regulations are applied to some uses. Too many prohibitions without alternatives kill the prohibition, according to a representative of the Afro-descendant community and for whom, "as it is forbidden to fish both over there and here, I have only one solution, to fish where I want": in his eyes, this sets illegality up to be the norm. Because it cannot be enforced, the regulation is a medium for negotiation, in this case, a highly conflictual one.

These three cases cover the diversity of the situations observed in our case studies, with regulations used either to inform and raise awareness, or to negotiate and, in this case, a tacit selection is made by the MPA between inviolable regulations and flexible regulations. The enforcement of regulations is then subject to conventions [41] that tacitly establish which rules are to be respected, which are negotiable and which infringements will be tolerated, with de facto prioritisation. Concerned with raising the awareness of its interlocutors and in the position of negotiator, the MPA endeavours to preserve an "area of potential agreement", not breaking the thread of dialogue with those responsible for the greatest pressures on the environmental assets it intends to protect. This is a pragmatic choice in light of uncontrollable uses (that are prolific and evolving and/or prior to the creation of the MPA, and thus enjoy significant legitimacy), which can only be limited by negotiation that relies on regulations that do not involve sanctions. Thus, the MPA does not dictate regulations in view of ensuring that they are all respected, but does so in order to have a medium for dialogue and to negotiate uses.

Nevertheless, as we will see, in certain cases, one eventuality can be sanctions for their own sake, with a deviation from the rule to the detriment of both the most vulnerable populations and conservation. Negotiation is then used to legitimise regular and illegal harvesting by the authorities responsible for inspections either for private ends and/or to fund the inspections. In India, for example, in the Gulf of Mannar, what should be a "no-go zone" is qualified as an "open access zone" by a local scientist, concealing a permanent negotiation between seaweed harvesters and National Marine Park rangers. Locally, the authorities responsible for offshore

[7] Playa Blanca receives more than 10,000 people per day, although its load capacity was assessed by the Park at 3,124 people per day. Located on the park's boundary, its tourist numbers lead to a proliferation of uncontrolled uses offshore and catering activities with no liquid waste management (Castaño C. A., 2016, La debacle ambiental y social de Playa Blanca, Opinión, 2016/12/29).

inspections evoke a "gentleman's agreement", which, if violated, leads to informal fines that are negotiated and adapted to the limited capacity to pay of groups that cannot be prevented from working in view of their vulnerability. Our interviews confirm the study conducted by Rajagopalan [42] that observed that, in a context of intensified inspections in the Biosphere Reserve, officials confiscated harvesting tools, which the offenders could retrieve if they offered "gifts". The ICSF, an NGO supporting fish-workers, made it clear to the community that it should demand a formal sanction rather than paying informal fines, but this has occurred only once since 2008. If the regulation is a medium for negotiation, the danger is that one of the results of its creation is to offer the possibility of negotiated and illegal harvesting of resources and taking from users, as Sundaresan [43] illustrated in a more urban context.

4.3.3 Recognising regulations' real functions: their advantages and disadvantages

In all our case studies where the MPA has as its disposal a legal arsenal, prioritising regulations and their different uses is part of the normal functioning of an MPA that attempts an effectiveness that is impossible to envisage if dialogue with certain stakeholders breaks down. However, this is not without risk. In addition to the possibility of the illegal use of the regulation, its use as a medium for negotiation gives rise to unequal treatment at the expense of populations with the lowest social capital. Indeed, on the one hand, the effort the MPA must make to enforce a rule is all the more considerable since the offenders have social capital that allows them to curb its application, and on the other hand, this social capital facilitates negotiation: consequently, the rules are often applied differently depending on the public concerned. In Colombia, for example, in the RSBCNNP, islanders are banned from erecting artificial defences on the foreshore to protect themselves from coastal erosion. However, while the rule is enforced for local afro-descendant communities, it is enforced far less strictly for holiday homes and tourist enterprises (this is easily observed in the landscape), which travel to Cartagena to engage in dialogue with an administration whose language and logics they understand. This is the source of a sense of injustice, which is sometimes expressed with violence, as in the case when the NNP's premises were vandalised on San Bernardo in 2018. The differentiated application of the rule is primarily the result of a differential of social capital from one public to another. It is crucial to acknowledge the reality of how these legal regulations are used, as much to counter the "regulatory illusion" as to control the eventual negative effects of these uses.

5. Discussion

5.1 Lessons learnt, scope and limitations of the comparative analysis

Lastly, it emerges from the case studies that MPA status does not offer *a priori* protection but constitutes a potentially activatable capital to aid conservation. Its activation depends on the MPA's inscription process in the local institutional landscape, often hampered by factors such as funding by project, social non-acceptance, and oppositions between public organisations. We then observe that, whatever the MPA's degree of institutionalisation, the actual scale of the protection initiative does not correspond to the classified perimeter. It is therefore necessary to take into consideration the area with effective protection rather than the area officially declared, with a level of active protection (where the MPA is proactive), which is a fraction of the classified perimeter, and then a level of passive protection, i.e. the classified perimeter broadened to include nearby areas (where actors use the existence of the MPA to oppose potential threats to the environment). The case studies enabled us to observe the way in which legislative tools at the MPA's disposal are used. What emerges is the fact that the effectiveness of an MPA is not correlated to its regulatory and that the absence of regulations specific to the MPA does not mean that the MPA is ineffective. Firstly because MPAs might not create regulations but instead activate rules that exist but are not known and/or not respected, and then because having a considerable legislative arsenal signifies nothing about the way in which it is used. The case studies have enabled us to specify the way in which regulations are drawn up and used by MPAs, primarily as a medium for information, awareness raising or negotiation, with the advantages and biases inherent in these practices.

The scope of these findings is linked to the fact that they call into question a representation of the triptych (a status, a perimeter, regulations) upon which MPAs are founded: these three elements are, in fact, less specific and decisive than they appear to be, and it is worth analysing their role and outlines in each case. This representation forms the basis for the international assessment of each country's conservation effort, which we propose to reconsider. But can these findings be generalised from thirteen case studies? There is nothing representative about our sample, but the recurrence of the mechanisms is instructive: it is true that regulations are not used by all in the same way, but all the MPAs prioritise regulations that they attempt to enforce while other rules are above all used as a medium for dialogue, information and negotiation. It is true that the ratio between the actively protected perimeter and the MPA perimeter is highly variable, but these two levels rarely coincide completely. With regard to passive protection, it is meaningful in every case. In addition to these recurrences, the precise analysis of the mechanisms studied leads us to state that they exist in many MPAs. However, it would be worth completing this analysis with the precise assessment of actively protected perimeters, the estimation of passively protected perimeters and a more precise identification of the self-regulation mechanisms constructed in certain MPAs.

5.2 Measuring the conservation effort: propositions

The quantitative "one-upmanship" that marks the creation of MPAs, although criticised [For example, see 3], remains championed by countries as well as many NGOs.[8] But it is based on assessment indicators of protected surfaces that our analysis contributes to challenging. In light of our analysis, we propose to measure them by taking into account mainly actively protected perimeters as defined on the basis of the analysis and which it is possible to quantify, and then passively protected perimeters, which could, by default, be assimilated with perimeters that are classified even if, in reality, they are slightly larger. Admittedly, this second level cannot be precisely circumscribed in that it depends on how stakeholders in environmental disputes seize hold of (or do not) the existence of the MPA – but it can be approximated. Furthermore, for a more detailed assessment, these perimeters should not be studied without examining the MPA's institutionalisation process, its degree of consolidation and what might constitute a threat to it: to do this, it would be useful to mobilise the indicators regarding the durability of the MPA's funding, its social acceptance, and the degree of adherence and convergence between public organisations with regard to this territorial conservation policy. Concerning the assessment of levels of protection within the perimeters under consideration, our

[8] During the Global Forum for MPAs in 2017, current international commitments were described as the "minimum goal to be attained", the ambition being to protect "at least 30% of the world's seas" according to the IUCN, while the WWF asserted the need to raise this objective to 40%.

analysis also leads us to challenge a vision championed in certain political discourse (the "fully natural zones" proposed by the French presidency) and scientific studies [17, 39], according to which, the more uses with potential impacts are covered by a legal framework, the more effective the protection is: our analysis indicates that it is important both not to ignore the significance of systems of rules that are not legally recognised but adopted by stakeholders, and to examine the way in which legal regulations are effectively used.

6. Conclusion

Our comparative analysis based on thirteen differentiated case studies brings to light recurrences for protection levels and the conditions of the effectiveness of the conservation policies embodied by MPAs as well as the way in which they make use of (or do not) the regulations associated with their implementation. These findings run counter to the received ideas that form the basis for political discourse and decisions and some scientific studies according to which: (a). A classified area is protected ... and can therefore be recorded as such; (b). The entire classified perimeter enjoys protection; (c). Protection relies upon regulations drawn up with the aim of being enforced. When countries pledge to increasing the surface area of their waters classified as MPAs, they consider that the MPA status is a guarantee of protection and that it will provide classified perimeter-wide protection – two ideas that are called into question by this analysis. In light of this, considering two levels of protection, active and passive, gives us a clearer idea of both the reality of the protection measures and the extent of the efforts made by each country. Even so, however, quantitative objectives should not relegate qualitative ones to the background. Furthermore, it would be worth examining the role that regulatory plays in an MPA's effectiveness and efficiency. This is what we attempted to do by illustrating that firstly, an MPA can act efficiently to limit the impact of uses without necessarily having specific regulations at its disposal, and then that regulations are used in relations with stakeholders more than merely as instruments to exercise authority: when used to serve this relationship, they can prove to be more useful than when they are used to confront actors whose contributions are crucial to the MPA's effectiveness.

More generally, the findings of our analysis are an encouragement to examine more closely the way in which MPAs work and the conditions of their effectiveness. The three simple ideas challenged here are all illusions that lead to confusion, particularly for decision-makers who congratulate themselves for reaching quantitative objectives for creating MPAs without paying sufficient attention to the means they need to devote for their effective management and the success of their institutionalisation. What's more, these illusions mask the reality of MPAs: if the MPA status does not guarantee the protection of the perimeter under consideration, nor even its institutionalisation, and if the regulations associated with the conservation policies embodied by an MPA are not necessarily devised to be enforced, what exactly is an MPA? It is a local public policy that is constantly under construction (to be considered as a process rather than a state), marked by a dual level of action (active protection and passive protection), mobilising regulations that are first and foremost a medium for dialogue and negotiation.

Author details

Jean-Eudes Beuret[1]* and Anne Cadoret[2]

1 Institut AGRO, UMR ESO 6590, Rennes, France

2 Aix Marseille University, CNRS, TELEMMe UMR 7303, Aix en Provence, France

*Address all correspondence to: beuret@agrocampus-ouest.fr

References

[1] World Wildlife Fund, 2014. Living Planet Report. Species and spaces, people and places, Gland, Switzerland. https://www.worldwildlife.org/publications/living-planet-report-2014, [Accessed 2020-01-13]

[2] UNEP-WCMC, 2020. World Database on Protected Areas, United Nations Environment World Conservation Monitoring Centre – IUCN, https://www.protectedplanet.net/[Accessed 2020-01-13]

[3] Féral, F., 2011. L'extension récente de la taille des aires marines protégées : une progression des surfaces inversement proportionnelle à leur normativité, *VertigO,* Hors-Série 9, http://vertigo.revues.org/10998; DOI: 10.4000/vertigo.10998

[4] Pasquaud, S., Lobry, J., 2010. Regard critique sur la mise en place d'indicateurs d'évaluation de l'efficacité des aires marines protégées, Sciences Eaux et Territoires, 3, 122-125. DOI : 10.14758/SET-REVUE.2010.3.23

[5] Stafford, R., Howlett, S.J., Clitherow, T.J., Spiers, E., 2016. An integrated evaluation of potential management processes on marine reserves in continental Ecuador based on a Bayesian belief network model, Ocean & Coastal Management, 121, 60-69. DOI: 10.1016/j.ocecoaman.2015.12.010

[6] Ramirez, L.F., 2016, Marine protected areas in Colombia: Advances in conservation and barriers for effective governance, Ocean and Coastal Management, 125, 49-62. DOI: 10.1016/j.ocecoaman.2016.03.005.

[7] D'Anna, G., Fernández, T. V., Pipitone, C., Garofalo, G., Badalamenti, F., 2016. Governance analysis in the Egadi Islands Marine Protected Area: A Mediterranean case study. Marine Policy, 71. DOI: 10.1016/j.marpol.2015.12.009.

[8] Araújo, J.L., Bernard, E., 2016. Management effectiveness of a large marine protected area in Northeastern Brazil, Ocean & Coastal Management, 130, 43-49. DOI: 10.1016/j.ocecoaman.2016.05.009

[9] Muallil, R.N., Deocadez, M.R., Martinez, R.J.S., Campos, W.L., Mamauag, S.S., Nañola, C.L., Aliño, P.M., 2019. Effectiveness of small locally-managed marine protected areas for coral reef fisheries management in the Philippines, Ocean & Coastal Management 179, DOI: 10.1016/j.ocecoaman.2019.104831

[10] Stafford R., 2018, Lack of evidence that governance structures provide real ecological benefits in marine protected areas. Ocean and Coastal Management, 152, 57-61. DOI: 10.1016/j.ocecoaman.2017.11.013.

[11] Féral, F., 2007. L'administration des aires marines protégées en Afrique de l'Ouest, Mondes en développement, 2, 138, 43-60. DOI: 10.3917/med.138.0043

[12] Chaboud, C., 2007. L'exploitation durable des ressources marines et côtières, In: Chaboud, C., Froger, G., Méral, P. (Dir.), Madagascar face aux enjeux du développement durable : des politiques environnementales à l'action collective locale. Karthala, Paris, 229-257. ISBN 978-2-84586-908-0.

[13] Cadoret, A., Beuret, J.-E., 2016. Aire Marine Protégée, intérêt général environnemental et territoire, un rendez-vous manqué ? Le cas de Mayotte. VertigO, 16(1), DOI: 10.4000/vertigo.17173.

[14] Depraz, S., 2008. Géographie des espaces naturels protégés : genèse,

principes et enjeux territoriaux. Armand Colin, Paris. EAN : 9782200347581

[15] Goeury, D, 2014, Les aires marines protégées », In : Woessner R. (Dir.), Mers et océans, Atlande, Clefs Concours, Paris. ISBN-13: 978-2350302751

[16] Héritier, S., Depraz, S., Peyvel, E., 2009, Des espaces naturels protégés : pour quoi faire ? Cafés géographiques de Lyon, http://cafe-geo.net/wp-content/uploads/espaces-naturels-proteges.pdf [Accessed 2019-07-04]

[17] Zupan, M., Bulleri, F., Evans, J., Fraschetti, S., Guidetti, P., Garcia-Rubies, A., Sostres, M., Asnaghi, V., Caro, A., Deudero, S., Goñi, R., Guarnieri, G., Guilhaumon, F., Kersting, D., Kokkali, A., Kruschel, C., Macic, V., Mangialajo, L., Mallol, S., Macpherson, E., Panucci, A., Radolovic, M., Ramdani, M., Schembri, P. J., Terlizzi, A., Villa, E., Claudet, J., 2018. How good is your marine protected area at curbing threats? Biological Conservation, 221, 237-245. DOI: 10.1016/J.BIOCON.2018.03.013.

[18] Dehens, L. A., Fanning, L. M.,2018. What counts in making marine protected areas (MPAs) count? The role of legitimacy in MPA success in Canada. Ecological Indicators 86, 45-57. DOI: 10.1016/J.ECOLIND.2017.12.026.

[19] Islam, G., Tai, S.Y., Kusairi, M.N., Ahmad, S., Aswani, F., Senan, M., Ahmad, A., 2017, Community perspectives of governance for effective management of marine protected areas in Malaysia, Ocean & Coastal Management, 135, 34-42. DOI: 10.1016/j.ocecoaman.2016.11.001

[20] Taylor, E., Baine, M., Killmer, A., Howard, M., 2013. Seaflower marine protected area: governance for sustainable development, Marine Policy 41, 57-64. DOI: 10.1016/J.MARPOL.2012.12.023

[21] Bennett, N.J., Dearden, P., 2014. Why local people do not support conservation : community perceptions of marine protected area livelihood impacts, governance and management in Thailand. Marine Policy, 44, 107-116. DOI: 10.1016/j.marpol.2013.08.017.

[22] Day, J., Dudley, N., Hockings, M., Holmes, G., Laffoley, D., Stolton, S., Wells, S., 2012. Application des catégories de gestion aux aires protégées : lignes directrices pour les aires marines. IUCN, Gland, Switzerland. ISBN : 978-2-8317-1525-4.

[23] Aubertin, C., Rodary, E., 2009. Aires protégées, espaces durables ? IRD Éditions, Objectifs Suds, Marseille. DOI: 10.4000/books.irdeditions.5655

[24] Kelleher, G., Bleakley, C., Wells, C., 1995. Priority Areas for a Global Representative System of Marine Protected Area, World Bank, Washington D.C. http://documents.worldbank.org/curated/en/936381468780944183/South-Pacific-Northeast-Pacific-Northwest-Pacific-Southeast-Pacific-and-Australia-New-Zealand, [Accessed 2020-01-13].

[25] Jameson, S.C., Tupper, M.H., Ridley, J.M., 2002, The three screen doors: can marine "protected" areas be effective? Marine Pollution Bulletin 44. 1117-1183. DOI: 10.1016/S0025-326X(02)00258-8.

[26] Kelleher, G., 1999 Guidelines for Marine Protected Areas, IUCN, Gland/Cambridge.https://portals.iucn.org/library/sites/library/files/documents/PAG-003.pdf [Accessed 2020-01-13].

[27] Dudley, N. (Ed.), 2008. Guidelines for applying protected area management categories. IUCN, Gland, Switzerland. ISBN: 978-2-8317-1636-7

[28] Humphreys, J., Herbert, R.J.H., 2018. Marine protected areas: Science, policy & management. Estuarine, Coastal and Shelf Science, 215, 215-218. DOI: 10.1016/J.ECSS.2018.10.014.

[29] Roy, S.N., 2009. L'étude de cas, In Gauthier, B., (Ed.) Recherche Sociale, de la problématique à la collecte de données, 5ième édition, pp. 199-225, Presses de l'Université du Québec, Québec.

[30] Dougherty, D., 2002. Grounded Theory Research Methods, In: Baum J.A.C. (Ed.) Companion to Organizations, Oxford : Blackwell, 849-866. DOI: 10.1002/9781405164061.ch37

[31] Yin, R.K., 2003. Case Study Research : Design and Methods, 3rd edition, Thousand Oaks, CA : Sage. DOI: 10.1177/1609406919862424

[32] Musca, G., 2006. Une stratégie de recherche processuelle : l'étude longitudinale de cas enchâssés, Revue Management, 2006/3, 9, 153-176. DOI: 10.3917/mana.093.0153

[33] Yin, R.K., 1981. The case study as a serious research strategy, Science Communication, Knowledge : Creation, Diffusion Utilization, Vol 3(1), 97-114. DOI: 10.1177/107554708100300106.

[34] Mucchielli, A., 2007. Les processus intellectuels fondamentaux sous-jacents aux techniques et méthodes qualitatives, Recherches Qualitatives, Hors-Série 3, 1-27. ISSN 1715-8702

[35] Alexandre, M., 2013. La rigueur scientifique du dispositif méthodologique d'une étude de cas multiples, In Recherches qualitatives, 32(1), 26-56. ISSN 1715-8702.

[36] Colletis, G., Pecqueur, B., 1993. Intégration des espaces et quasi-intégration des firmes : vers de nouvelles rencontres productives ? *Revue d'Economie Régionale et Urbaine*, 3, 489-508. DOI : 10.4000/ei.900.

[37] Pecqueur, B., Zimmermann J.-B., 2004. Economie de proximités, Editions Hermès-Lavoisier, Paris. halshs-00280083.

[38] Fouqueray, M., Papyrakis, E., 2019, An empirical analysis of the cross-national determinants of marine protected areas, Marine Policy 99, 87-93. DOI: 10.1016/j.marpol.2018.10.017.

[39] Horta e Costa, B., Claudet, J., Franco, G., Erzini, K., Caro, A., Gonçalves, E. J., 2016. A regulation-based classification system for Marine Protected Areas (MPAs). Marine Policy 72, 192-198. DOI: 10.1016/J.MARPOL.2016.06.021.

[40] Cadoret, A., 2021. Conflicts and acceptability of visitation management measures for a marine protected area: The case of Porquerolles, Port-Cros National Park, Ocean & Coastal Management, 204, DOI: 10.1016/j.ocecoaman.2021.105547

[41] Thévenot, L., Eymard Duvernay, F., Favereau, O., Orléan, A., Salais R., 2005. Values, Coordination and Rationality : the Economics of Conventions, In Oleinik A. (Ed.) The institutionnal Economics of Russia's Transformation, Aldershot (UK), Ashgate.

[42] Rajagopalan, R., 2008. Les ramasseuses d'algues dans le Golfe de Mannar, au Tamil Nadu, sont confrontées à un avenir incertain, HINA, les femmes et la pêche - Bulletin de la CPS n°17. ISSN 1029-0230

[43] Sundaresan, J., 2017. Urban planning in vernacular governance: land use planning and violations in Bangalore, India, Progress In Planning, 127, 1-23. DOI: 10.1016/j.progress.2017.10.001.

Chapter 8

Galician Atlantic Islands National Park: Challenges for the Conservation and Management of a Maritime-Terrestrial Protected Area

Javier Ferreiro da Costa, Pablo Ramil-Rego, Manuel A. Rodríguez Guitián, Hugo López Castro, Carlos Oreiro Rey, Luis Gómez-Orellana and José Antonio Fernández Bouzas

Abstract

At present, biodiversity conservation and management in Spanish National Parks in Spain must respond to a series of regulations at a European, national and regional level, also adapting to scientific-technical progress. The availability of increasingly precise data on the values to be conserved (ecosystems, habitats, species, geodiversity) in these protected areas enables more detailed management, but also requires more rigorous, powerful, and multidisciplinary tools. Maritime-terrestrial national parks are highly sensitive areas to public use, so their impact must be one of the most important factors to take into account when planning their management. This work evaluates the past and present challenges for conservation in Galician Atlantic Islands National Park (NW Spain), where biodiversity conservation and management has evolved over time in a significant way, providing a valid case study applicable to other national parks worldwide, as well as similar situations in other contexts and scenarios. Future challenges are arising in the National Park to improve the conservation status of natural habitats and wildlife, mainly through new European initiatives that may establish important synergies with other countries.

Keywords: biodiversity, conservation, management, National Park, Natura 2000

1. Introduction

Replace the entirety of this text with the introduction to your chapter. The introduction section should provide a context for your manuscript and should be numbered as a first heading. When preparing the introduction, please bear in mind that some readers will not be experts in your field of research.

In the 18th century, in the middle of the Enlightenment, there was still the conception that natural resources were inexhaustible and providence was trusted as a generator of new resources, but the impact on the territory and the demographic increase originated the first initiatives to protect the natural areas in the world, linked to hunting management measures and preventing the disappearance of some species, due to its ecological [1] or esthetic [2] importance. The current concept of protection of Natural Areas has traditionally been attributed to the declaration of Yellowstone National Park, on March 1, 1872, in the United States of America.

Years later, between the late 19th and early 20th centuries, other countries imitated USA, widespread the declaration of National Parks all over the world: Australia (1879), Canada (1885), New Zealand (1891), Mexico (1898), South Africa (1898), Argentina (1903), Sweden (1909), Latvia (1911), Georgia (1912), Switzerland (1914), Italy (1916), and Spain (1918). During the 20th century, the concept of protection of Natural Areas has undergone a visible evolution in different stages [3], ranging from the protection of emblematic and singular spots at the beginning of the century, which would be increased in number and levels of protection under a great diversity of legal categories, and finally, it was intended to integrate the conservation of Natural Areas with sectoral policies and land use planning under the framework formed after 1992 "Earth Summit".

In Spain, National Parks law was approved in August 1916, which can even be considered as the first law of national parks in the World, and under this regulation, the two first National Parks appeared in 1918: Covadonga and Ordesa. After that, Franco's dictatorship (1939–1975) was a "dark period" when the protected areas regulation (and therefore new area declarations) was completely subordinated to development policies (forestry, intensive agriculture, reservoir construction, industry, gaming, fishing, tourism). The joint arrival of the democracy in 1977 along with the Spanish Constitution in 1978, paved Spain's way towards Europe, which culminated in 1986, when Spain was officially integrated into the European Economic Community (nowadays the European Union). All these changes brought new and modern regulations about protected areas and biodiversity conservation, although the biggest step was the creation Natura 2000 network, based on Directives 92/43/EEC and 79/409/EEC (now replaced by 2009/147/EC). In fact, Spain is the EU member that holds the largest area occupied by Natura 2000 sites, including 27.3% of its terrestrial area into this network [4].

Nowadays, Spain has Law 42/2007, of Natural Heritage and Biodiversity, which establishes the basic legal regime for the conservation, sustainable use, improvement, and restoration of natural heritage and Spanish biodiversity, including the regulation of natural protected areas, and obviously including the National Parks among all of them. These have been endowed with their own and specific regulatory framework, constituted by Law 30/2014, in which National Parks are considered as models for nature conservation and as examples of participatory management, and by Royal Decree 389/2016, in which the Master Plan of Spanish National Parks Network was approved, as well as the strategic objectives of the National Parks in terms of conservation, public use, research, training, awareness, cooperation, planning, coordination, monitoring, and evaluation. All this Spanish regulatory framework around protected areas and National Parks is complemented by the regional protected areas regulation, as foreseen in the Spanish Constitution.

So, Spanish Government has been consolidated for coordination function of the National Parks Network through the National Parks Autonomous Agency (OAPN the acronym in Spanish), establishing their own instruments for management, planning, social participation, as well as their own image as a brand that identifies them highlighting their value and social appreciation. On the other hand, the management and organization of National Parks correspond directly to the Spanish autonomous regions in whose territories they are located, including the maritime-terrestrial ones when there is an ecological continuity between terrestrial and marine ecosystems, which must be supported by the best existing scientific evidence and be thus expressly recognized in the declarative law.

In this paper, as one of the first planning and dissemination tasks developed by LIFE INSULAR project (LIFE20 NAT/ES/001007), we evaluate the past, current, and future challenges for conservation in one of the most unique National Parks in Spain, Maritime-Terrestrial Galician Atlantic Islands National Park (hereinafter PNG), located in NW Spain, considering the present regulatory framework around it, as well as the intrinsic characteristics of this specific natural protected area. The document assesses the huge progress in biodiversity conservation and management that has been made in the PNG over time, including the legal designations of the archipelagos, the trends in public use, the established protection measures, the limiting of visitors, the statutory instruments for planning and management, and the consequences of all of them to halt the biodiversity loss in island natural ecosystems.

2. The Maritime-Terrestrial Galician Atlantic Islands National Park

This National Park was declared under Spanish Law 15/2002. Located on the Atlantic coast of Galician region (NW Spain), PNG is made up of four archipelagos (Cíes, Ons, Sálvora, and Cortegada) and the marine waters that surround them (**Figure 1**). PNG comprises a total area of 8480 ha (**Table 1**), corresponding the

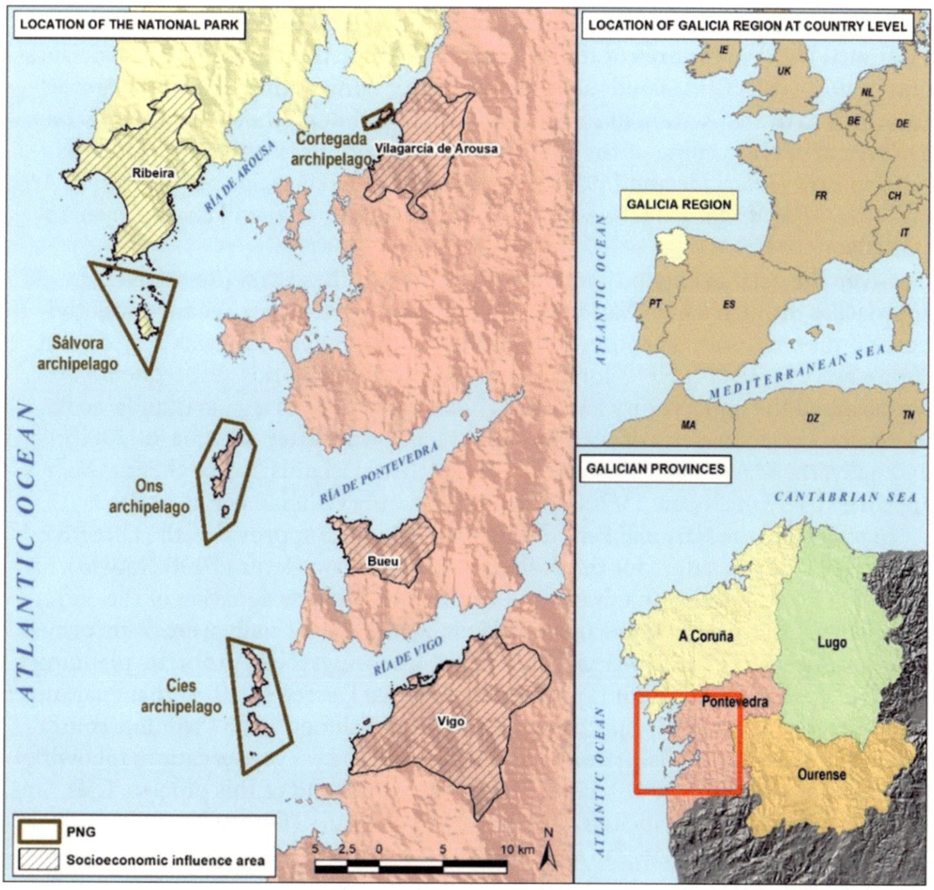

Figure 1.
Location map of PNG.

	Cíes	Ons	Sálvora	Cortegada	Total
Terrestrial area (ha)	433	470	248	44	1195
Marine area (ha)	2658	2171	2309	147	7285
Total	3091	2641	2557	191	8480

Table 1.
Marine and terrestrial area are occupied by the archipelagos of the National Park.

86% to marine waters. According to Law 15/2002, the biggest archipelago in the National Park is Cíes, with 3091 ha, of which 433 ha are terrestrial and 2658 ha are marine waters. Ons archipelago occupies 2641 ha, with 470 ha of terrestrial land (the largest terrestrial archipelago) and 2171 ha of marine waters. Sálvora Archipelago holds 2309 ha of marine waters and 248 ha of terrestrial area. Cortegada Archipelago is the smallest of the four, assuming a total area of 191 ha in the National Park, of which 43.8 ha are terrestrial lands.

Each archipelago is located in four different municipalities that are spread in two Galician provinces (**Figure 1**). Archipelagos located in Pontevedra province are Cíes (belonging to Vigo municipality), Ons (belonging to Bueu municipality), and Cortegada (belonging to Vilanova de Arousa municipality). In A Coruña province, the resting archipelago of PNG is Sálvora (belonging to Ribeira municipality). According to Laws 15/2002 and 42/2007, the whole four municipalities configure the socioeconomic influence area of PNG (**Figure 2**).

The environmental values of this territory (unique in Galicia with the status of a National Park) in the area of the Atlantic coast led to the proposal for the declaration in 1975 of the Cíes Islands as a "Natural Area" (one of the existing categories according to the legislation in force at that time), although they would finally be declared at the beginning of the 1980s under other of the existing categories, as Natural Park (Royal Decree 2497/1980), and later in 1988 as Special Protection Area for Birds (SPA) under Directive 79/409/EEC, a category that will also be given to Ons Islands in 1990.

Given the need to expand the protection scope of these first preventive approaches through a new National Park, in order to harmonize economic activities and the environment conservation, a first management plan was drafted and approved (Decree 274/1999) for Cíes, Ons, and Sálvora islands. Subsequently, it was assessed the opportunity and need to also integrate Cortegada islands, so its corresponding management plan was approved 3 years after (Decree 88/2002). So finally, the Maritime-Terrestrial Galician Atlantic Islands National Park was approved by Spanish Law 15/2002, the first and only Galician National Park.

In parallel to the National Park designation, after the approval of the Directive 92/43/EEC, the procedure for the designation of Galician Natura 2000 Network had started in 1999 with the first drafts and finished with the designation of the definitive Sites of Community Importance (SCI) in 2004. These finally were transformed to Special Areas of Conservation (SAC) in 2014, through the appropriate planning and management instrument (approved by regional Decree 37/2014) that guarantees the maintenance or, where appropriate, the reestablishment, of a favorable conservation status of natural habitats and species interesting for conservation, following the foreseen procedure by Directive 92/43/EEC. At the end of this process, Cíes, Ons, and Sálvora islands were included in three different SACs (**Figure 2**), which are managed by the autonomous region of Galicia.

Considering the high importance of the marine biodiversity of the National Park, this has two additional protection categories by international instruments. The first one was conferred in 2008, as it was integrated in OSPAR network, which is focused

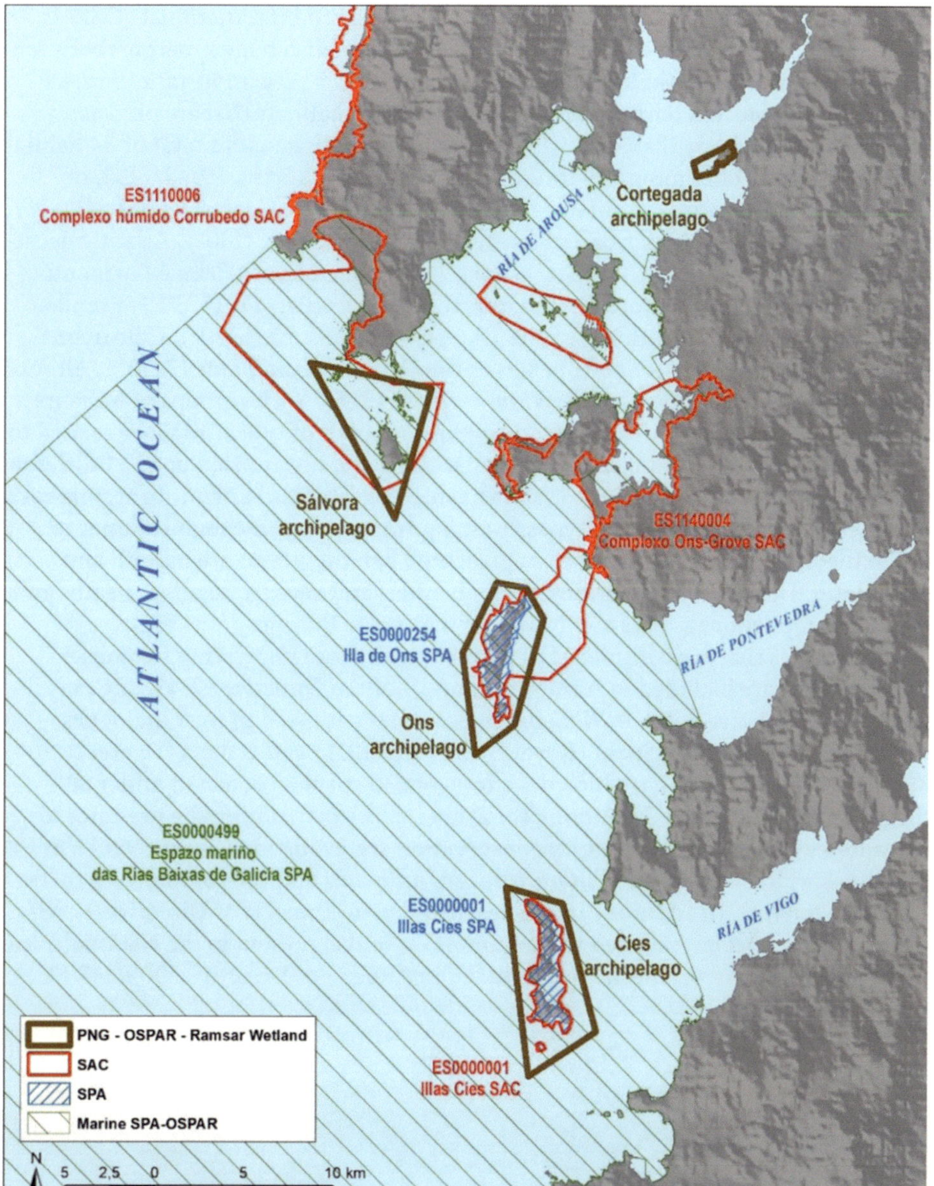

Figure 2.
Overlapping of the different categories of natural protected areas in the territorial scope of PNG.

on the protection and conservation of marine ecosystems and diversity of North-East Atlantic, becoming the first Spanish protected area under OSPAR Convention. Subsequently, the Spanish Government declared a series of SPAs in Spanish marine waters in 2014, including in one of them the Cíes, Ons, and Sálvora archipelagos of the PNG (**Figure 2**), and that was also integrated in OSPAR network. The second additional protection category by international instruments is very recent, as in May 2021 the National Park has definitely been included in the List of wetlands of international importance, as defined by the Ramsar Convention (**Figure 2**).

So Maritime-Terrestrial Galician Atlantic Islands National Park is a very important protected area into the Galician territory, and also at a Spanish level. It holds several types of protected areas (**Figure 2**), from a national (National Park), regional (Protected Wetland), European (SAC, SPA), and international (OSPAR, Ramsar) point of view, which are overlapped and establish huge synergies between them. The biodiversity sheltered by the National Park is very important, both in terms of the protected harbored habitats and species habitats that are present.

According to the available data [5], the National Park houses a total of 34 habitat types considered of community interest in Annex I of Directive 92/43/EEC, of which eight habitats are classified as priority conservation: Coastal lagoons (1150*), Fixed coastal dunes with herbaceous vegetation-grey dunes (2130*), Atlantic decalcified fixed dunes (2150*), Temperate Atlantic wet heaths with *Erica ciliaris* and *Erica tetralix* (4020*); Arborescent matorral with *Laurus nobilis* (5230*), Pseudo-steppe with grasses and annuals of the *Thero-Brachypodietea* (6220*), Calcareous fens with *Cladium mariscus* and species of the *Caricion davallianae* (7210*), Alluvial forests with *Alnus glutinosa* and *Fraxinus excelsior* (91E0*). The groups of habitats of community interest that register the highest number of types are those related to coastal environments and halophilic vegetation, being inventoried up to 11 different types of habitats, as well as dune systems, which register also a significant number of dune system habitats (seven types) ranging from embryonic mobile dunes to decalcified fixed dunes. Other well-represented habitat groups are natural and semi-natural grassland formations, rocky habitats and caves, temperate heaths and scrub, sclerophyllous scrubs and forests of temperate Europe.

The species that are considered as protected are those interesting for conservation as they are included in Annexes II and IV of the Directive 92/43/EEC, in Annex I of Directive 2009/147/EC, together with those listed in the Catalogues of Threatened Species in Spain (Royal Decree 139/2011) and Galicia (Decree 88/2007). The National Park includes a total of 530 species protected under all these regulations (**Table 2**). Among these taxa, it is worth noting the presence of two species considered for priority conservation according to Directive 92/43/EEC: the flora species **Omphalodes littoralis* subsp. *gallaecica* and the common sea turtle (**Caretta caretta*). In addition, 15 species included in Annex II of Directive 92/43/EEC and 16 in species included in its Annex IV, are also present in the National Park. Regarding the birds, 25 species are included in Annex I of Directive 2009/147/EC.

	HD				BD	SCTS			GCTS		Total
	P	II	IV	V	I	En	Vu	SP	E	V	
Plants	1	2	2	4	—	1	—	2	7	2	12
Invertebrates	—	4	2	1	—	—	1	6	1	4	11
Fishes	—	3	—	2	—	—	—	2	—	1	5
Amphibians	—	1	1	—	—	—	—	2	—	3	3
Terr. reptiles	—	—	1	—	—	—	—	8	1	4	8
Marine reptiles	1	1	2	—	—	—	—	1	1	1	2
Birds	—	—	—	—	25	2	4	88	1	7	123
Terr. mammals	—	2	4	—	—	—	1	3	—	1	4
Marine mammals	—	2	4	—	—	—	2	2	—	2	4
Total	2	15	16	7	25	3	9	112	11	25	530

[HD]: Habitats Directive (92/43/EEC); [P]: priority species; [II, IV, IV]: Annex where the species is included; [BD]: Birds Directive (2009/147/EC); [I]: Annex I; [SCTS]: Spanish Catalog of Threatened Species; [En]: Endangered; [Vu]: Vulnerable; [SP]: List of Wild Species under Special Protection; [GCTS]: Galician Catalog of Threatened Species; [E]: Endangered; [V]: Vulnerable.

Table 2.
Protected species richness in PNG.

The cataloged species include 3 species considered as Endangered by the Spanish Catalog of Threatened Species, and 11 species considered as Endangered by the Galician Catalog of Threatened Species.

3. Conservation problems and management issues previous to PNG

Available information confirms the human presence on the islands from the Mesolithic to the Roman empire [6–11]. Throughout the Middle Ages, the islands depended on different monastic orders, maintaining a feudal regime. Between the 16th and 18th centuries, the islands maintained a system similar to the previous monastic regime, administered by the nobles of the towns located on the continent [12–14]. During this period the archipelagos will be witnesses and victims of numerous naval warlike conflicts against other nations, or against pirate invasions of various kinds [9], which led to the construction of fortifications and defensive bastions (batteries, barracks, arsenal, etc.).

During the 19th century, residents from the nearby coasts moved to the islands to attend to them and carry out various labors. The four archipelagos supported a population that ranged from 30 to 550 neighbors per island (depending on the island size). The insular inhabitants developed agriculture (potatoes, corn, vegetables), intense livestock activity, fishing, and shellfish. The islands were also used as a hunting ground, as well as different industries and facilities were installed (salting, lighthouses, etc.) in which the inhabitants of the islands worked [15–20].

During the 20th century, these infrastructures and facilities declined or were automated, as the quality of life in the continent was improved notably. This caused the archipelagos to gradually lose their resident population. The loss of population motivated the reduction (even total loss) of the crop areas due to the abandonment of cultivated lands, which were replaced by natural ecosystems (coastal heaths, sand dunes, etc.). This process was more evident in Sálvora and Cortegada islands, where agricultural activity was abandoned earlier, and this was not replaced by other actions, as happened in Cíes and Ons. In this way, Sálvora and Cortegada recovered their naturalness as they were subjected to natural dynamic processes, and the crop fields were replaced by dune habitats, coastal heaths, and native forests so that they currently have most of their occupied surface by natural ecosystems.

The depopulation of the islands progressed in parallel with growing activity in Cíes and Ons of the Spanish Forest Heritage (PFE the acronym in Spanish), created in 1935 but whose activity was definitely boosted from 1940, after the Spanish Civil War. The activity of the PFE focused for more than 20 years on the transformation of the natural habitats in the island territories (coastal heaths, fixed dunes, humid dune slacks) through productive afforestation [21–24] with exotic species (*Pinus* spp., *Eucalyptus* spp., *Acacia* spp.), many of which have the invasive capacity, following a methodology that was used in the rest of the Spanish coastal ecosystems [25]. So, natural communities of great biodiversity value were replaced by very low value synanthropic invasive formations that were potentially harmful to the surrounding ecosystems, although the high natural elements hosted by these islands had motivated them to be previously proposed in 1917 to be declared a "Notable Site" under the 1916 National Parks Law and the provisions that developed it [26], and there were also previous available scientific works that highlighted the relevant role of the insular natural environment [27, 28]. The activity of the PFE was especially relevant in Cíes, where it passed from a scenario characterized by natural herbaceous and shrub island habitats to a landscape in which exotic wooded formations occupy more than half of the surface [29]. In Ons, the afforestation was carried out in a smaller proportion, although it was also established at the expense of natural coastal habitats.

The afforestation of these island territories was used as an indoctrination measure of a country under an authoritarian regime in a context of political isolation and economic autarchy [30], to show and evaluate the "patriotic and lucrative work" that the PFE forestation works constituted. However, the high surface area that they reached in Cíes was a source of conflict with the few residents who still lived in the archipelagos, which led to them causing the uprooting of forest plants and even causing intentional fires with the purpose of destroying plantations. But the presence of inhabitants on the islands was in a regressive phase so that in the 1970s there were hardly any inhabitants left on the islands. Afforestation, on the contrary, was consolidated, so that even in the first Spanish vegetation cartographies, the forest plantations were already represented as the dominant coverage in the island territories [31]. The replacement of the PFE by Conservation Nature Institute (ICONA the acronym in Spanish) in 1971, abandoned the "lucrative" argument of forestation, but its vision of insular ecosystems was that of "arid, rugged, harsh, sterile and bare lands", and whose afforestation was necessary to improve its appearance so any visitor who arrived by sea to the Cíes could find a Spain "warm, fertile, forested, rich, industrial and peaceful". So, afforestation in this archipelago continued to be carried out by ICONA at the expense of natural habitats and still using exotic species, which is why they continued to win criticism from environmental sectors [17, 20, 32].

Starting in the 1960s, an unusual interest in tourism began to grow on the coast in general, and on the island territories in particular. The Cíes Islands played an important role in this new phenomenon, derived from the accelerated abandonment of its inhabitants and the growing promotion as a destination for touristic excursions. In this way, organized boat visits to the islands began to be promoted, which attracted numerous groups of people who uncontrollably accessed places of high ecological fragility, such as cliffs, fixed dunes, coastal heaths, rocky slopes, and humid dune slacks, causing a high negative impact on natural ecosystems due to the promotion of garbage, the production of fires due to uncontrolled bonfires, the erosion and loss of natural habitats, the collection of wild flora species, or the capture and nuisance of wild fauna population species.

In many of these cases, the high attraction that the Cíes islands had for the enjoyment and recreation of visitors, motivated an excessive profusion of free and uncontrolled camping, which was carried out in a completely unsustainable way. Due to the increasing interest in the island lands, a new trend began by the former inhabitants towards the sale of their few private properties. As a result of this process, the new buyers proceeded to build chalets, sheds, shacks, additions, etc., in an uncontrolled way and without any kind of permit or authorization, accompanied in many cases by the introduction of ornamental non-native plant species to decorate the properties of the new owners, although many of these species over time showed a high invasive potential, negatively affecting the surrounding natural ecosystems. So much so, that in the late 1970s it was possible to identify the presence of more than 300 uncontrolled shacks, and as many tents, dispersed throughout the island territory, without any type of environmental criteria or caution [17]. At the end of every summer season, a Dantesque spectacle of garbage and waste covered the islands: several boats were necessary to eliminate the kitchens, refrigerators and even ping-pong tables that were scattered throughout the island, as well as an intense smell of latrine invaded the area permanently.

In addition to the afforestation of PFE and ICONA, as well as uncontrolled buildings and public use, during the summers of the 1970s, some of the islands (the South Island of Cíes, mainly) were occupied by companies of Special Operational Commands (COES the acronym in Spanish) of the Spanish Army, who remained in the island territories performing survival practices. For this, they did not hesitate to build cabins using trunks and branches that were cut down or removed from the trees present on the islands, as well as they fed on everything that could be edible for humans, eggs, and chickens of seabirds, mainly [20]).

4. The first considerations of the islands as a natural protected area

All the described impacts generated several conservation problems on the natural habitats and on flora and fauna species of the island territories, constituting a threat to their long-term maintenance. These consequences, and fundamentally those derived from the presence of uncontrolled people in the archipelagos, were detected and denounced in the local media and several publications, so since 1975 Vigo municipality started various proposals for the declaration of the Cíes Islands as a natural protected area. ICONA was also developing at that time an inventory of the areas that could receive the declaration of one of the categories of natural protected areas established in the then current Law 15/1975 [33], and Cíes islands were the first proposed area in Pontevedra province. So the situation was completely favorable, and finally, in 1980, the Cíes Islands were considered as a Natural Park, in order to "preserve such an exceptional place and the need to properly condition it so that it could be enjoyed and admired by present and future generations. Its beauty was joined to the presence of very interesting colonies of seabirds that nested in the islands, such as the European herring gull, the European shag, and the common guillemot".

The protection regime of the new Natural Park was established in 1982, protecting all-natural values, as these were considered the geological, botanical, faunal, and landscape values, as well as archaeological and historical remains. To preserve all these, every user should be properly authorized by ICONA, establishing some expressly prohibited activities such as the access of visitors through unauthorized places or in a greater number than authorized by ICONA, the free camping, the garbage disposal, or the bonfires, among many others. According to this regulation, the Managing Board of the National Park established and approved in the same 1982 year a limit of 3000 persons per day in Cíes islands. Regarding urban uses, these did not appear expressly forbidden, but a procedure was established to eliminate the clandestine invasions, occupations, exploitation, and installations, as well as the mandatory coordination with the urban policy, including the elaboration of a Special Plan for Protection of the Natural Park. Even this plan was never approved, the declaration of the Natural Park would serve to paralyze the irregular construction activity in the archipelago, accompanied by dismantling and demolition of illegal constructions [20].

The transfer of the competencies in nature conservation to Galician regional government (Xunta de Galicia) were initially established Royal Decree 167/1981, which were finally consolidated by Royal Decrees 1706/1982 and 1234/1983, a new regulation was established in the Natural Park in 1983, in order to regulate the access of visitors to the islands. Joining to the 3000 persons per day limit, the visits to the Natural Park started to be prohibited when, through collective, public, or private boats of transport, a number greater than 10 people disembarked, or when they were disembarked by a place not expressly authorized by Xunta de Galicia. The visits of less than 10 people should be provided by the appropriate authorities of the National Park staff.

Regarding the forest plantations, the declaration of the Natural Park did not contribute to their environmental suitability. Although the new afforestation with exotic species would be partially replaced by native species plantations [34], ICONA still kept planting some areas with non-native species (*Quercus rubra*, *Fraxinus ornus*) between 1979 and 1994, or they even made mixed plantations (with native and non-native species). The basis of these actions does not differ much from the plantations made by PFE during the mid-20th century, as they lacked a prior assessment and they were carried out without a proper species selection criterion.

So, as a consequence of the Natural Park declaration, the wooden area in the islands continued to be increased, as a result of the forest plantations, but also it could be ascertained the invasive potential of some of the introduced species (*Eucalyptus globulus*, *Acacia dealbata*, *Acacia melanoxylon*), as they formed new several stands surrounding the previously planted plots, or even they formed new ones away from them. The resultant forest formations from all these plantations and the subsequent stands formed by invasive alien trees were represented in the vegetation maps at the end of the 20th century [35, 36], so high natural value habitats (coastal heaths, fixed dunes, humid slacks) were lost or negatively affected in these areas, although to a lesser extent than during the PFE period.

On the opposite, the definitive abandonment of the islands by the irregular residents and the elimination of their illegal constructions, allowed in these areas to increase the fixed dunes and coastal heaths in the archipelago, so natural values of the islands could partially recover. But as a result of the previous continued presence of these people, as well as by the continuation of transit of external visitors to the islands (even their presence was controlled by the Natural Park staff), the colonization of herbaceous invasive alien species was identified [34]. Perhaps the most worrying case is *Arcthoteca calendula*, although there are other species that also occupy significant areas, like *Zantedeschia aethiopica*, *Cortaderia selloana*, *Tritonia x crocosmiiflora*, *Carpobrotus edulis*, *Vinca difformis*, *Yucca gloriosa*, *Arundo donax*, *Tropaeolum majus*, etc.

In any case, all the efforts made in Cíes were evident and also favorable for the seabird colonies that were the main reason for the Natural Park declaration, so all those reached benefits allowed Cíes to be designated as a Special Protection Area for Birds (SPA) under Directive 79/409/EEC, as SPA "Illas Cíes" (ES0000001).

Although the other archipelagos (Ons, Sálvora, Cortegada) were not foreseen to be designated as natural protected areas in short term after Cíes Natural Park declaration, they were suffering some kind of similar conservation problems, so specific regulations were necessary to be approved, but outside the legal framework for natural protected areas.

The Ons situation during the 1980s and 1990s was very similar to Cíes. The new afforestation made from 1984 used native and exotic species [34], some of them with invasive potential, and mostly over high natural and seminatural value habitats (coastal heathland, hay meadows). The forest plantations were accompanied by a high density of new paths and roads, causing a high fragmentation of natural habitats, reducing their conservation status, and decreasing their permeability for the migration and genetic exchange of wild species. The basis of these actions was very similar to those made by PFE during the mid-twentieth century, as they lacked a prior assessment and they were carried out without a proper species selection criterion.

The unregulated access of visitors to Ons, especially during the summer, was motivating uncontrolled camping and shaft establishment in unsuitable locations and causing ecological, landscape, and social damages, so conservation measures were necessary. As Ons island territories were completely owned and managed by Xunta de Galicia, this regional public administration established in 1985 a first regulation for the access of visitors to the islands, so new permanent and non-permanent constructions were forbidden, only allowing to camp in the area enabled for this purpose, and including the prohibition of bonfires. This first regulation was later updated in 1994, in order to prevent new impacts over high fragility insular areas, establishing a limit of 80 people staying overnight in the campsite. However, new studies of insular carrying capacity were being drafted at that time, so when they were available, a new regulation was approved in 1995, in order to update the applicable rules to visitors and adapt the management, considering the improvement of the camping area, as it was prepared to receive 200 people. But the demand for new visits to the islands continued to be increasing, so the campsite was enlarged again, and 4 years later the limit raised up to 400 people in 1999, including the prohibition of circulation for motor vehicles.

As a result of all these regulations, Ons islands were allowed to be designated as a Special Protection Area for Birds (SPA) under Directive 79/409/EEC, as SPA "Illa de Ons" (ES0000254), because of the importance of the seabird colonies that were present in the insular territories.

Sálvora island did not suffer any significant change during the 1980s and 1990s, because it had been depopulated earlier than Cíes and Ons, it still was private property, and it was not so interesting for touristic purposes as this island is located in the outer part of Ulla river estuary, surrounded by very rough sea and many rocky reefs, making very difficult the approaching for boats. In fact, Sálvora has a long history of shipwrecks, which has led it to be described as a "boat cemetery" [37].

In the opposite, Cortegada island is located in the inner part of Ulla river estuary, surrounded by very calm waters and sandy flats, so the touristic interest was very high during the 1980s and the 1990s. It was privately owned by a real estate company [38, 39] whose main goal was the urbanization of the island, including hotels, chalets, casino, sporting marina, etc. This even motivated that a proper plan was drafted and approved by Vilagarcía de Arousa council, in order to adapt the municipality planning to allow Cortegada urbanization. But the lack of funding, coupled with the discovery of archaeological remains on the island, delayed this initiative, which was increasingly finding opposition from local environmental groups [40]. The natural and archaeological values, as well as the awareness rising from the society, led Xunta de Galicia in 1991 to establish a preventive protection regime for Cortegada island (through Decree 193/1991), according to the then current Law 4/1989 (that had substituted Law 15/1975), especially regarding the probable urban uses that were being planned at that time and that could potentially constitute a disturbance factor. This meant that any authorization or activity license to be developed in the archipelago that could transform its natural reality should be submitted to a mandatory and binding report from the regional public body responsible for urban and land use planning. This protection regime would be reinforced by the Complementary and Subsidiary Urban Norms of Pontevedra province, which in the same 1991 year would finally include the whole archipelago like "Natural Area", which meant that it was excluded from any possible urbanistic development.

At the end of 1990s, Cíes Natural Park was 10 years working, the rest of the islands had specific regulations to guarantee their conservation, and the efforts were starting to be successful at European level, with two SPAs in Cíes and Ons. In this scenario, it was necessary to expand the protection scope from Cíes to other Galician Atlantic islands in order to harmonize their economic activities and environment conservation, so a first joint management plan was drafted and approved (Decree 274/1999) for Cíes, Ons and Sálvora islands. The singularity and faunal richness of all these islands, as well as their variety of plant communities, high-value landscape, and geomorphology, justified the general interest of their conservation and met the criteria to become a National Park. Subsequently, it was assessed the opportunity and need to also integrate Cortegada islands, as they harbored similar natural, cultural and ethnographic values, so its corresponding management plan was approved 3 years after (Decree 88/2002). In accordance with the unique and fragile ecosystems and landscapes that deserve special protection in the four archipelagos, both management plans designed a zonation for them. But it was established that this was considered as an indicative or preliminary zonation to serve as guidance until the moment it should be definitely established in their corresponding Master Plan for Use and Management (mandatory Plan according to the legal framework on protected areas).

So, all the necessary steps had been made, and the Maritime-Terrestrial Galician Atlantic Islands National Park was finally approved by Spanish Law 15/2002, first and only Galician National Park. The situation was also favorable to new natural protected areas initiatives, because the approval of the Directive 92/43/EEC had started, in parallel to the National Park declaration, the procedure for the designation of Galician Natura 2000 Network in 1999 with the first drafts, continued with the designation of the definitive Sites of Community Importance (SCI) in 2004, and the definitive transformation to Special Areas of Conservation (SAC) in 2014 through regional Decree 37/2014 approval. At the end of this process, these islands were included into three different SACs: Cíes was included in SAC "Illas Cïes" (ES0000001), Ons was included in "Complexo Ons-O Grove" (ES1140004), and Sálvora was included in SAC "Complexo húmido de Corrubedo" (ES1110006).

5. Actual challenges and strategies for biodiversity conservation

Five months after the declaration of PNG, an ecological catastrophe occurred in the archipelagos, as they received the impact of the oil spill from the Prestige oil tanker, which sank 130 miles off the Galician coast. Sálvora and Ons were the most exposed islands, although Cíes also received a significant amount of oil. This discharge caused negative effects on marine and terrestrial ecosystems, including their habitats and species [41–46]. The clean-up work to remove the fuel from the coasts of Galicia lasted for 20 months. However, the importance of the islands of PNG in the coastal dynamics was confirmed during this great tragedy, since these archipelagos acted as a natural barrier against fuel, preventing a large part of it from reaching the neighboring coasts, especially the estuaries of Vigo, Pontevedra and Arousa, one of the most productive marine territories in the Atlantic Ocean.

In any case, the PNG declaration would entail a change in the management criteria of the terrestrial ecosystems of the archipelagos, beginning to carry out works and actions aimed at improving the conservation status of natural ecosystems, by reducing their impact factors fundamentally.

In Cíes islands, small experiments were beginning to be carried out on the elimination of non-native wooded formations planted by PFE and ICONA. The elimination of these formations responded to biodiversity conservation criteria since the occupation of the territory by forest plantations of non-native species caused a decrease of conservation status for natural habitats, as well as a break in the connectivity of ecosystems and therefore a reduction of their permeability for the present species. The elimination of exotic tree plantations also met safety criteria: sometimes, the size achieved by planted trees for more than 60 years caused these stands to reach a state of senescence, with high sizes that pose a risk of falling trees that could cause damage to natural components, people or real estate. Moreover, the presence of herbaceous invasive alien species still continued after the PNG declaration [47], so several initiatives for their removal started. Besides the herbaceous invasive alien species (*C. edulis*, *Arcthoteca calendula*, *Z. aethiopica*, *C. selloana*, *Tritonia x crocosmiiflora*, etc.), they started elimination of *E. globulus* and *A. melanoxylon* that previously invaded and encroached natural habitats. Nowadays, the invasive alien species conservation problem is one of the main concerns in NW Spain [48].

The elimination of forest plantations and invasive alien species formations was complemented by several actions for the recovery of fixed dunes. In Cíes, at various points of the islands, small formations of *E. globulus* and *A. melanoxylon* were eliminated on coastal scrub and dune habitats, and in parallel a series of visitor access control devices were installed for several years, preventing the transit of people in the dune system of Rodas beach. In Ons, a few forest plantations were carried out with native species, with a purpose to restore degraded areas, and dune regeneration actions also started on the beaches of Melide and Canexol, consisting in the elimination of tree and herbaceous invasive species (*E. globulus*, *A. melanoxylon*, *C. edulis*) and followed by the establishment of exclusion zones from people transit, to favor the regeneration of dune habitats. In Sálvora, the prohibition of transit through dune systems since PNG declaration led to the recovery of the degree of coverage of its characteristic herbaceous formations. In Cortegada, small areas of *Pinus* and *Eucalyptus* formations started to be removed, since the degree of senescence of these formations advised their elimination due to the high risk and danger to the surrounding habitats, as well as to the visitors of PNG.

All these measures allowed the recovery of habitats degraded by uncontrolled public use prior to the declaration of PNG, which caused sand erosion and area loss of the different types of natural habitats. The implementation of these measures made it possible to stop the area loss, and even achieve an increase in the occupied area by the dunes. An increase in the degree of coverage of the characteristic species of dune habitats was also achieved, as well as a recovery of their natural structure, which results in an improvement of the functionality of the dune ecosystem, and ultimately in an improvement of its future prospects.

The success of these initiatives would result in the recognition of these archipelagos as a new protected area under the European legal framework, which would join the SACs and SPAs that had already been previously designated by the regional government in Cíes, Ons, and Sálvora. In this way, the Spanish Government declared in 2014 a series of new SPAs in Spanish marine waters (Order AAA/1260/2014), including in one of them the maritime waters under Spanish sovereignty or jurisdiction that are surrounding these same three archipelagos (Cíes, Ons and Sálvora): SPA "Espacio Marino de las Rías Baixas de Galicia" (ES0000499).

Futhermore, considering the high importance of the marine biodiversity of PNG, this received two additional protection categories by international instruments. The first one was conferred in 2008, as it was integrated in OSPAR network, which is focused on the protection and conservation of marine ecosystems and diversity of North-East Atlantic, becoming the first Spanish protected area under OSPAR Convention. The second one is very recent, as in May 2021 PNG has definitely been included in the List of wetlands of international importance, as defined by the Ramsar Convention. This declaration implies its automatic consideration as "Protected Wetland", a regional category of protected area that is specifically designed for Galician wetlands that fulfill a function of international importance for natural resources conservation and especially as a habitat for waterfowl.

Paradoxically, the improvement of the conservation status of island ecosystems, and the promotion of new protected areas at a European and international level, would increase the demand for visits to the PNG archipelagos, in order to know and enjoy their landscape, their coasts, and their beaches. Every summer more and more people would visit the island territories, both in organized groups and on a discretionary basis, reaching the islands in collective or individual transport boats. The archipelagos that receive the highest number of visits are those with the largest continental area (see **Table 3**): Cíes often exceeds 300,000 annual visitors, Ons easily exceeds 140,000 visitors a year, Sálvora can reach more than 20,000 visits/year, and Cortegada often exceeds 10,000 visitors in the busiest years. In total, PNG is easily over 400,000 visits/year, touching 490,000 visitors during the peak years.

	Cies	Ons	Sálvora	Cortegada	Total
Area					
Terrestrial	433 ha	470 ha	248 ha	44 ha	1195 ha
Marine	2658 ha	2171 ha	2309 ha	147 ha	7285 ha
Visitors/year					
2021	270,798	139,734	13,048	5390	428,970
2020	208,404	95,918	10,183	4065	318,570
2019	296,205	150,684	14,908	10,477	472,274
2018	291,283	160,468	21,380	13,092	489,953
2017	303,516	102,178	14,243	10,354	440,661

Source: National Parks Autonomous Agency.

Table 3.
Visitors per year in PNG archipelagos during 2017–2021 period (until September 15, 2021).

In comparison with the Spanish Network of National Parks, PNG is among the ones with the lowest number of visitors per year, especially compared to some of the large mountain National Parks (**Table 4**), such as Guadarrama (33,960 ha) or Picos de Europa (67,128 ha), which can receive up to 2–3 million visitors a year. However, taking into account the territorial dimensions of PNG, with barely 1200 ha of land area, it receives a significant annual number of visitors, not negligible, similar or even higher than that of other large terrestrial National Parks, such as Aigüestortes (14,119 ha), Doñana (54,252 ha), or Monfragüe (18,396 ha).

Comparing the island National Parks (**Table 5**), the number of visits is directly proportional to those that occupy a greater land area. Such is the case of the Canarian National Parks such as Garajonay (3986 ha), Taburiente (4387 ha), Teide (18,990 ha), and Timanfaya (5107 ha), which range from more than 500,000 visitors a year in Taburiente, to more than 4 million annual visitors in Teide. The smallest insular National Parks, PNG, and Cabrera, both with a land area of just over 1000 ha, reach more than 400,000 visits per year in PNG, and just over 120,000 visits per year in Cabrera. Not inconsiderable figures, especially in the case of the Galician islands, taking into account that both are the only National Parks that have the consideration of "Maritime-Terrestrial" within the Spanish Network of National Parks, since their land area is a minority and they are mostly occupied by marine waters (7285 ha in the Galician islands representing 86% of the National Park, and 89,478 ha in Cabrera representing 99% of the National Park), making them difficult for visitors to arrive from the nearest coasts because the access is only possible by boat as there are no airports within these island territories.

The gradual increase of visitors to PNG (**Table 3**), given the condition of a National Park, motivated the preparation of a study of the carrying capacity of these archipelagos [49]. The concept of carrying capacity is a term widely used in the study of ecology, tourism, or sustainable use of resources [50–53], trying to approximate the maximum number of visitors that can use an area without significant alterations to the conservation status of the vulnerable elements or to the quality of the visitor experience. Traditional management approaches based on the strict application of the carrying capacity principle are suboptimal, so an adaptive management framework has been demanded, but it has been scarcely explored [54].

The results of the study of the carrying capacity [49], after taking into account the physical, psychological, ecological-environmental, global, and seasonal carrying capacity, allow establishing the maximum thresholds of visitors per day

National parks	2013	2014	2015	2016	2017	2018	2019
Aigüestortes	433,529	485,935	525,067	586,334	560,086	552,014	560,723
Cabañeros	84,616	88,196	100,993	104,565	112,760	108,561	100,493
Cabrera	93,291	108,038	120,505	121,189	126,143	118,232	82,007
Doñana	277,173	296,777	300,287	288,637	288,759	258,683	388,325
Garajonay	817,220	865,493	828,758	870,486	907,277	1,245,480	1,016,324
Guadarrama	1,140,910	2,815,024	2,989,556	2,440,128	2,691,890	2,284,293	1,519,039
Monfragüe	278,400	253,153	288,644	280,319	288,589	263,036	457,555
Ordesa	589,400	590,050	598,950	608,950	566,950	578,850	589,450
Picos Europa	1,545,830	1,842,272	1,913,858	2,101,293	2,047,956	1,958,240	1,791,411
PNG	318,034	363,121	399,890	400,465	440,661	489,953	472,274
S. Nevada	611,095	690,150	780,702	734,539	732,657	655,259	789,756
T. Daimiel	250,295	155,755	192,025	181,106	170,098	196,623	157,424
Taburiente	375,180	392,990	445,084	509,183	525,961	510,600	487,060
Teide	3,292,247	3,212,632	3,289,444	4,079,823	4,327,527	4,330,994	4,443,628
Timanfaya	1,452,365	1,575,029	1,655,772	1,703,258	1,723,276	1,692,339	1,626,970
Total	11,559,585	13,734,615	14,429,535	15,010,275	15,510,590	15,243,157	14,482,439

Source: OAPN visitor count data.

Table 4.
Number of visitors per year in the Spanish Network of National Parks during the 2013–2019 period.

	PNG	Cabrera	Garajonay	Taburiente	Teide	Timanfaya
Area						
Terrestrial	1195 ha	1.316 ha	3.986 ha	4.387 ha	18.990 ha	5.107 ha
Marine	7285 ha	89.478 ha	—	—	—	—
Visitors/year						
2019	472,274	82,007	1,016,324	487,060	4,443,628	1,626,970
2018	489,953	118,232	1,245,480	510,600	4,330,994	1,692,339
2017	440,661	126,143	907,277	525,961	4,327,527	1,723,276
2016	400,465	121,189	870,486	509,183	4,079,823	1,703,258
2015	399,890	120,505	828,758	445,084	3,289,444	1,655,772
2014	363,121	108,038	865,493	392,990	3,212,632	1,575,029
2013	318,034	93,291	817,220	375,180	3,292,247	1,452,365
2012	280,798	104,499	752,095	354,901	2,660,854	1,474,383
2011	322,396	185,358	825,638	424,832	2,731,484	1,549,003
2010	292,374	160,306	610,254	387,805	2,407,480	1,434,705
2009	274,716	60,662	625,801	377,349	3,052,830	1,371,349
2008	254,000	60,804	860,000	408,088	2,866,057	1,600,175
2007	238,939	76,541	884,858	389,024	3,142,418	1,748,149
2006	213,897	71,987	854,824	377,582	3,349,204	1,778,882
2005	213,897	71,987	854,824	377,582	3,349,204	1,778,882
2004	182,394	73,540	859,860	367,938	3,540,195	1,815,186
2003	171,999	66,535	641,754	395,264	3,364,873	1,841,431

Source: OAPN visitor count data.

Table 5.
Number of visitors per year in the insular National Parks during the 2003–2019 period.

(**Table 6**). This study was used as a scientific basis for the thresholds that would be established by the Master Plan of Use and Management (MPUM) of PNG, approved by Decree 177/2018, constituting the first protected area in Galicia to implement a study of these characteristics within its regulatory scope. The overall objective of this plan was the maintenance or, where appropriate, the reestablishment, in a favorable conservation status, of natural habitats and flora and fauna species of interest for conservation, taking into account economic, social, and cultural requirements, as well as regional and local particularities. So, MPUM included the provisions of the PNG Declaration Law (Law 15/2002) and of its initial planning instruments (Decree 274/1999, Decree 88/2002), as well as those established by Law 30/2014 on National Parks and by the Master Plan of the National Parks Spanish Network (Royal Decree 389/2016). Obviously, MPUM incorporated the guiding principles of European (Directives 92/43/EEC and 2009/147/EC), Spanish (Law 42/2007), and regional (Law 9/2001, currently replaced by Law 5/2019; Decree 37/2014) regulations of natural heritage, biodiversity, and protected areas.

In addition to the establishment of a carrying capacity limit for PNG archipelagos based on scientific-technical criteria, the MPUM (Decree 177/2018) defined definitive zoning, taking as orientation the preliminary zonation of the initial management plans (Decree 274/1999, Decree 88/2002), and following the criteria established in the Master Plan of the National Parks Spanish Network (Royal Decree

	Cíes		Ons		Sálvora		Cortegada	
	PS	**LS**	**PS**	**LS**	**PS**	**LS**	**PS**	**LS**
A	1600–1800	0	1200–1300	0	0	0	0	0
B	100–200	250–450	100–200	250–450	150–250	150–250	150–250	150–250
C	2000		—		—		—	
D	500–600	0	250–300	0	—	0	—	0
E	75–125	0	60–70	0	15–20	0	15–20	0
F	250–450		250–450		150–250		150–250	

[PS]: Peak Season; [LS]: Low Season; A: límite de acceso diario en las navieras autorizadas para la realización de transporte colectivo; B: límite de acceso diario en grupos organizados y autorizados; C: límite de acceso diario por transporte marítimo (A + B); D: límite de personas en el camping; E: límite de fondeos diarios. Temporada baja; F: límite de acceso diario en grupos organizados y autorizados.

Table 6.
Maximum thresholds of visitors per day in PNG archipelagos, according to carrying capacity study [49].

389/2016) for the zoning of maritime-terrestrial national parks. These criteria determine that vertical dimension has to be taken into account to adapt the delimitation of the marine zonation to the different depths and ecosystems, considering the water column, the seabed, and the isobaths. This aspect was incorporated into the zoning of Decree 177/2018, in which various marine zones of moderate use are contemplated on the surface of marine waters, while the seabed is included in another category of zoning (reserve marine zone, restricted-use marine zone). So, PNG has become the first Spanish National Park that has implemented this three-dimensional methodology in its zoning scope and therefore has integrated it into its measure regime for management and conservation of natural heritage and biodiversity, when it comes to establish certain limitations of use for the different zoning categories that are defined in MPUM. In addition, taking into account the Natura 2000 consideration of PNG archipelagos, the zonation of the MPUM also kept a direct correspondence with the zoning units of the Master Plan of the Galician Natura 2000 Network (Decree 37/2014), in accordance with Law 42/2007, which provides that the Spanish categories of protected areas must be assigned to those internationally recognized, for the purposes of homologation and compliance with international commitments.

At present, after the initial conservation actions that followed the declaration of the National Park, and the regulation of visitor access according to the load capacity established with scientific-technical criteria, new challenges are being posed in the Atlantic islands of Galicia. In September 2021 has started a new LIFE project entitled "Integrated strategy for sustainable management of insular habitats in Natura 2000 islands of the Atlantic Ocean", whose acronym is LIFE INSULAR (LIFE20 NAT/ES/001007). It's a project that targets a favorable conservation status of fixed grey-dunes habitat (2130*) and its contact habitat (4030) in Atlantic Ocean islands, spread across Atlantic and Macaronesian biogeographical regions. The project has a transnational scope, so eight Spanish and Irish Natura 2000 SACs have been selected to develop conservation actions, addressing common conservation problems and threats to increase the area and improve the structure and future prospects of targeted insular habitats in five different islands from both Member states. Three of them will be islands from PNG: Cíes, Ons, and Sálvora. Best practices of proven effectiveness will be applied, from September 2021 to December 2026, on targeted insular habitats to address common conservation problems and threats from a transnational approach. The covered area by the targeted habitats will be increased by elimination of old senescent forest plantations established by PFE and ICONA, cultivation of characteristic plant species of insular habitats, and their restoration. Their structure and function will be improved through the control of competition against plant invasive alien species, as well as their future prospects through the improvement of habitat knowledge and protection measures against anthropogenic pressures. The project will be complemented by a transnational strategy to inform and raising public awareness to the general public about the relevance, natural values, and ecosystem services provided by insular ecosystems, as well as transferring the measures developed in the project for their replicability at the EU level through specific replication and networking strategies. LIFE INSULAR is expected to have a great demonstrative character, allowing high replicability and transferability to other European island territories, or even worldwide, so it is considered that the selected insular territories will be representative from two biogeographical regions where European Natura 2000 islands in the Atlantic Ocean are located.

6. Conclusions

Biodiversity conservation and management in Spanish-protected areas have evolved over time in a significant way, and especially the Galician Atlantic Islands National Park, one of the two maritime-terrestrial National Parks in Spain. Prior to its declaration as a National Park, during the 20th century the islands that form it were gradually depopulated, which caused the abandonment of agrosystems and their substitution by natural habitats recovery. But from the 1950s the PFE first, and ICONA second, transformed coastal scrubs and dune systems by afforestation with exotic species (some of them invasive species), constituting a decrease in the conservation value of the islands, as high-natural value habitats are substituted by low-natural value forest formations. During the 1960s and 1970s the uncontrolled visitors caused a lot of damages to the natural heritage of the archipelagos. The Cíes Natural Park declaration in 1980, and subsequently the establishment of several protection measures in the rest of the archipelagos, helped to halt the biodiversity loss in these islands.

Finally, the declaration of the Maritime-Terrestrial Galician Atlantic Islands National Park introduced a new way of management under scientific-technical criteria, that was executed in these four archipelagos through developing conservation actions to restore habitats, assessing the conservation problems, and halting the impacts. This change of perspective made possible a significant improvement of the conservation status of natural ecosystems, allowing new declarations of a huge number of protected areas at regional, national, European, and international level, overlapping and reaching important synergies between them.

So this National Park has become a reference in Galician and Spanish conservation scheme, as a lot of visitors travel to the islands in order to know first-hand the natural values that have motivated the declaration of all those different categories of protected areas. This has led to establish the National Park planning several limits of number of visitors depending on the island, the season, or the type of tourism they are developing.

Nowadays, new challenges arise in the National Park, such as the removal and control of plant invasive alien species, the elimination of senescent forest formations, or the restoration of natural ecosystems using characteristic plant species of insular habitats employing local and compatible genetic material for plant production. The genetic characterization of the insular plant reproduction material, versus the continental one, appears as one of the future fields for further research in the archipelagos. The start of new European initiatives to achieve these goals within the islands, establishing important synergies with other countries, is a valid alternative and powerful for reaching success in improving the conservation status of natural habitats and wildlife.

Acknowledgements

The authors thank National Parks Autonomous Agency (OAPN the acronym in Spanish) for providing the data regarding visitors per year of the Spanish Network of National Parks.

Author details

Javier Ferreiro da Costa[1*], Pablo Ramil-Rego[1], Manuel A. Rodríguez Guitián[1], Hugo López Castro[1], Carlos Oreiro Rey[1], Luis Gómez-Orellana[1] and José Antonio Fernández Bouzas[2]

1 GI-1934-TB, Institute of Agricultural Biodiversity and Rural Development (IBADER), University of Santiago de Compostela, Campus Terra, Lugo, Spain

2 Maritime-Terrestrial Galician Atlantic Islands National Park, Vigo, Spain

*Address all correspondence to: javier.ferreiro.dacosta@gmail.com

References

[1] Alba Alonso J. Potencialidad de los espacios naturales protegidos como impulsores del desarrollo socioeconómico de su área de influencia el caso del Parque Natural de Somiedo (Asturias) 1988-98. PhD tesis. Madrid: Universidad Autónoma de Madrid; 1999

[2] Paluzie i Mir L. Los espacios naturales protegibles, su conservación, regulación legal e incidencia en la ordenación del territorio. Barcelona: Universitat Politécnica de Catalunya; 1990. p. 231

[3] Mallarach JP. Evaluación de la gestión de los espacios naturales protegidos: criterios, métodos y retos. En: Gómez Limón J and Múgica de la Guerra M. (Coords.): Actas del 5° Congreso Europarc-España: evaluación de la gestión de los espacios naturales protegidos. ESPARC 99. Fundación Fundación Fernando González Bernáldez: Madrid; 2000

[4] Ferreiro da Costa J, Ramil-Rego P. Biological conservation and nature protection strategies in Spanish Atlantic region. In: Sen B, editor. Selected Studies in Biodiversity. Rijeka, Croacia: InTech; 2018. pp. 25-44

[5] Ramil Rego P, Rodríguez Guitián MA, Ferreiro da Costa J, Rubinos M, Gómez-Orellana L, de Nóvoa Fernández B, et al. Os Hábitats de Interese Comunitario en Galicia. Lugo: Universidade de Santiago de Compostela; 2008

[6] Díaz Álvarez P. Carta y noticia arqueológica de las Islas Cíes. Vigo: Tipografía Faro de Vigo; 1958. p. 22

[7] Díaz Álvarez P. Las islas de los Dioses. Vigo: Banco Simeón; 1981. p. 121

[8] Díaz Álvarez P. Ánforas romanas en el eje atlántico Galaico-lusitano. Vigo: El Autor; 1984. p. 94

[9] Patiño Gómez R, González Fernández M. Historia de las Islas Cíes. Vigo: RP Edicións; 1989. p. 96

[10] Monteagudo L. Localizaçao das Cassiterides o Oestrymnides. Revista de Guimaraes. 1957;**67**(3-4):372-416

[11] Casal García R. Pedras do anelo do Noroeste peninsular. Gallaecia. 1980;**6**:101-110

[12] Taboada y Leal N. Descripción topográfica-histórica de la ciudad de Vigo, su ría y alrededores. Santiago de Compostela: Imprenta de la Viuda e Hijos de Compañel; 1840. p. 230

[13] Santiago y Gómez J. Historia de Vigo y su comarca. Madrid: Imp. y Lit. del Asilo de Huérfanos del Sagrado Corazón de Jesús; 1896. p. 604

[14] López Ferreiro A. Historia de la S.A.M. Santiago de Compostela: Iglesia de Santiago de Compostela; 1898

[15] Álvarez Limeses G. Pontevedra. In: Carreras y Candi F. (dir.), Geografía General del Reino de Galicia. Barcelona: Casa Editorial Alberto Martín; 1936

[16] Otero Pedrayo R. Guía de Galicia. Santiago de Compostela: Sucesores de Galí; 1945. p. 589

[17] Fernández de la Cigoña E. Islas Cíes, Parque Natural de Galicia. Asociación Gallega para la Cultura y la Ecología. Viana do Castelo: Gráfica da Casa dos Rapazes; 1986

[18] Fernández de la Cigoña E. Illas Ons e Sálvora. Historia Natural e humana e outros traballos. Vigo: Edicións Xerais; 1989

[19] Fernández de la Cigoña E. Illas de Galicia: Cíes, Ons, Sálvora, Tambo, San Simón e Cortegada. Vigo: Montes e Fontes. Edicións Xerais; 1991a. p. 229

[20] Fernández de la Cigoña E. Cíes y Ons. La ruta de las islas. Naturaleza Gallega. Vol. Vol. IV. Vigo: Asociación Galega para a Cultura e a Ecoloxía (AGCE); 1991b

[21] Marcide Odriozola I, Basanta P. Plan de intensificación inmediata para el desarrollo de la repoblación forestal hasta conseguir un mínimo de 5.000 hectáreas anuales pro provincia. La Coruña: Desarrollo de la industria de la madera. Congreso Agrícola de Galicia. Imprenta Moret; 1944

[22] Areses R. Las repoblaciones forestales en Galicia, como negocio. En: Instituto de Ingenieros Civiles de España: II Congreso Nacional de Ingeniería. Tomo IV. Agricultura, Montes e Industrias derivadas. Madrid: Instituto Nacional de Investigaciones Agronómicas; 1951a. pp. 139-156

[23] Areses R. Algunas plantas de adorno o utilidad, como complemento de las repoblaciones forestales en Galicia. En: Instituto de Ingenieros Civiles de España: II Congreso Nacional de Ingeniería. Tomo IV. Agricultura, Montes e Industrias derivadas. Madrid: Instituto Nacional de Investigaciones Agronómicas; 1951b. pp. 171-217

[24] Areses R. La provincia de Pontevedra y la restauración forestal de sus montes. Montes. 1953;**50**:95-107

[25] Kith Tassara M. Correcciones y repoblaciones de dunas. En: Instituto de Ingenieros Civiles de España: II Congreso Nacional de Ingeniería. Tomo IV. Agricultura, Montes e Industrias derivadas: Madrid: Instituto Nacional de Investigaciones Agronómicas; 1951. pp. 237-245

[26] Mulero Mendigorri A. La protección de espacios naturales en España: antecedentes, contrastes territoriales, conflictos y perspectivas. Madrid: Mundi-Prensa; 2002. p. 309

[27] Merino B. Flora descriptiva de Galicia. Vol. 3 Tomos. Santiago de Compostela: Tipografía Galaica; 1905-1909

[28] Losa España D. Datos para el estudio de la flora gallega. Plantas de las islas Cíes. Anales del Jardin Botanico de Madrid. 1943;**4**:357-391

[29] Rico Boquete E. Política forestal e repoboacións en Galicia (1941-1971). Santiago de Compostela: Monografías da Universidade de Santiago de Compostela, n° 187. Servicio de Publicacións e Intercambio Científico; 1995

[30] Gómez Mendoza, J., Mata Olmo, R. (1992): Repoblaciones forestales públicas desde 1940. Objetivos, criterios y resultados. Agricultura y Sociedad 65, págs. 15-64.

[31] Ceballos L. Mapa forestal de España. Escala 1:400.000. Madrid: Dirección General de Montes, Caza y Pesca Fluvial. Ministerio de Agricultura; 1966. p. 50

[32] Fernández Sánchez J, Pradas Regel R. Los Parques Nacionales Españoles. Una aproximación histórica. Madrid: Red de Parques Nacionales. Organismo Autónomo de Parques Nacionales; 1996. p. 482

[33] ICONA. Inventario Abierto de Espacios Naturales de Protección Especial. Pontevedra. Pontevedra: Instituto Nacional para la Conservación de la Naturaleza. Dirección General de Urbanismo; 1978

[34] Fernández Alonso JI, Blanco-Dios JB, Bernárdez Villegas JG, Rigueiro Rodríguez A. Flora y vegetación de las Islas Cíes (Parque Nacional Marítimo-Terrestre de las Islas Atlánticas de Galicia). Madrid: Organismo Autónomo Parques Nacionales; 2011. p. 750

[35] Ruiz de la Torre J. Mapa forestal de España. Escala 1:200.000. Pontevedra.

Madrid: Fundación General de la Universidad Politécnica de Madrid. Escuela Técnica Superior de Ingenieros de Montes. ICONA. Ministerio de Agriculutura, Pesca y Alimentación; 1991. p. 108

[36] ICONA. Segundo Inventario Forestal Nacional. 1986-1995. Galicia. Pontevedra. Madrid: Área de Banco de Datos e Inventario Patrimonial. Servicio de Inventario Forestal. Ministerio de Agricultura, Pesca y Alimentación; 1993. p. 233

[37] Díaz Guerrero F. Naufragios célebres (I). En: Historia de las Rías: 269-284. Faro de Vigo SA. Valencia: Artes Gráficas del Mediterráneo; 2000

[38] Pazos Pérez LJ. La isla de Cortegada en su historia. Gráficas Salnés Cambados, Pontevedra: Apuntes sobre su donación; 2002

[39] Pazos Pérez LJ. Cortegada, unha aldea abandonada dende 1907. Aunios. 2003;7:50-52

[40] Garrido Castromán D. Cortegada. Una isla real. Cambados: Servicio de Publicaciones de la Diputación de Pontevedra. Gráficas Salnés; 2007. p. 345

[41] Cajaraville MP, Garmendia L, Orbea A, Werding R, Gomez-Mendikute A, Izagirre U, et al. Signs of recovery of mussels health two years after the Prestige oil spill. Marine Environmental Research. 2006; **62**:S337-S341

[42] Viñas L, Franco MA, Soriano JA, Gonzalez JJ, Ortiz L, Bayona JM, et al. Accumulation trends of petroleum hydrocarbons in commercial shellfish from the Galician coast (NW Spain) affected by the Prestige oil spill. Chemosphere. 2009; **75**(4):534-541

[43] Morales-Caselles C, Jimenez-Tenorio N, Riba I, Sarasquete C, y DelValls TA. Kinetic of biomarker responses in juveniles of the fish Sparus aurata exposed to contaminated sediments. Environmental Monitoring and Assessment. 2007;**131**(1-3):211-220

[44] Morales-Caselles, C, Kalman, J, Micaelo C, Ferreira AM, Vale C, Riba I, et al. Sediment contamination, bioavailability and toxicity of sediments affected by an acute oil spill: Four years after the sinking of the tanker Prestige (2002). Chemosphere. 2008;**71**(7): 1207-1213

[45] Camphuysen CJ, Bao R, Fortin M, Roselaar CS, Heubeck M. Post-mortem examination of Great Northern Divers Gavia immer killed in the Prestige oil spill, Galicia, Spain, 2002/03. Seabird. 2010;**23**:53-65

[46] Villares R, Real C, Fernández JÁ, Aboal J, Carballeira A. Use of an environmental specimen bank for evaluating the impact of the Prestige oil spill on the levels of trace elements in two species of Fucus on the coast of Galicia (NW Spain). Science of the Total Environment. 2007;**374**(2-3):379-387

[47] Fernández-Bouzas JA, Díaz Gato D, González Baz M, Rodríguez Guitián MA, Ferreiro da Costa J, López Castro H, et al. Medios Insulares: antropización y expansión de especies exóticas invasoras. Situación en el Parque Nacional Marítimo-Terrestre das Illas Atlánticas. En: Ramil-Rego P, Vales C, editor. Especies Exóticas Invasoras: situación e propostas de mitigación. Lugo: Monografías do Ibader, Serie Biodiversidade; 2019. pp. 79-107

[48] Ramil-Rego P, Rodríguez Guitián MA, Gómez Orellana L, Ferreiro da Costa J and López Castro H. Especies Exóticas Invasoras en Galicia: Un problema preocupante en la protección de la Biodiversidad. En: Ramil-Rego P, Vales C, editor. Especies

Exóticas Invasoras: situación e propostas de mitigación. Lugo: Monografías do Ibader, Serie Biodiversidade; 2019. pp. 11-37

[49] Ramil-Rego P, Fernández Bouzas JA, Gómez-Orellana Rodríguez L, Ferreiro da Costa J, Rodríguez Guitián MA. Evaluación de la Capacidad de Carga del Parque Nacional Marítimo-Terrestre das Illas Atlánticas de Galicia. Lugo: GI-1934-TTB. IBADER; 2011. p. 255

[50] Echamendi P. La capacidad de carga turística. Aspectos conceptuales y normas de aplicación. Anales de Geografía de la Universidad Complutense. 2001;**21**:11-30

[51] Getz D. Capacity to absorb tourism: Concepts aud implications for strategie planning. Annals of Tourism Research. 1983;**10**:239-263

[52] Heberlein TA, Shelby B. Carrying capacity, values and the satisfaction model: A reply to greist. Journal of Leisure Research. 1977;**9**(2):142-148

[53] McCool S, Lime D. Tourism carrying capacity: tempting fantasy or useful reality? Journal of Sustainable Tourism. 2001;**9**(5):372-388

[54] Piñeiro-Chousa J, López-Cabarcos MÁ, Romero-Castro N, Vázquez-Rodríguez P. Sustainable tourism entrepreneurship in protected areas. A real options assessment of alternative management options. Entrepreneurship and Regional Development. 2021;**33**(3-4):249-272

Section 4

Community, Ecotourism and Protected Areas

Chapter 9

Towards Sustainable Community Conservation in Tropical Savanna Landscapes: A Management Intervention Framework for Ecotourism Projects in a Changing Global Climate

Boycen Mudzengi

Abstract

Community-based conservation embedded in the Communal Areas Management Programme for Indigenous Resources, as exemplified by the Mahenye ecotourism project, faces numerous challenges due to climate change. It, therefore, becomes imperative to adopt community-based conservation models for the changing global climatic dynamics. The specific objectives of the research were to do the following: (i) identify the shocks emanating from climate change at Mahenye ecotourism project, (ii) indicate adaptations to make the ecotourism model at Mahenye more resilient in the face of shocks emanating from climate change, and (iii) develop a management intervention framework for ecotourism projects in a changing global climate. We approached the research from a qualitative perspective. The shocks emanating from climate change at the Mahenye ecotourism project included a shortage of water and forage for wild animals during drought years, flood-induced damage of buildings and roads due to increased incidence and severity of tropical cyclones, reduced bioclimatic comfort due to temperature rises, and increased theft of flora and fauna due to climate change-related socio-economic deprivation. The adaptations include recalibrating variables ranging from amenities, income streams, marketing, and linkages. The research results could inform environmental planners on strategies for ensuring the sustainability of community ecotourism in a changing climate.

Keywords: adaptations, climate change, community-based conservation, ecotourism, sustainability

1. Introduction

Tropical savanna landscapes in Sub-Saharan Africa are endowed with enormous biological and cultural diversity, however, this endowment is under increasing ecological and social challenges as exemplified by the case of the Mahenye community ecotourism project, southeast Zimbabwe [1]. The Mahenye community

IntechOpen

ecotourism project is facing innumerable challenges ranging from climate change, increasing resource demands, gender imbalances, cultural changes, socio-economic decay, global uncertainties, and health shocks such as the coronavirus (COVID-19) pandemic [1–5]. Community ecotourism is a subtype of Community Based Natural Resources Management (CBNRM) and involves having fun while supporting the protection of natural and cultural resources. It also involves maintaining a low visitor impact and providing the local community with socio-economic benefits [6]. The community ecotourism industry in Sub-Saharan Africa tropical savanna landscapes is mostly nature-based, with wildlife and other natural and cultural attractions being fundamental to its development. Community ecotourism is embedded in the Communal Areas Management Programme for Indigenous Resources (CAMPFIRE) in Zimbabwe and has the potential to significantly contribute to local community development and environmental sustainability [7].

Global climate change presents an extensive existential shock to the resources upon which community ecotourism relies on to thrive in tropical savanna landscapes [8–11]. Global climate change is epitomized by increasing incidences of droughts, heatwaves, tropical cyclones, and floods as well as changing weather patterns. Some of these extreme weather events that are characteristic of climate change have a negative effect on the bioclimatic comfort of ecotourists. Further, climate variability and change pose significant threats to the biophysical environment, which in turn undermines the ecological capability to ensure sustainable wildlife survival in the form of flora and fauna [12]). This is fundamental as wildlife is the drawcard of the ecotourism economy in the African tropical savannas. Climate change also leads to socio-economic deprivation as agricultural yields fall leading to increasing overexploitation of natural resources upon which ecotourism relies [5].

Lindsey et al. [13] highlighted systemic flaws in current conservation models in Africa in the face of COVID-19 pandemic and socio-economic shocks and suggested opportunities to restructure for greater resilience. It is therefore imperative to adopt community-based conservation models to make them more robust and resilient in the face of global climate change. A research gap also exists as few research has been done on the shocks emanating from climate change and possible adaptations to the stresses at CAMPFIRE projects across Zimbabwe. The specific objectives of the research were to: (i) identify shocks emanating from climate change at the Mahenye community ecotourism project, (ii) indicate adaptations to make the ecotourism model at Mahenye more resilient in the face of shocks emanating from climate change, and (iii) develop a management intervention framework for community-based ecotourism ventures in a changing global climate.

2. Methods

2.1 Study area

The Mahenye community ecotourism project is situated in Chipinge District (**Figure 1**), in the remote but biologically and culturally diverse southeast boundary area of Zimbabwe with Mozambique. Mahenye Ward had a total population of 3671 and the number of households was 707 in 2012 [14]. Mahenye is the land of the Shangaan people, also known as the Tsonga or Hlengwe [15]. The tradition and culture of the Mahenye community have remained strong as the Shangaan are ethnically discrete within the Chipinge District. All the other wards of Chipinge District comprise primarily Shona-speaking Ndau peoples [2]. Thus, Mahenye is characterized by discreteness and isolation.

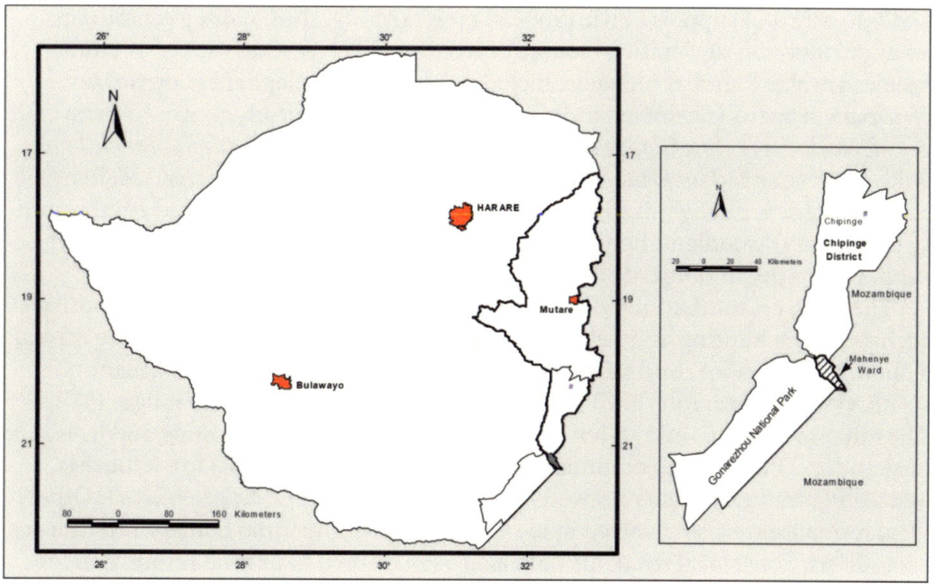

Figure 1.
Location of Mahenye in Southeast Zimbabwe (Source: Authors).

The Mahenye ecotourism is a community-based project initiated in 1982 to promote both biophysical conservation and socio-economic development in the Ward. The ecotourism project is owned by the whole Mahenye community and its secretariat is under the Jamanda Community Conservancy and Trust, whose office bearers are elected by the community. The Mahenye ecotourism project started as a community-driven initiative in partnership with the then Department of National Parks and Wildlife Management, now Zimbabwe Parks and Wildlife Management Authority in 1982 thereby making it the birthplace of CBNRM programmes in Sub-Sahara Africa. This arrangement was officially endorsed when the central government granted appropriate authority over wildlife to Chipinge Rural District Council in 1991 [2]. The Mahenye ecotourism project receives substantial donor funding, however from the year 2003 to 2015 some donor fatigue was experienced. Currently, the Mahenye ecotourism is a community private partnership between the community and a business corporate, River Lodges of Africa which runs Chilo Lodge for the purpose of accommodating visitors to the Ward [16]. River Lodges of Africa has a lease agreement with the Mahenye community and it makes direct financial payments to the Jamanda Community Conservancy and Trust.

The Mahenye community ecotourism project was chosen as it served as one of the early models for the development of the CAMPFIRE in Zimbabwe and the Community Based Natural Resource Management (CBNRM) approach in Sub-Saharan Africa [2, 17], and has been able to remain resilient in the face of significant challenges throughout its operational phase [1]. Mahenye is also within the Great Limpopo Transfrontier Conservation Area (GLTFCA). The Transfrontier Conservation Areas (TFCAs) initiatives seek to promote and facilitate regional peace, tourism, biophysical conservation, cooperation, and socio-economic development in Southern Africa [18, 19].

Mahenye is characterized by teeming tropical savanna biogeography. The average monthly maximum temperatures are 25.9°C in July and 36°C in January. The average monthly minimum temperatures range between 9°C in June and 24°C in January [20]. The annual average rainfall is low ranging between 400 and 600 mm and supports little rain-fed crop farming, thus making ecotourism an important non-agricultural source of livelihood [2]. A wide variety of animal species are also found at Mahenye including the African elephant (*Loxodonta africana*), leopard (*Panthera pardus*), buffalo (*Syncerus caffer*), giraffe (*Giraffa camelopardalis*), zebra (*Equus quagga*), hippopotamus (*Hippopotamus amphibius*), wildebeest (*Connochaetes taurinus*) and kudu (*Tragelaphus strepsiceros*). Mahenye is also mainly covered by mixed mopane (*Colophospermum mopane*) and combretum (*Combrertum*) woodland but a dense riverine forest is found along the Save River supporting a broad range of floral, fish, and avian species [2].

The main ecotourism activities or products at Mahenye are both consumptive such as trophy hunting and fishing and non-consumptive comprising game drives, a photographic safari, birdwatching and identification, canoeing, village tours, scenic views for example, at Chivilila Falls along the Save River and lodges [21]. The other main economic activities in the study area are crop farming and livestock husbandry. There is also community gardening, marketing crafts to ecotourists, and selling traditional beer brewed from Ilala palms (*Hyphenea petersiana*). Other local residents are also involved in low-paying jobs at the Chilo Lodge which houses ecotourists. Some local residents have also been trained as natural resource monitors and game scouts. Further, some local residents also perform traditional dances to ecotourists at cultural festivals [21].

2.2 Data collection and analysis

The research was approached from a qualitative perspective basing the study on data mining and key informant interviews using cellphone calls, e-mailing, and social media platforms. The study was also based on field observations undertaken in May 2019 and the researcher's prior knowledge about Mahenye having carried out research in the area from 2008 to the present. The study also used the case-study approach.

Key information was sought from an environmental management professional at the Environmental Management Agency (EMA) headquarters in Harare, hereafter referred to as Expert 1. A key informant interview was carried out with a senior official at the Zimbabwe Parks and Wildlife Management Authority (ZPWMA), hereafter referred to as Key Informant 1. Other key informant interviews were carried with officials at the Zimbabwe Tourism Authority (ZTA) in Harare and Chipinge Rural District Council (RDC), hereafter referred to as Key Informant 2 and Key Informant 3. Further, key informant interviews were carried with 3 key informants with experience in the Mahenye community ecotourism project, hereafter referred to as Key Informant 4, Key Informant 5, and Key Informant 6. These 3 key informants are residents of Mahenye Ward and have experience in working for the community-based ecotourism project as CAMPFIRE committee members and employees of the Jamanda Community Conservancy. The research engaged 7 key informants as qualitative aspects work with saturation. Engaging more research participants from Mahenye was also problematic given poor access to virtual technologies in the remote community. The key informant interviews were conducted virtually between June and December 2020. The interviews were carried out virtually due to human movement restrictions and social distancing requirements being used as COVID-19 health containment measures. The respondents were informed of the academic purposes of the study and gave their informed consent to participate.

A combination of content and thematic analysis was used to sort the large volumes of collected data into focused and meaningful information for the purpose of addressing the research objectives [22–24]. Data analysis also included identifying and documenting the shocks emanating from climate change at the Mahenye community ecotourism project. The adaptations that can make the ecotourism model at Mahenye more resilient in the face of climate change were determined from the author's field experiences, strategies mentioned in the academic documents, and from key informants. The information obtained from data analysis and authors' field experiences was then used to develop the management intervention framework for community-based ecotourism ventures in a changing global climate. The management intervention framework was adapted from the Sustainable Livelihoods Approach (SLA) [25, 26]. The SLA has been successfully used to understand and promote sustainable rural development [27]. The SLA is a diagnostic tool that provides a framework for understanding and improving the sustainability of livelihood in the face of biophysical, socio-economic, and health shocks.

3. Results and discussion

3.1 Shocks emanating from climate change at the Mahenye community ecotourism project

The shocks emanating from climate change at the Mahenye community ecotourism project were found to be a shortage of water and forage for the wild animals during drought years, flood-induced damage of buildings and roads due to increased occurrence and severity of tropical cyclones, reduced bioclimatic comfort due to temperature rises and increased theft of flora and fauna due to climate change-related socio-economic deprivation.

3.1.1 Shortage of water and forage for the wild animals during drought years

Climate change has resulted in increased incidences of drought leading to shortages of water and forage for the wild animals at Mahenye. According to Key Informant 4 drought negatively impact the survival of wild animals and vegetation upon which community ecotourism relies on. This is worsened by the challenges encountered in pumping underground water for the wild animals such as fuel shortages and poorly serviced pumps due to failure to acquire spare parts as noted by Key Informant 1. Key Informant 3 also noted that the trophy hunting quotas were not achieved during drought years due to wild animal decimation resulting in lower revenues accruing to communities, for example, from trophy elephants. Further, Expert 1 noted that due to drought-induced nutritional shortages some wild animals in southeastern Zimbabwe were wasting away in terms of appearance for most of the year thereby reducing their esthetic value for ecotourists.

Other studies have also noted the negative effect of increasing drought incidences on vegetation production and wildlife populations in tropical savanna ecosystems [28–32]. Increasing drought incidences also have potentially adverse impacts on ecosystem integrity as wild animals concentrate on dwindling perennial watering points triggering erosion and siltation [33]. Further, drought conditions and aridification have altered the migratory patterns of game in the Kgalagai Transfrontier Park in Southern Africa and the Etosha National Park in Namibia as animals have to travel considerably further between wetland and dryland grazing areas [34]. As Namibian tourism is heavily reliant on game viewing and trophy hunting, such migratory behavior has negative impacts on the industry [34]. However, some rangers argue that low forage during drought years allows ecotourists to view game animals with ease [12].

3.1.2 Flood-induced damage of buildings and roads

Climate change has resulted in increased occurrence and severity of tropical cyclone events leading to heavy rainfall downpours and flood-induced damage of buildings and roads at Mahenye. Key Informant 5 noted that the Mahenye Safari Lodge, which is located on Gayiseni Island in the middle of the Save River, has not been operating since 2008 as it was extensively damaged by flooding after a cyclone hit the area leaving only Chilo Lodge operating. Six (6) key informants also attributed the poor state of the roads in the study area to the damage caused by heavy rainfall downpours. Observations also showed that the gravel road linking Mahenye to the main Birchenough Bridge-Chiredzi- Ngundu Highway has been badly damaged in some sections partly due to heavy downpours that have been exacerbated by poor maintenance. Further, some sections of the highways linking Mahenye to major cities such as Harare and Bulawayo and other touristic centers such as Victoria Falls, Kariba, Nyanga, and Great Zimbabwe were damaged by heavy rainfall experienced during the 2020–2021 crop growing season. This makes driving difficult and a risk for ecotourists visiting Mahenye on their travel itineraries. This has the potential to negatively impact the attractiveness, marketability, and income streams of the Mahenye community ecotourism.

Climate change-induced flooding has also damaged ecotourism infrastructure in Botswana [35], South Africa [12, 36, 37], and Nigeria [38]. Darkoh et al. [35] noted that climate change-induced floods resulted in the complete isolation of the Moremi Game Reserve in the Okavango Delta, Botswana thereby leading to business losses and retrenchment of tourism employees. Other researches in the Okavango Delta of Botswana have also shown that climate change-induced flooding has the potential to result in crocodiles and hippopotamus moving into human-populated areas and increase the prevalence of malaria and cholera [35, 39, 40]. However, high rainfall totals associated with tropical cyclone events can boost primary production thereby making more forage available for some wild animals such as grazers like zebra and wildebeest.

3.1.3 Reduced bioclimatic comfort due to temperature rises

Temperature rises at Mahenye have resulted in reduced bioclimatic comfort for ecotourists. According to Expert 1 incidences of heatwaves are increasing across Zimbabwe including Mahenye in the southeast lowveld. Further, extremely high temperatures often lead to violent summer thunderstorms in the tropical savannas. These temperature rises are associated with an increase in greenhouse gases. Key Informant 5 noted that ecotourists at Mahenye have experienced thermal discomfort during incidences of heatwaves in 2016 and 2019. Thermal discomfort results in ecotourists not feeling relaxed and satisfied. Key Informant 1 also noted that the increase in extremely hot days in southeast Zimbabwe including Mahenye has made it difficult to conduct some of the slotted afternoon game drives during some summer months as wild animals would be hiding under bushes to avoid the intense heat and ecotourists would not be eager to do outside activities due to fear of sunburn.

Temperature rises have also negatively affected the hospitality industry in other touristic areas in Zimbabwe such as Victoria Falls [10] and Hwange National Park [41]. Similarly, other studies have shown the negative impacts of temperature rises on bioclimatic comfort in the Okavango Delta, Botswana [11], Namibia [34], Lesotho, Swaziland and Zambia [42], Kruger National Park, South Africa which is also part of the GLTFCA [12] and Nigeria [38]. Further, the temperature rises at a time when rainfall totals are decreasing have potential to result in quick grass vegetation loss in tropical savanna touristic landscapes of the GLTFCA [12]. This is so as persistent droughts reduce the grass fuel load for natural wildfires that are essential for the growth of fresh grass vegetation and curbing bush encroachment [12]. Furthermore, due to increased temperature and diminishing rainfall, the sour veld was reportedly translocating nutrients to the roots faster. This potentially negatively affects the health of grazing animals such as the buffalo and rhinoceros in the GLTFCA [12]. However, the influence of high temperatures on tourists' discomfort is often relative to the perceived temperatures of a destination [43]. Further, temperature rises can potentially attract adventure and extreme sport tourism as some people are either keen to conquer or watch athletes competing in the blistering heat.

3.1.4 Increased theft of flora and fauna due to climate change-related socio-economic deprivation

Climate change-related socio-economic deprivation is resulting in increased theft of flora and fauna at the Mahenye community ecotourism project. Key Informant 5 and Key Informant 6 noted that Mahenye is facing climate change-related issues such as increased poaching of wild animals and timber during drought years. Drought and flooding events associated with climate change have led to increased food insecurity and poverty, leading to the Mahenye community relying more on natural resources. This poses a heightened threat to biodiversity as communities engage in illegal hunting for bushmeat and tree cutting for wood fuel. Cases of human-wildlife conflicts also increase as communities go out into the wilderness in search of natural resources. Other studies have also noted food shortages and poverty as the drivers of illegal hunting in Zimbabwe [44, 45] and Tanzania [46].

3.2 Adaptations to make the ecotourism model at Mahenye more resilient in the face of shocks emanating from climate change

Adaptations to make the ecotourism model at Mahenye more resilient in the face of climate change include promoting climate change compatible ecotourism development and applying as well as lobbying for funds to mitigate climate change effects from international conservation agencies. Climate-proof ecotourism can be achieved by pumping underground water to ensure sufficient water for the wild animals during drought years and constructing climate-smart buildings and roads. Broadening sustainable livelihood options is also another way of mitigating climate change-induced deprivation. Expert 1 noted other livelihood options that can be promoted at Mahenye as sustainable beekeeping, fisheries, selling of traditional wild foods and beverages, crafts, oils, and natural healthcare products. Therefore, resources need to be channeled towards the diversification of livelihood options and community empowerment. This can be kick-started by creating a Livelihood Diversification Fund using proceeds from ecotourism.

Further, adaptations may include the forging of partnerships between Mahenye community ecotourism project and universities to undertake research on climate change mitigation and developing strategies that will ensure sufficient water and forage for the wild animals so important to ecotourism during periods of extreme climatic events. The research can also explore the development of other options for ecotourists in case wildlife declines due to climate change. Climate change mitigation research can also integrate indigenous knowledge systems with modern scientific climate know-how taking advantage of Mahenye community's strong adherence to traditional customs. The utility of local ecological knowledge in adapting and coping with climate change has been shown by some research in the Middle Zambezi Biosphere Reserve [47].

The other adaptation strategies to drought include rainwater collecting [48], rainwater rituals and prayers, and seeking drought relief from external agencies [49]. Consumption adjustment strategies such as turning to drought-resistant traditional foods are another way of adapting to drought. At Mahenye the community can turn to the *hoka* meal. *Hoka* is a drought-resistant long grass with seeds that are soaked, dried, and then ground into a meal. Further, to adapt to climate change in the most advantageous way Németh [50] suggests policymakers should increase local power over resources through the devolution of authority over natural property to grassroots structures at the sub-district level, ensure fair ecotourism benefit distribution, safeguard the inevitable livelihood transition and channel financial capital into reducing vulnerability.

Further, adaptation strategies to temperature rises that have resulted in reduced bioclimatic comfort for ecotourists include constructing more swimming pools and shaded areas as well as planting more trees. In addition, the bioclimatic comfort of ecotourists during periods of intense heat can be improved by providing a greater number of air conditioners and ensuring that cool drinks and refreshments are always available. These adaptation strategies were being implemented in Botswana [51, 52]. Ecotourism businesses in Botswana had also readjusted the timing of game drives and walking trails to cooler times of the day to ensure that human comfort levels are maintained [52]. The readjustment of the timing of outdoor ecotourist activities to ensure human bioclimatic comfort has also been suggested for the Kruger National Park in South Africa [12].

3.3 Management intervention framework for community-based ecotourism projects in a changing global climate

Given the scenario of a changing global climate, it becomes imperative to develop a management intervention framework for community-based ecotourism projects such as Mahenye in order to ensure that the ventures are more resilient to the shock. Our proposed management intervention framework for community-based ecotourism projects in a changing global climate (**Table 1**) shows the shocks affecting ventures and possible adaptations at local and higher levels. The management model also shows the livelihood outcomes that may result from undertaking various adaptations in response to shocks.

Climate change shock	Management adaptations at the local level	Management adaptations at higher levels	Livelihood outcomes
Shortage of water and forage for the wild animals during drought years	Short-term • Rainmaking rituals and prayers • Rainwater harvesting • Selling natural resources • Applying and lobbying for funds to mitigate climate change effects from international conservation agencies Long-term • Pumping underground water to ensure sufficient water for the wild animals during drought years • Broadening sustainable livelihood options	Short-term • Seeking drought relief • Channeling financial capital into reducing vulnerability Long-term • Increasing local power over resources by devolving authority over natural property to grassroots structures • Ensuring fair ecotourism benefit distribution • Safeguarding the inevitable livelihood transition	• Reduced vulnerability to shock • Increased community well-being

Climate change shock	Management adaptations at the local level	Management adaptations at higher levels	Livelihood outcomes
Flood-induced damage of buildings and roads	Short-term • Applying and lobbying for funds to mitigate climate change effects from international conservation agencies • Renovating damaged ecotourism infrastructure Long-term • Constructing climate-smart ecotourism infrastructure • Broadening sustainable livelihood options	Short-term • Channeling financial capital into reducing vulnerability Long-term • Increasing local power over resources by devolving authority over natural property to grassroots structures	• Reduced vulnerability to shock • Increased community well-being
Reduced bioclimatic comfort	Short-term • Readjusting of the timing of game drives and walking trails to cooler times of the day • providing a greater number of air conditioners • Ensuring availability of cool drinks and refreshments • Applying and lobbying for funds to mitigate climate change effects from international conservation agencies Long-term • Applying and lobbying for funds to mitigate climate change effects from international conservation agencies • Ensuring climate-compatible accommodation is available • Retrofitting of old buildings and infrastructure to ensure they are climate compatible • constructing more swimming pools • Providing a greater number of shaded areas • Planting more trees	Short-term • Channeling financial capital into reducing vulnerability Long-term • Promoting research on strategies to ensure bioclimatic comfort for ecotourists at all times through the provision of climate-compatible accommodation	• Reduced vulnerability to shock • Increased attractiveness to ecotourists • Increased community well-being
Increased theft of flora and fauna	Short-term • Applying and lobbying for funds to mitigate climate change effects from international conservation agencies Long-term • Broadening sustainable livelihood options	Short-term • Channeling financial capital into reducing vulnerability Long-term • Increasing local power over resources by devolving authority over natural property to grassroots structures • Ensuring fair ecotourism benefit distribution • Safeguarding the inevitable livelihood transition	• Reduced vulnerability to shock • More sustainable use of natural resource base • Increased community well-being

Source: Authors.

Table 1.
Proposed management intervention framework for community-based ecotourism projects in a changing global climate.

4. Conclusions

Based on this study, the shocks emanating from climate change at the Mahenye community ecotourism project are not substantially different from those found in ventures of a similar nature and circumstances as revealed by literature. The shocks range from shortage of water and forage for wild animals during drought years, flood-induced damage of buildings and roads due to increased incidence and severity of tropical cyclones, reduced bioclimatic comfort due to temperature rises, and increased theft of flora and fauna due to climate change-related socio-economic hardships. The adaptations to make the Mahenye community ecotourism project more resilient in the face of these shocks include recalibrating variables ranging from amenities, income streams, marketing, and linkages. However, the effective restructuring of these variables at Mahenye may be negated by the constrained macro-economic situation in Zimbabwe. It is hoped that the proposed management intervention framework may enable similar ecotourism projects to continue benefiting humans and wildlife for generations in the face of climate change.

Conflict of interest

The author declares no conflict of interest.

Author details

Boycen Mudzengi
Great Zimbabwe University, Masvingo, Zimbabwe

*Address all correspondence to: boycenmudzengi@gmail.com

References

[1] Mudzengi BK, Gandiwa E, Muboko N, Mutanga CN. Towards sustainable community conservation in tropical savanna ecosystem: A management framework for ecotourism ventures in a changing environment. Environment, Development and Sustainability. 2020. DOI: 10.1007/s10668-020-00772-4

[2] Murphree M. Community, council & client: A case study in ecotourism development from Mahenye, Zimbabwe. In: Hulme D, Murphree M, editors. African Wildlife & Livelihoods: The Promise and Performance of Community Conservation. Oxford: James Currey; 2001. pp. 177-193

[3] Balint PJ, Mashinya J. The decline of a model community-based conservation project: Governance, capacity, and devolution in Mahenye, Zimbabwe. Geoforum. 2006;**37**(5):805-815

[4] Byington S. Lowveld practices in Mahenye, Zimbabwe: A critical analysis of resilience in a marginalized Southern African community [thesis]. Missoula, MT: University of Montana; 2014

[5] Mudzengi BK, Gandiwa E, Muboko N, Mutanga CN. Ecotourism and sustainable development in rural communities bordering protected areas. A case study of opportunities and challenges for Mahenye, Chipinge District, Southeast Zimbabwe. Nature and Faune. 2020;**33**(1):70-77

[6] Cetin M, Sevik H. In: Butowski L, editor. Assessing Potential Areas of Ecotourism through a Case Study in Ilgaz Mountain National Park. Rijeka: InTech; 2006. pp. 81-110

[7] Snyman S. The role of private sector ecotourism in local socio-economic development in southern Africa. Journal of Ecotourism. 2017;**16**(3): 247-268

[8] Midgley GF, Bond WJ. Future of African terrestrial biodiversity and ecosystems under anthropogenic climate change. National Climate Change. 2015;**5**(9):823

[9] Kilungu H, Leemans R, Munishi P, Nicholls S, Amelung B. Forty years of climate and land-cover change and its effects on tourism resources in the kilimanjalo national park. Tourism Planning and Development. 2019:1-19. DOI: 10.1080/21568316.2019.1569121

[10] Dube K, Nhamo G. Climate variability, change and potential impacts on tourism: Evidence from the Zambian side of the Victoria Falls. Environmental Science and Policy. 2018;**84**:113-123

[11] Dube K, Mearns K, Mini S, Chapungu L. Tourists' knowledge and perceptions on the impacts of climate change on tourism in Okavango Delta, Botswana. African Journal of Hospitality, Tourism and Leisure. 2018;7(4):1-18

[12] Dube K, Nhamo G. Evidence and impact of climate change on South African national parks. Potential implications for tourism in the Kruger National Park. Environmental Development. 2019. DOI: 10.1016/j.envdev.2019.100485

[13] Lindsey P, Allan J, Brehony P, Dickman A, Robson A, Begg C, et al. Conserving Africa's wildlife and wildlands through the COVID-19 crisis and beyond. Nature Ecology and Evolution. 2020. DOI: 10.1038/s41559-020-1275-6

[14] Zimbabwe National Statistical Agency. Census 2012 Preliminary Report. Harare: Zimbabwe National Statistical Agency; 2012. p. 114

[15] Mombeshora S, Le Bel S. Parks-people conflicts: The case of

Gonarezhou National Park and the Chitsa community in south-east Zimbabwe. Biodiversity and Conservation. 2009;**18**(10):2601-2623

[16] Machena C, Mwakiwa E, Gandiwa E. Desktop Review of the Communal Areas Management Programme for Indigenous Resources (CAMPFIRE) and other Community-Based Natural Resources Management Models in the Region. Harare, Zimbabwe: Government of Zimbabwe and European Union; 2017. p. 136

[17] Wildlife in Livelihood Development. Jamanda Community Conservancy [Internet] 2015. Available from: www.wild-africa.org/jamanda-conservancy/4588094811 [Accessed: June 13, 2021]

[18] Duuton S, Archer F. Transfrontier parks in South Africa. Cultural Survival Quarterly. 2004;**28**(1):1-3

[19] Ferreira S. Problems associated with tourism development in Southern Africa: The case of Transfrontier Conservation Areas. Geographical Journal. 2004;**60**:301-310

[20] Gandiwa E, Chikorowondo G, Zisadza-Gandiwa P, Muvengwi J. Structure and composition of Androstachys johnsonii woodland across various strata in Gonarezhou National Park, south-east Zimbabwe. Tropical Conservation Science. 2011;**4**(2):218-229

[21] Gohori O. Towards a Tourism and Community Development Framework: An African Perspective [thesis]. Potchefstroom: North-West University; 2020

[22] Braun V, Clarke V. Using thematic analysis in psychology. Qualitative Research in Psychology. 2006;**3**(2):77-101

[23] Woodhouse LD. Thematic analysis of documents from the SOPHE health disparity elimination research agenda summit: Illuminating competencies for future researchers and practitioners. Health Promotion Practice. 2006;**7**(3):346-353

[24] Gupta V, Levenburg N. A thematic analysis of cultural variations in family businesses: The CASE project. Family Business Review. 2010;**23**(2):155-169

[25] Scoones I. Sustainable Rural Livelihoods: A Framework for Analysis. Brighton: Institute of Development Studies, University of Sussex; 1998. p. 22

[26] Morse S, McNamara N. The theory behind the sustainable livelihood approach. Sustainable Livelihood Approach. 2013. DOI: 10.1007/978-94-007-6268-8_2

[27] Baker DM, Murray G, Agyare A. Governance and the making and breaking of social-ecological traps. Ecology and Society. 2018;**23**(1):38-45

[28] Wessels KJ, Dwyer PC. Impact of ENSO Events on the Kruger National Park's Vegetation. Stellenbosch: Sun Media; 2011. p. 5

[29] Gandiwa E. Population dynamics of large herbivores and the framing of wildlife conservation in Zimbabwe. Open Journal of Ecology. 2014;**4**:411-420. DOI: 10.4236/oje.2014.47036

[30] Case MF, Staver AC. Fire prevents woody encroachment only at higher than historical frequencies in a South African savanna. Journal of Applied Ecology. 2017;**54**(3):955-962

[31] Kupika OI, Gandiwa E, Kativu S, Nhamo G. Impacts of climate change and climate variability on wildlife resources in southern Africa: Experience from selected protected areas in Zimbabwe. In: Selected Studies in Biodiversity. Rijeka: IntechOpen; 2017. DOI: 10.5772/intechopen.70470

[32] Urban M, Berger C, Mudau T, Heckel K, Truckenbrodt J, Onyango Odipo V, et al. Surface moisture and vegetation cover analysis for drought monitoring in the southern Kruger National Park using sentinel 1, sentinel 2, and landsat 8. Remote Sensing. 2018;**10**(9):1482

[33] Sutherland K, Ndlovu M, Pérez-Rodriguez A. Use of artificial waterholes by animals in the southern region of the Kruger National Park, South Africa. South African Journal of Wildlife Resources. 2018;**2**(48):023003

[34] Reid H, Sahlén L, MacGregor J, Stage J. The economic impact of climate change in Namibia: How climate change will effect the contribution of Namibia's natural resources to it's economy. In: Discussion Paper 07-02. London: International Institute for Environment and Development; 2007. p. 54

[35] Darkoh MBK, Khayesi M, Mbaiwa JE. Impacts and responses to climate change at the micro-spatial scale in Malawi, Botswana and Kenya. In: Salih MAM, editor. Local Climate Change and Society. London: Routledge; 2014. pp. 109-124

[36] Steyn JN, Spencer JP. Climate change and tourism: Mitigating and adaptive strategies with special reference to the Western Cape Province of South Africa. African Journal of Physical Health Education, Recreation and Dance. 2012;**18**(1):1-19

[37] Fitchett JM, Hoogendoorn G, Swemmer AW. Economic costs of the 2012 floods on tourism in the Mopani district municipality, South Africa. South Africa: Transactions of the Royal Society of South Africa. DOI: 10.1080/0035919X.2016.1167788

[38] Dillimono HD, Dickinson JE. Travel, tourism, climate change, and behavioural change: Travellers' persepectives from a developing country, Nigeria. Journal of Sustainable Tourism. 2015;**23**(3):437-454

[39] Kgathi DL, Kniveton D, Ringrose S, Turton AR, Vanderpost CHM, Lunqvist J, et al. The Okavango: A river supporting its people, environment and economic development. Journal of Hydrology. 2006;**331**:3-17

[40] Hambira WL, Saarinen J. Policy-makers' perceptions of the tourism-climate change nexus: Policy needs and constraints in Botswana. Development Southern Africa. 2015;**32**(3):350-362

[41] Mushawemhuka W, Rogerson JM, Saarinen J. Nature-based tourism operators' perceptions and adaptation to climate change in Hwange National Park, Zimbabwe. Bulletin of Geography. 2018;**42**(42):115-127

[42] Preston-Whyte RA, Watson HK. Nature tourism and climate change in Southern Africa. In: Hall CM, Higham J, editors. Tourism, Recreation and Climate Change. Bristol: Channelview Publications; 2005. pp. 130-142

[43] Gössling S, Bredberg M, Randow A, Sandström E, Svensson P. Tourist perceptions of climate change: A study of international tourists in Zanzibar. Current Issues in Tourism. 2006;**9**(4/5): 419-435

[44] Gandiwa E. Preliminary assessment of illegal hunting by communities adjacent to the northern Gonarezhou National Park, Zimbabwe. Tropical Conservation Science. 2011;**4**(4): 445-467

[45] Lindsey PA, Romanãch SS, Matema C, Mupamhadzi I, Muvengwi J. Dynamics and underlying causes of illegal bushmeat trade in Zimbabwe. Oryx. 2011;**45**(1):84-95

[46] Loibooki M, Hofer H, Campbell KLI, East ML. Bushmeat hunting by communities adjacent to the

Serengeti National Park, Tanzania: The importance of livestock ownership and alternative sources of protein and income. Environmental Conservation. 2002;**29**:391-398

[47] Kupika OL, Gandiwa E, Nhamo G, Kativu S. Local ecological knowledge on climate change and ecosystem-based adaptation strategies promote resilience in the middle Zambezi Biosphere Reserve, Zimbabwe. Scientifica. 2019;**2019**. DOI: 10.1155/2019/3069254

[48] Katanha A, Chigunwe G. Climate change adaptation challenges in semi arid regions of Dande Valley in Zimbabwe. International Journal of Science and Research. 2014;**3**(7): 633-640

[49] Kamara JK, Akombi BJ, Agho K, Renzaho A. Resilience to climate-induced disasters and its overall relationship to well-being in southern Africa: A mixed-methods systematic review. International Journal of Environmental Research and Public Health. 2018;**15**(11):2375

[50] Németh D. Southern African conservation projects and their impacts on local people's drought adaptation strategies: A meta-synthesis [thesis]. Utrecht: Utrecht University; 2020

[51] Saarinen J, Hambira WL, Atlhopheng J, Manwa H. Tourism industry reaction to climate change in Kgalagadi South District, Botswana. Development Southern Africa. 2012;**29**(2):273-285

[52] Hambira WL, Saarinen J, Manwa H, Atlhopheng JR. Climate change adaptation practices in nature-based tourism in Maun in the Okavango Delta area Botswana: How prepared are tourism businesses? Tourism Review International. 2013;**17**:19-29

Chapter 10

The Role of Community Reserved Forests in the Conservation of Anuran Amphibians in Meghalaya, North-East India

Ronald Kupar Lyngdoh Tron, Duwaki Rangad, Wankitlang Shangpliang, Baiakmenlang Manners and Iasyllok Rynjah

Abstract

The state of Meghalaya is situated in the north-eastern India and it comprises three major regions, namely, the Khasi Hills, the Jaintia Hills and the Garo Hills inhabited by three main tribal groups, the Khasis, the Jaintias and the Garos respectively. The tribal communities of Meghalaya protect and nurture the forests located close to their habitation and consider them as sacred. These Community reserved forests are managed by the community for their benefits and they comprise almost about 90% of the total forest cover in Meghalaya. With the recent trends of development and construction in the state many habitats are getting destroyed at an alarming rate. These community reserve forests have been seen to provide the maximum number of existing and stable habitats for many amphibian (anuran) species. In addition, they served as suitable sites for the breeding activities and oviposition by anurans. Discovery of many new anuran species have also been reported from such reserved forests.

Keywords: Anurans, Amphibians, Conservation, Community reserved forests, Meghalaya, India

1. Introduction: Meghalaya- the people, the forests and conservation

Meghalaya (in sanskrit, Meghalaya meaning "abode of clouds") is one of the seven states that are popularly known as the seven-sisters, located in the North Eastern part of India. Lying between 25° 47′ and 26° 10' N latitude, and 89° 45′ and 92° 47′ E longitude the state of Meghalaya is represented by an irregular terrain in the western and northern regions, and steep slopes to the south and west sharing a 496 km long international border with Bangladesh (**Figure 1**). It has a wide range of altitudinal variation ranging from about 50–1950 m [1], with Shillong peak as the highest peak. With a geographical area of 22 429 square km. [2], the diverse topography of the state provides for a variety of unique vegetation types at different levels of altitude accompanied by varied climatic conditions and edaphic composition. In general, the forests types in Meghalaya can be broadly classified into temperate and tropical mainly based on the rainfall, altitude and composition of dominant species [3].

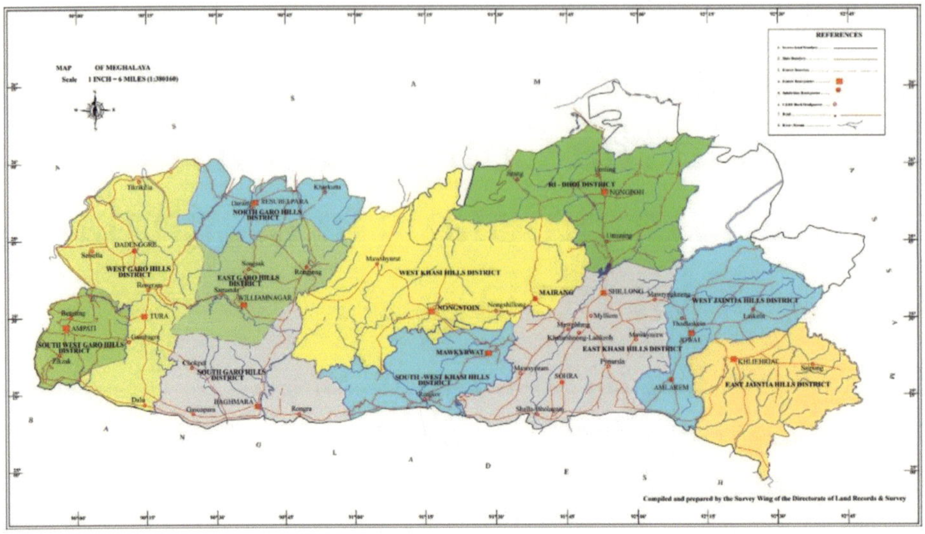

Figure 1.
Map of Meghalaya [source: http://megipr.gov.in].

The variation in elevation and physical relief affects the climate of Meghalaya. The geographical area of Meghalaya is divided into three major regions, namely, the Khasi Hills, the Jaintia Hills and the Garo Hills. Garo Hills is relatively lower in elevation as compared to Khasi and Jaintia Hills and therefore experiences higher temperature conditions and humidity. The Khasi and Jaintia Hills experience a moderate climate because of higher elevation. Rainfall in the state is also influenced by the difference in elevation and topography. The average annual rainfall varies from place to place, about 2600 mm in western Meghalaya, between 2500 and 3000 mm over the northern parts and about 4000 mm over south-eastern Meghalaya [4]. The southern parts of the Meghalaya plateau have the Cherrapunji -Mawsynram region which receives the heaviest rainfall, an annual average of 14,000 mm. In fact, the two places Mawsynram and Cherrapunjee (also Sohra) in Meghalaya are famous for receiving the highest rainfall in the World. A combination of all these factors results in a variety of unique habitats [5, 6].

Most tribal populations in India have a close interaction with nature and especially with the forests. They live close to the forests and depend on them for obtaining various resources for their needs. Meghalaya is a very diverse state having a sizeable number of indigenous groups that includes the Khasis, the Garos and Jaintias (in higher numbers); and other groups like the Karbis, Mikirs, Hajongs, Kochs and Rabhas in smaller numbers [7, 8]. The three major regions, namely, the Khasi Hills, the Jaintia Hills and the Garo Hills are inhabited by three main indigenous communities, the Khasi tribe, the Jaintias and the Garos respectively. The tribal people of Meghalaya nurture tracts of forests that are located close to their settlement as reserved forests as part of their culture, or religious belief or for different community benefits like water sources, forest resources, etc. The Khasi and Jaintia hills are home to a large number of forests that are held as sacred by the indigenous people and remain undisturbed by any human activity. The forests are believed to be the dwelling place of the deities and hence considered as spiritual places similar to a place of worship as in any religion. It is therefore forbidden to collect or gather anything that belongs to the forest, even as small as leaves, wood, water, etc. without the knowledge of the elders of the community or the local people in charge of the forest or care takers of these forests. These practices are

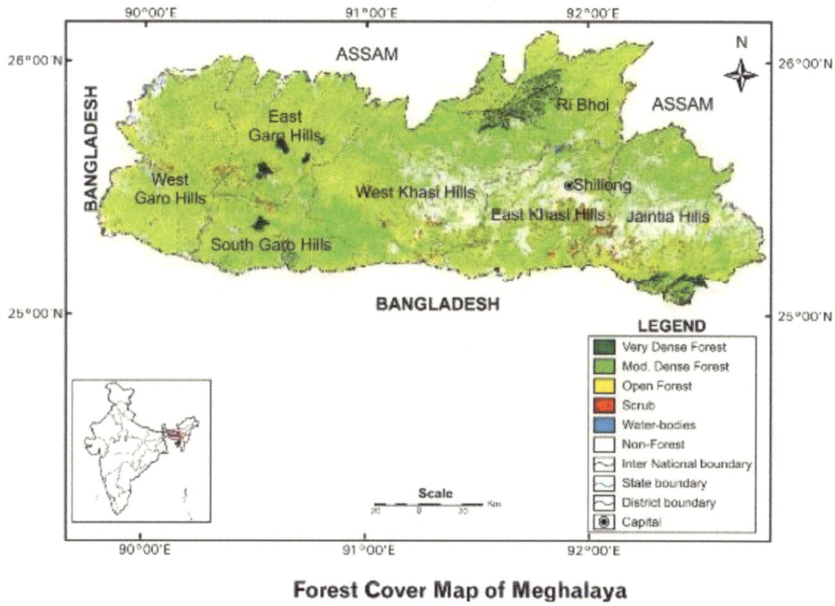

Figure 2.
Map of Meghalaya showing forest cover [source: https://megbiodiversity.nic.in/].

passed on from one generation to another like some traditional or customary law. Such practices of the tribal communities have been seen to promote conservation by their cultural beliefs, religious beliefs and even their customary laws [9–11].

The pattern of land ownership plays an important role in determining the type of land use in Meghalaya and thus, the amount of forest cover. In the three regions of Meghalaya the land ownership and land tenure system vary according to administration and religious beliefs of the people. According to the 6th schedule of Indian constitution, land ownership systems in Meghalaya and other parts of North Eastern India are imparted with a special status. Tiwari and Shahi broadly classify the land ownership system in Meghalaya into two types, i.e. riotwary and customary land system [12]. In the riotwary system the government deals directly with the land owners without interference of intermediaries.

The State Forest Department has classified the forests of Meghalaya (**Figure 2**) into the following six categories, see in [13].

i. Reserved forests (including government forests, national parks and sanctuaries) cover 993.0 sq. km and are owned and controlled by the State Forest Department. Local communities have very few rights over these forests.

ii. Unclassified forests, which cover 7146.5 sq. km, are forests where local communities have all the rights and de facto control. Most of these forests are used for shifting cultivation.

iii. Private forests cover 384.0 sq. km and belong to individuals, who use them primarily for personal consumption.

iv. Protected forests cover 129.0 sq. km and are used by local communities, primarily for personal consumption. Local communities have rights to these forests, but they are controlled by the State Forest Department, which considers the status of protected forest as an interim measure; the department intends to convert these forests into reserved forests.

v. Village forests, which cover 25.9 sq. km, were demarcated and registered by the village community under the United Khasi–Jaintia Management of Forests Act 1958. Most of these forests are used for subsistence purposes.

vi. Community (Raid) forests, which cover 768.0 sq. km, are large community forests (Raid means commune) that are managed by the Raid or commune head under the local administrative head.

This type of forest management according to traditions and rituals in Meghalaya existed before British occupation. And although they cannot be considered to be scientific in approach but were seen to be very effective in protection and conservation of forests. According to the tribal customary laws of Meghalaya, the forests can be further divided into different types according to their intended use. These include- sacred forests (Law Kyntang), village forests (Law Shnong), village restricted forests (Law Adong), forests belonging to a group of villages (Law Raid), private forest or community land (Law Ri-Sumar), private forests or private land (Law Ri-Kynti), clan forest (Law Kur) and cemetery forests (Law Lum Jingtep). These forests are currently called community reserved forests or community conserved areas and they serve a number of ecosystem services to the communities including serving as catchment areas for water sources, conservation of flora and faunal biodiversity, and sanctuary to a variety of medicinal plants [14].

The present study aims to highlight the impact of community reserved forests on the conservation of anuran amphibian biodiversity in the state of Meghalaya. Our study is based on review of an extensive survey of literature. In addition, our aim is to establish the importance of the reserved forests in preservation of pristine habitats for both floral and faunal diversity in the state.

2. Conservation of Amphibian biodiversity by community reserved forests of Meghalaya

The state of Meghalaya is blessed with a rich assemblage of diverse flora and fauna. Being part of the North east India, which falls under the Eastern Himalayas as well as Indo-Myanmar Biodiversity Hotspots, the state supports some of the rich and endemic species of both flora and fauna. Further, owing to its unique biogeographic position, Meghalaya serves as a corridor zone for the occurrence of flora and fauna of both Southeast Asia and Peninsular India. Some of the unique animals found in the forests of Meghalaya include the endangered Western Hoolock Gibbon (*Hoolock hoolock*) whose distribution is restricted to the closed-canopy rainforests of North East India, Bangladesh and Myanmar. In addition, the Capped Langur (*Trachypithecus pileatus*), Macaques (Rhesus *Macaca mulatta*, Assamese *Macaca assamensis*, Northern Pig-tailed *Macaca leonina* and Stumped-tailed *Macaca arctoides*) are also found in the forest canopies of Meghalaya. Among the carnivores, the Clouded Leopard (*Neofelis nebulosa*) is Meghalaya's state animal and other big cats such as Tiger (*Panthera tigris*) and Leopard (*Panthera pardus*) are found in the deep jungles of Meghalaya. Threatened and rare ungulates include the Himalayan Serow (*Capricornis thar*), Hog Deer (*Hyelaphus porcinus*), Sambar Deer (*Rusa unicolour*) and the globally endangered Indian Wild Water Buffalo (*Babalus arnee*). The endangered Chinese Pangolin (*Manis pentadactyla*) is also found in forest covers of Meghalaya. Adding to the list of wild animals is the endangered

Asian Elephant (*Elephas maximus*) which inhabit the wild forests of Meghalaya. In terms of the herpetofauna, the state of Meghalaya is a home to a diverse group of animals ranging from venomous snakes such as the King Cobra (*Ophiophagus Hannah*), MacClelland's Coral Snake (*Sinomicrurus macclellandi*), the White-lipped Pit Viper (*Cryptelytrops albolabris*) to the less venomous and non-venomous ones such as the Khasi Earth Snake (*Stoliczkaia khasiensis*) and the Khasi Keelback (*Amphiesma khasiensis*). In addition to snakes, the state also has records of lizards such as the Khasi Hills Bent-toed Gecko (*Crytodactylus khasiensis*), the Tokay Gecko (*Gekko gecko*) and some of the recently discovered Karst-dwelling bent-toed geckos (*Cyrtodactylus jaintiaensis, Cyrtodactylus karsticola* and *Cyrtodactylus agarwali*) [15] includinga skink *Spenomorphus apalpebratus* [16] from Mawphlang Sacred Grove. Apart from the wild animals, Meghalaya is also a home to wide variety of invertebrates such as spiders, colourful butterflies, moths, leeches, ants, giant earthworms, millipedes, centipedes, beetles as well as crickets and praying mantis.

2.1 Amphibian records from Meghalaya (old records to new discoveries)

Meghalaya, North East India is evident to have the richest expression of amphibians in North East India. The hilly terrain of the state with its numerous hills, valleys, streams, rivers, drainages along with cascading waterfalls, rainfed pools and grasslands all of which serve as congenial or compatible habitats that harbor a wide variety of amphibians with high level of endemism. The amphibians include anurans (tailless amphibians such as frogs and toads), salamanders (tailed amphibians) and caecilians (limbless amphibians). Among amphibians, anurans are the major and diverse component of many terrestrial and freshwater ecosystems. The pioneering studies relevant to diversity of amphibians in Meghalaya, North East India is evident from the accounts made by some workers such as Boulenger [17, 18], Yazdani and Chanda [19], Pillai and Yazdani [20], Pillai and Chanda [21–24], Sahu and Khare [25] and Hooroo [26]. Earlier records on the amphibian species of Meghalaya include descriptions contributed by Boulenger [18] who described a new frog *Rana garoensis* from Garo Hills while Roonwal and Kripalani [27] described *Philautus cherrapunjiae* from Cherrapunjee. Further, Yazdani and Chanda [19] described the Khasi Hills Rock Toad, *Ansonia meghalayana* from Mawblang near Cherrapunjee and this species was later reallocated to the genus *Bufoides* by Pillai and Yazdani [20]. This endemic Rock Toad (*Bufoides meghalayana*) (**Figure 3**) which was thought to be extinct from the wild was rediscovered again after 30 years from the same locality by Das *et al.* [28]. Pillai and Chanda [29] reported and described a new species of *Philautus* (*Raorchestes*) from Shillong, viz. *Raorchestes shillongensis* (**Figure 4**). Pillai and Chanda [22] also described two new frogs from

Figure 3.
Bufoides meghalayana.

Figure 4.
Raorchestes shillongensis.

Figure 5.
Hylarana danielli.

Mawphlang, Meghalaya viz. *Rana danieli* (**Figure 5**) and *Rana mawphlangensis*. In addition, Chanda [30] described a new frog *Rana mawlyndipi* (Ranidae) from Khasi hills, Meghalaya, India. The limbless amphibian, *Ichthyophis garoensis* was described as a new species by Pillai and Ravichandran [31] from Garo Hills.

Amphibians currently include 8352 recognized species with representatives found virtually in all temperate and tropical lands except for Arctic and Antarctic latitudes and in many oceanic islands. At present, 445 species of amphibians (composed of three orders- Anura, Gymnophiona and Caudata) are known from India. Of these, 404 species belong to Anura, 39 species belong to Gymnophiona and 2 species belongs to Caudata [32]. The seven sister states of North-Eastern India that comprised of Arunachal Pradesh, Assam, Nagaland, Manipur, Mizoram, Tripura, Meghalaya and Sikkim harbours 146 species of amphibian assemblages out of which 53 are endemic [33].

In Meghalaya, no comprehensive studies have been made on the endemic amphibian fauna and their distribution. However, Hooroo *et al.* [26] reported for the first time the Painted Balloon Frog, *Kaloula pulchra* from Cherrapunjee, East Khasi Hills district, Meghalaya. Sen [34], reported that there are 49 species of amphibians in the state of Meghalaya. Mahony [35] redescribed *R. mawphlangensis* and reallocated the generic name and placed it in the genus *Odorrana* (**Figure 6**) based on morphological characters of the holotype. Since then, some more reports have been made on the documentation of new amphibian fauna in the state of Meghalaya. Mathew and Sen [36] described three new species of caecilians, *Ichthyophis nokrekensis, Ichthyophis alfredi and Ichthyophis daribokensis* from Nokrek Biosphere Reserve. *Pterorana khare* (Ranidae) was also reported as a new state record form Meghalaya

Figure 6.
Odorrana mawphlangensis.

Figure 7.
Leptolalax khasiorum.

by Rangad *et al.* [37]. A new species of megophryid frog of the genus *Leptolalax*, viz. *Leptolalax khasiorum* (**Figure 7**) was described by Das *et al.* [38] from the sacred groves of Mawphlang, East Khasi Hills, North-eastern India. Another new species of *Leptolalax* was discovered from Nokrek Biosphere Reserve viz. *Leptolalax nokrekensis* by Mathew and Sen [39]. In addition, a new Dicroglossid species was described from the same forest (Mawphlang Sacred Grove), Meghalaya by Purkayastha and Matsui [40] viz. *Fejervarya sengupti*. The discovery of a new genus of the limbless amphibian from Tura, Garo Hills namely *Chikila gaiduwani* by Kamei *et al.* [41] is also noteworthy to mention. Adding to the new discoveries of amphibian species are the new species records of four megophryid frogs namely *Xenophrys megacephala* [42] from Ri Bhoi district, *Xenophrys oropedion* [43] (**Figure 8**) from Malki forest (Riat Laban Reserved Forest) Shillong, *Xenophrys falvipunctata* [44] from Mawphlang Sacred Grove and *Xenophrys oreocrypta* [44] from Tura, Garo Hills. Hence till date there are 61 species of amphibians in the state of Meghalaya. Further, the list of amphibian species belonging from different families that have been recorded throughout our surveys (2015 till date) from the diversified habitats of different forest areas of Meghalaya (sacred groves, reserved and protected forests) have been listed in **Table 1**.

Endemic species have a generally restricted distribution and potential threats to these species can carry more risk of extinction than for broadly distributed species. Since, these species are highly adapted to their home range, any alterations

Figure 8.
Xenophrys oropedion.

in the prevailing environmental factors caused due to anthropogenic or natural causes within their range, their adaptations can function as a source of competitive strength or weakness.

Thus, endemic species are a focus for the conservation of biological diversity, or biodiversity. The first comprehensive attempt to document the endemic species of amphibians in the state of Meghalaya was made by Saikia and Kharkongor [45], who reported in their checklist that there are 19 amphibian species which are considered endemic to the state of Meghalaya (**Table 2**).

Amphibians are a group of organisms that are highly selective about their breeding habitat and choice of suitable oviposition sites. Therefore, they are highly sensitive to changes of the variables that govern an amphibian habitat. Relatively low vagility [46, 47] and narrow habitat tolerance [48–50] seems to amplify the effect of habitat degradation, fragmentation and habitat loss on amphibians. Amphibians are among the planet's most threatened taxa and about one-third of the world's species are threatened with extinction [51]. Habitat loss and fragmentation appear to contribute directly to most of these threats [50, 52, 53]. Anthropogenic activities have brought about different degree of threats towards amphibian community [54]. Amphibian habitats in these forested areas are becoming smaller day by day. Water and moisture conservation abilities are reduced due to decrease in forest coverage. Landslides and soil erosion are covering forest creeks and thus, reducing the number of water sources in these forests. Such threats eventually hamper the breeding and breeding sites of many amphibian species. However, in the state of Meghalaya, these protected areas serve as important ground not only for amphibian diversity and abundance but also for their breeding and development. The protected forest areas seem to harbour a large number of microhabitats to many amphibian species. This is evident from the recent works by various scholars. For instance, *L. khasiorum* is reported to be one of the earliest breeders of amphibians in the forest stream of Mawphlang sacred groove [55]. Other amphibian species occurring in sympatry at this sacred groove include: *X. oropedion, Sylvirana danieli, Hylarana leptoglossa, Philautus sp., Polypedates himalayensis, Rhacophorus bipunctatus, Euphlyctis cyanophlyctis, Duttaphrynus melanostictus* (**Figure 9**), *Amolops gerbillus, Amolops formosus, and Fejervarya sengupti* [35, 38, 40, 43]. Similarly, Khongwir *et al.*, [56] studied the breeding and nesting behavior of *Rhacophorus maximus* (**Figure 10**) in a Mawsynram and Sohra, the regions which lie in the southern slopes of the State and receive exceptionally high levels of rainfall. Multiple amplecting pairs are seen in the temporary rainfed pond at under the forested cover which appeared to be a congenial breeding habitat during the peak of the breeding period. Further,

DOI: http://dx.doi.org/10.5772/intechopen.99750

Forests	Families	Species	Total no. of Families & Species
1. Riat Laban Reserve Forest	i. Bufonidae	*Duttaphrynus melanostictus*	5 Families 16 Species
	ii. Dicroglossidae	*Fejervarya nepalensis, Fejervarya pierrie, Fejervarya syhadrensis, Fejervarya teraiensis.*	
	iii. Ranidae	*Pterorana khare, Odorrana livida, Odorrana mawphlangensis, Hylarana nicobariensis, Sylvirana danieli.*	
	iv. Rhacophoridae	*Philautus andersoni, Roarchestes shillongensis, Polypedates himalayensis, Polypedates teraiensis, Rhacophorus bipunctatus.*	
	v. Megophryidae	*Xenophrys oropedion*	
2. Upper Shillong Protected Forest	i. Bufonidae	*D. melanostictus.*	5 Families 17 species
	ii. Dicroglossidae	*F. nepalensis, Fejervarya pierrie, F. syhadrensis, F. teraiensis.*	
	iii. Ranidae	*Amolops gerbillus, Clinotarsus alticola, Odorrana chloronota, O. livida, Hylarana nicobariensis, S. danieli.*	
	iv. Rhacophoridae	*P. andersoni, Raorchestes shillongensis, Polypedates himalayensis, P. teraiensis, R. bipunctatus.*	
	v. Ichthyophiidae	*Icthyophis garoensis.*	
3. Laitkor Protected Forest	i. Bufonidae	*D. melanostictus.*	4 families 12 species
	ii. Dicroglossidae	*F. nepalensis, Fejervarya pierrie, F. syhadrensis, F. teraiensis.*	
	iii. Ranidae	*O. livida, S. danieli.*	
	iv. Rhacophoridae	*P. andersoni, R. shillongensis, Polypedates himalayensis, P. teraiensis, R. bipunctatus.*	
4. Mylliem Community Forest	i. Bufonidae	*D. melanostictus.*	4 Families 11 species
	ii. Dicroglossidae	*Euphlyctis cyanophlyctis, F. nepalensis, Fejervarya pierrie, F. syhadrensis, F. teraiensis.*	
	iii. Hylidae	*Hyla annectans.*	
	iv. Rhacophoridae	*R. shillongensis, Polypedates himalayensis, P. teraiensis, R. bipunctatus.*	

Forests	Families	Species	Total no. of Families & Species
5. Laitkroh Community Forest	i. Bufonidae	*D. melanostictus.*	4 Families 9 species
	ii. Dicroglossidae	*E. cyanophlyctis, F. nepalensis, Fejervarya pierrie, F. syhadrensis, F. teraiensis.*	
	iii. Hylidae	*H. annectans.*	
	iv. Rhacophoridae	*P. teraiensis, R. bipunctatus.*	
6. Cherrapunjee (Community Forest & Sacred grove)	i. Bufonidae	*Bufoides meghalayana, D. melanostictus.*	6 Families 16 species
	ii. Dicroglossidae	*E. cyanophlyctis, Fejervarya asmati, F. nepalensis, Fejervarya pierrie, F. syhadrensis, F. teraiensis.*	
	iii. Hylidae	*H. annectans.*	
	iv. Microhylidae	*Kaloula pulchra.*	
	v. Ranidae	*A. gerbillus, C. alticola.*	
	vi. Rhacophoridae	*Kurixalus naso, P. teraiensis, R. bipunctatus, Rhacophorus maximus.*	
7. Laitkynsew Village Community Forest	i. Bufonidae	*Bufoides meghalayana, D. melanostictus.*	6 Families 16 species
	ii. Dicroglossidae	*E. cyanophlyctis, Ingerana borealis, F. nepalensis, Fejervarya pierrie, F. syhadrensis, F. teraiensis.*	
	iii. Megophryidae	*Xenophrys parva.*	
	iv. Microhylidae	*K. pulchra, Microhyla ornata.*	
	v. Ranidae	*C. alticola.*	
	vi. Rhacophoridae	*K. naso, P. teraiensis, R. bipunctatus, R. maximus.*	
8.Pynursla Community Forest	i. Bufonidae	*D. melanostictus.*	3 Families 9 species
	ii. Dicroglossidae	*E. cyanophlyctis, F. nepalensis, Fejervarya pierrie, F. syhadrensis, F. teraiensis.*	
	iii. Rhacophoridae	*P. teraiensis, R. bipunctatus, R. maximus.*	

Forests	Families	Species	Total no. of Families & Species
9.Mawsynram Forest & Sacred Grove	i. Bufonidae	*D. melanostictus.*	5 Families 13 Species
	ii. Dicroglossidae	*E. cyanophlyctis, F. nepalensis, Fejervarya pierrie, F. syhadrensis, F. teraiensis.*	
	iii. Hylidae	*H. annectans.*	
	iv. Ranidae	*O. chloronota, O. livida.*	
	v. Rhacophoridae	*Polypedates himalayensis, P. teraiensis, R. bipunctatus, R. maximus.*	
10.SyntuKsiar, Jowai (Community Forest & Sacred grove)	i. Bufonidae	*D. melanostictus.*	4 Families 10 Species
	ii. Dicroglossidae	*E. cyanophlyctis, F. nepalensis, Fejervarya pierrie, F. syhadrensis, F. teraiensis.*	
	iii. Ranidae	*C. alticola.*	
	iv. Rhacophoridae	*P. teraiensis, R. bipunctatus, R. maximus.*	
11.Ummulong Community Forest	i. Bufonidae	*D. melanostictus.*	4 Families 11 Species
	ii. Dicroglossidae	*E. cyanophlyctis, F. nepalensis, Fejervarya pierrie, F. syhadrensis, F. teraiensis.*	
	iii. Ranidae	*C. alticola, O. chloronota.*	
	iv. Rhacophoridae	*P. teraiensis, R. bipunctatus, R. maximus.*	
12.Nongpoh Forest Areas	i. Bufonidae	*D. melanostictus.*	6 Families 11 Species
	ii. Dicroglossidae	*E. cyanophlyctis, F. nepalensis, Fejervarya pierrie, F. syhadrensis, F. teraiensis.*	
	iii. Megophryidae	*Leptobrachium smithi.*	
	iv. Microhylidae	*Microhyla berdmorei.*	
	v. Ranidae	*Hylarana leptoglossa.*	
	vi. Rhacophoridae	*P. teraiensis, R. bipunctatus.*	

Forests	Families	Species	Total no. of Families & Species
13. Mawpun Forest Area	i. Bufonidae	*D. melanostictus.*	4 Families 8 Species
	ii. Dicroglossidae	*E. cyanophlyctis, F. nepalensis, Fejervarya pierrie, F. syhadrensis, F. teraiensis.*	
	iii. Rhacophoridae	*P. teraiensis.*	
	iv. Ichthyophiidae	*Icthyophis garoensis.*	
14. Mawphlang Sacred Grove	i. Bufonidae	*D. melanostictus.*	6 Families 15 species
	ii. Dicroglossidae	*E. cyanophlyctis, F. nepalensis, Fejervarya pierrie, F. syhadrensis, F. teraiensis, Fejervarya sengupti*	
	iii. Megophryidae	*Xenophrys falvipunctata*	
	iv. Microhylidae	*Leptolalax khasiorum*	
	v. Rhacophoridae	*P. teraiensis, R. bipunctatus.*	
	vi. Ranidae	*O. livida, S. danieli, O. mawphlangensis, Hylarana leptoglossa*	

Table 1.
List of amphibian species belonging to different families documented from different forest areas highlighting the diversity of anuran amphibians in community reserved forests and sacred forests of Meghalaya. Majority of the amphibian species that have been recorded were found in the diversified habitats of protected and reserved forests as well as sacred groves present in different regions of the state of Meghalaya (recorded by our team from 2015 till date).

Order	Family	Sl. No.	Species
Gymnophiona	Chikilidae	1	*Chikila fulleri* (Alcock, 1904)
		2	*Chikila gaiduwani* (Kamei, Gower, Wilkinson & Biju, 2013)
	Ichthyophiidae	3	*Ichthyophis alfredii* (Mathew & Sen, 2009)
		4	*Ichthyophis daribokensis* (Mathew & Sen, 2009)
		5	*Ichthyophis garoensis* (Mathew & Sen, 2009)
Anura	Bufonidae	6	*Bufoides meghalayanus* (Yazdani & Chanda, 1971)
		7	*Pedostibes kempi* (Boulenger, 1919)
	Dicroglossidae	8	*Fejervarya sengupti* (Purkyastha & Matsui, 2012)
	Megophryidae	9	*Leptolalax khasiorum* (Das, Tron, Rangad & Hooroo, 2010)
		10	*Leptolalax nokrekensis* (Mathew & Sen, 2009)
		11	*Megophrys megacephela* (Mahony, Sengupta, Kamei & Biju, 2011)
		12	*Megophrys oropedion* (Mahony, Teeling & Biju, 2013)
	Ranidae	13	*Hylarana garoensis* (Boulenger, 1920)
	Rhacophoridae	14	*Chiromantis cherrapunjee* (Roonwal & Kripalani, 1966)
		15	*Philautus garo* (Boulenger, 1919)
		16	*Philautus kempiae* (Boulenger, 1919)
		17	*Philautus namdaphaensis* (Sarkar & Sanyal, 1985)
		18	*Polypedates assamensis* (Mathew & Sen, 2009)
		19	*Raorchestes shillongensis* (Pillai & Chanda, 1973

Table 2.
Checklist of the endemic amphibians of the state of Meghalaya.

Figure 9.
Duttaphrynus melostictus.

Shangpliang *et al.*, [57] studied and observed the unique characteristics of the breeding activity and oviposition of Annandale's high-altitude tree frog, *Kurixalus naso* (Annandale, 1912) (**Figure 11**) at a study site located under the protected area (Law Adong) Mawsynram, Meghalaya, North East India. The amplecting females lay scattered seed-like eggs inside the excavated burrows and the males, using their hind limbs, expose the eggs by pushing them to the mouth of the burrowing hole. This breeding strategy revealed by the tree frog *K. naso,* without foam formation is

Figure 10.
Rhacophorus maximus.

Figure 11.
Kurixalus naso.

unique among frogs belonging to the family Rhacophoridae. Therefore, identification of such breeding habitats may help to understand the ecological requirements of the species and to further create more realistic conservation strategies for the long-term persistence of the amphibian community [57].

2.2 Contribution of community forests to conservation of biodiversity

Community forest have long since served as a means to protect and conserve the rich biological diversity not only because biodiversity has tangible benefits but the fact that these forests have significant religious connotations to the tribal societies and makes a significant contribution to their wellbeing and livelihood. This dependence on forests and forest resources has led the communities residing in close proximities to forest areas to understand the need for conservation and sustainable use of resources [14]. The two main factors that have contributed to the sustainable utilization of natural resources and management of forests are traditional ecological knowledge and traditional institutions prevailing in the indigenous communities worldwide [58]. The khasi communities in Meghalaya, through an age old tradition have been protecting nature and their natural resources thus reflecting great awareness for the need of conservation and management [59]. Community Forestry is successful in decreasing resource degradation and helpful in conservation of Biodiversity [60]. It has been stated that community forests have improved the overall forest conditions including biodiversity [61]. In Meghalaya, a Sixth Schedule state in NE India, over 90% of the forests are under direct or de-facto control of the communities [62]. They are managed by 'traditional institutions' (TIs), organized at village level and recognized by the Indian Constitution [63]. The forests provide livelihood and are also culturally important for the communities [64]. The local people develop various types of traditional forest management practices which contribute to the conservation of biodiversity and addressing the livelihood needs of the rural people [14, 65, 66]. Many species such as hoolock gibbon and elephant find place in tribal stories and mythologies, and people living close to forest have a very high tolerance for these species [14].

The state harbours about 850 species of medicinal plants, of which 377 species are used by 70–80% population [67]. Similarly, there are 249 wild species of edible plants belonging to 153 genera which are still consumed by people in Meghalaya [68]. Some of the medicinally important species reported are *Acorus calamus, Asparagus racemosus, Garcinia cowa, Myrica esculenta, Panax pseudoginseng and Rauvlfia serpentina*, etc. [2]. Medicinal plants are a vital resource for the traditional health care systems, as well as for modern medicines. It was observed that density of Community Forests and abundance of herbal practitioners are often correlated. Tiwari *et al.* [69] found that community forests in the form of sacred groves were homes to many medicinal plants. It is found that the traditional management practices not only help in conserving the resource as evident from the presence of large patches of well protected forests (for example 700 ha village protected forest in Pynursla) and ensuring its sustainable use, but at the same time serve as a common good and 'safety net' for the communities as seen in the village Nongpyndeng, where a large proportion of forest is being managed by the village council for the benefit of all inhabitants of the village. Often, more than one category of forest is found within the boundary of a single village or a group of villages. Over time, these communities have evolved a system of combining forest conservation and sustainable use at a micro level [70], unlike much of national and international efforts which are aimed at meeting these requirements at national or global scales.

In 2012, the Government of India under the Ministry of Environment and Climate Change (MoEFCC) in partnership with the United Nations Development Program (UNDP), constituted the India Biodiversity Awards (IBA) to recognise and honour the outstanding role played and the success achieved by communities to protect and conserve biodiversity [71]. The Yaongyimchen Community Biodiversity Conservation Area (YCBCA) about 10sq km in area, in Longleng district of Nagaland is a community-owned forest now a safe haven for 85 species of birds, including Amur falcons, 15 species of frogs, as well as leopards, barking deers, serows and otters. This transition did not happen overnight. Credit ought to be given to the local community who halted all hunting activities; even traditional traps were completely stopped in the community-conserved area. Another example can be taken from the Land of the Rising Sun - Arunachal Pradesh where the local villagers have used traditional knowledge to to protect the Bugun Liocichla in Arunachal Pradesh. The Singchung Bugun Village Community Reserve (SBVCR) in west Kameng district, Arunachal Pradesh was officially declared in 2017. The Committee was honoured with the IBA 2018 for using its "traditional knowledge to protect the bird and its habitat" threatened by activities like timber extraction, forest clearance and infrastructure development. It is to be noted that The Singchung Bugun Village Community Reserve Management Committee has members from the indigenous community as well as from the forest department.

In the year 2018, the Umru Biodiversity Management Committee in Ri-Bhoi, Meghalaya was specially recognized for their efforts in conserving the Amur Falcons. It is worth mentioning that in the same year, a Certificate of Appreciation was awarded to Ka Khloo Kongwasan Chyrmang Community Reserve in Jaintia Hills District, Meghalaya. Altogether there are 5451 BMCs in Meghalaya and each of these committees are working tirelessly to conserve the biodiversity in their area [71]. Besides, two National Parks and four Wildlife sanctuaries the state of Meghalaya, the Govt. of Meghalaya has vide Section 36 C of the Wild Life (Protection) Act, 1972 declared 74 private and community lands/forests as Community Reserves, the maximum in the country [72]. The Forests and Environment Department in consultation with local communities have prepared management plan for scientific management and conservation of the notified community reserves. Section 36D inserted in the Wildlife (Protection) Act, 1972 in 2003 provides for preparation and implementation of management plan and to take steps to ensure protection of wildlife and its habitat in the community reserve by a Community Reserve Management Committee [73]. This will come a long way in conserving biodiversity in CFs. In Jaintia Hills, the Forests and Environment Department is taking all possible measures to persuade the communities to allow the state government to notify rich habitats of wildlife as community reserves. As we have seen, progress so far is encouraging. The department with proactive cooperation and participation of local communities is taking measures for protection, conservation and scientific management of areas which have already been notified as community reserves [74].

3. Measures to protect community forests of Meghalaya

Forests play a major role in the sustenance of well-being in humans as they provide services such as water purification, provision of oxygen, and spiritual and cultural benefits. Thus, forests and forest resources prove to be not only a source of income but also are an important source of food, freshwater, medicines, firewood and materials for construction. It has been reported that many indigenous communities, forest biodiversity is fundamental to their culture and identity [75]. This dependence on forests and forest resources has led the communities residing in close proximities to forest areas to understand the need for conservation and sustainable use of resources [14]. These forests known as community forests are managed and controlled either by the clans, individuals, groups or traditional institutions according to the prevailing customary laws and practices [14]. The sacred groves which is the most famous amongst the community forests has been regarded as one of the best means for ecosystem conservation. This is due to the fact that it possesses higher species diversity as compared to its surrounding areas [76]. This class of community forest houses many rare, threatened and economically important species [69]. Khiewtam and Ramakrishnan [77] stated that due to human activities much of the vegetation are disturbed and it is only in these sacred forests that thick vegetation is prevalent. Human activities comprising of agricultural activities namely clearing of native vegetation, grazing of livestock, logging and construction has led to the alteration of vast areas on earth. These activities have had intense bearings not only on biodiversity but have also affected significant ecosystem processes such as pollination and nutrient cycling, habitat loss eventually leading to habitat fragmentation [78].

The common strategies adopted by most forest management institutions are penalty which could be monetary or nonmonetary like confiscation of equipment or fines in other forms strictly governed by customary practices of the respective community. Conflicts, whether intra–village or inter–village or inter–community are resolved by institutional mechanisms. Intervention of government agencies, in particular the Forest Department is sought when matters are not settled at the community level [79]. Thus, such collective efforts have contributed to the conservation of forest resources of the country. The establishment of Conservation and Community reserves has led the Govt. of Meghalaya vide Section 36 C of the Wild Life (Protection) Act, 1972 to declare 64 private and community lands/forests as Community Reserves. These reserves are not located within a National Park or Wildlife Sanctuary and are focused mainly on the preservation and protection of flora, fauna and traditional or cultural conservation values and practices. It is also believed that these will provide a safe habitat for the animals inhabiting such areas [80]. The concept of Conservation- and Community- reserves is although new yet it is believed to contribute to conservation of biodiversity.

The World Conservation Union in the year 1999 reported that in spite of forests being protected, yet they continue to be under the pressures of human activities and conservation is not ensured even if it involves legal designations. Further, the International Union for Conservation of Nature and Natural Resources (IUCN) reported that many protected areas lack institutional infrastructure necessary to regulate agriculture, grazing, forestry, mining, hunting, civil conflict, and tourism, financial and human resources and a reassuring legal framework [81]. The United Nations Conference on Environment and Development (UNCED) in Rio de Jenairo, Brazil, in an attempt to promote the management, conservation and sustainable development of all types of forests, introduced the non-legally Binding Authoritative Statement of Principles for a Global Consensus on the Management, Conservation and Sustainable Development of all Types of Forests, also known as the Forest Principles. By the year 2007, the United Nations General Assembly adopted the Forest Instrument which is an important step towards achieving sustainable forest management globally. The concept of sustainable forest management has influenced many new initiatives, prompted revisions to forest policies and practices and been widely accepted around the world by forestry organizations at all levels. It continues to evolve through implementation of criteria and indicators processes at the national, regional and ecoregional levels [82].

The FAO in 2000 stated that in order to meet the demands for food in a growing population, sustainable development of agriculture, fisheries as well as forests can be achieved through biotechnology [83]. The term biotechnology has been defined as the development or use of living organisms to produce, alter or improve a product or a living organism for a specified purpose. It comprises of not only conventional breeding, including domestication plant and animals, but rather, modern innovations emphasizing on biological systems [84]. Although this field has provided grounds for rapid development of new technologies, yet there is limited availability of studies on its role in forest plantation, and ecological benefits arising from genetically modified tree species. This basically arises due to the fact that trees possess the known characteristics of being sessile, having a longer lifespan, is outcrossing and can disperse pollen and seeds across very long distances, and would likely be established in environments with potential mating proximity populations of other species. These issues all of which have been overlooked [85]. However, the use of biotechnology has played a pivotal role in the processing sector, such as pulp and paper production. It also plays a significant role in various stages of the production, starting from planting to harvesting. The first ever application of biotechnology in forestry in order to increase seedling growth in tissue culture was the inoculation of seedlings with symbiotic organisms (specifically mycorrhizae). Owing to a better understanding of genetics as well as development of new techniques, forest biotechnology, now focuses on areas such as propagation, genetic transformation and genetic diversity studies which would ultimately lead to conservation of species that are not only economically but ecologically important as well [86].

4. Conclusion

Herpetofauna is currently facing a major decline on a global scale, resulting from various factors, such as climate change [87, 88], habitat loss, spread of invasive species, overcollection [89] and are therefore in an urgent need of intensive conservation effort. Natural forests are rapidly being replaced by agricultural developments and other human dominated land use types [90, 91]. The species rich tropical regions are quickly losing a large number of species presenting a big threat to global biodiversity loss. The community reserved forests now remain as the only areas that have remain untouched by drastic anthropogenic activities that destroy the breeding habitats of the anuran amphibians. Habitat destruction affects anurans drastically because of the fact that anurans are highly selective for breeding and oviposition habitats, have low vagility and narrow habitat tolerance. Measures that promote restoration of the forests cover and protection of the core habitat for amphibian diversity and abundance and preservation of their sheltered breeding and oviposition sites needs to be focused and implemented at the earliest.

Some important suggestive measures in this regard are: (i) restoration of temporary pools and different water bodies with diverse array of hydro-periods inside the forested area; (ii) minimizing the excessive use of pesticides and fertilizers in the agricultural fields adjacent to the forested area as water run-off might disturbed the survival of the herpeto-fauna; (iii) avoid utilization of the forested area as dumping site. In addition, this finding may provide platform to evaluate the relationship between diversity of amphibians and the diversity of the plant species within these forests. Further, evaluation of amphibian adaptive mechanism in these forests, comparative embryological and developmental processes and analysis of diverse reproductive strategies exhibited by the amphibians housed inside these forests may be taken into consideration for future research. This will help to establish the long-term persistence of the amphibian species and the sustainability of those populations at risk.

Acknowledgements

We would like the thank our host institution, St. Edmund's College, Shillong, Meghalaya, for the support. We would also like to thank our guide and mentor, Prof. (Mrs.) R.N.K. Hooroo, North Eastern Hill University, for the guidance during our research. We are grateful to the Chief Conservator of Forest, Wildlife Circle, Government of Meghalaya and Chief Wildlife Warden, Meghalaya, Shillong for permission to carry out our work. We also thank the Department of Forest and Environment, Government of Meghalaya for the valuable resources.

Conflict of interest

"The authors declare no conflict of interests."

Author details

Ronald Kupar Lyngdoh Tron[1*], Duwaki Rangad[1], Wankitlang Shangpliang[1], Baiakmenlang Manners[2] and Iasyllok Rynjah[3]

1 Department of Zoology, St. Edmund's College, Shillong, Meghalaya, India

2 Department of Biotechnology, St. Edmund's College, Shillong, Meghalaya, India

3 Department of Environmental Science, St. Edmund's College, Shillong, Meghalaya, India

*Address all correspondence to: xyloptz@gmail.com

References

[1] Talukdar G, Ghosh S, Roy PS. Landscape dynamics in north east region of India (Meghalaya State) using Spatial Decision Tree Model. Geo. Inter. 2004; 19(1): 11-18.

[2] Government of Meghalaya. Meghalaya State Development Report 2008-09. Department of Planning, Shillong. 5. Government of Meghalaya (2006). 2009.

[3] Kanjilal UN, Kanjilal PC, Das A, De RN, Bor NL. Flora of Assam, 5 Vols. Government Press, Shillong, India. 1934-1940.

[4] State of the Environment Report Meghalaya. Department of Environment and Forests, Government of Meghalaya and North Eastern Hill University, Shillong. 2005.

[5] Champion HG, Seth SK. A Revised Survey of the Forest Types of India. The Manager of Publications, Delhi-6. 1968.

[6] Puri GS, Gupta RK, Meher-Homji VM, Puri S. Forest Ecology. 2 Vol. Oxford and IBH Publishing, New Delhi. 1989; 582 pages.

[7] Singh BP, Dhyani SK. An appraisal of indigenous landuse systems in Meghalaya for developing sustainable rainfed hill agriculture. In: Kohli RK, Arya KS (eds) In: Proceedings of IUFRO-DNAES international meet on resource inventory techniques to support agroforestry and environemntal activities. DNAES, Chandigarh. 1996; 1-3 Oct 1996, pp 61-72.

[8] Sangma MS: Balpakram and its myths. In: Heritage of Meghalaya Vol II. Publ by Directorate of Art and Culture, Shillong, Meghalaya, India. 1998; pp. 4 – 8.

[9] Berkes, F, Colding, J. & Folke, C. Rediscovery of traditional ecological knowledge as adaptive management. Ecological Applications, 2000;10, 1251-1262.

[10] Berkes, F. Rethinking community-based conservation. Conservation Biology, 2004; 18, 621-630.

[11] Sasaki, K., Sasaki, Y. & Fox, S.F. Endangered traditional beliefs in Japan: influences on snake conservation. Herpetological Conservation and Biology, 2010; 5, 474-485.

[12] Tiwari B K, Shahi K. "Land ownership pattern in Meghalaya" in B K Tiwari, S Singh - eds., Ecorestoration of Degraded Hills, Kaushal Publication, Shillong. 1995.

[13] Tiwari BK, Barik SK, Tripathi R S. Sacred Forests of Meghalaya: Biological and Cultural Diversity. National Afforestation and Eco-Development Board, Shillong; 1999. pp 120.

[14] Tiwari BK, Tynsong H, Lynser MB. Forest management practices of the tribal people of Meghalaya, North-East India. Journal of Tropical Forest Science. 2010; 329-342.

[15] Agarwal I, Mahony S, Giri VB, Chaitanya R, Bauer AM. Six new *Cyrtodactylus* (Squamata: Gekkonidae) from northeast India. Zootaxa. 2018; 4524 (5): 501-535

[16] Dutta AR, Das I, Bauer AM, Tron RKL, Praveen K. Lizard Wears Shades. A Spectacled Sphenomorphus (Squamata: Scincidae), from the Sacred Forests of Mawphlang, Meghalaya, North-east India. Zootaxa. 2013; 3701 (2): 257-276.

[17] Boulenger GA. The fauna of British India, including Ceylon and Burma. Reptilia and Batrachia. Taylor and Francis. 1890. xviii + 541 pp.

[18] Boulenger G A. A monograph of the south Asian, Papuan, Melanesian and

Australian frogs of the genus *Rana*. Rec. Indian Mus. Calcutta. 1920. 20: 1-226.

[19] Yazdani GM, Chanda S K. A new toad *Ansonia meghalayana* (Family: Bufonidae) from Meghalaya (Assam) India, with observations on its breeding 332 on *Pandanus furcatus* Roxb. (Pandanales: Pandanaceae). J. Assam Sci. Soc. 1971; 14(1): 76-80.

[20] Pillai RS, Yazdani GM. *Bufoides*, a new genus for the rock toad, *Ansonia meghalayana* Yazdani and Chanda with notes on its ecology and breeding habits. J. Zool. Soc. India. 1973; 25(1-2): 65-70.

[21] Pillai RS, Chanda SK. The distribution pattern of Amphibia in northeast India. J. Assam Sci. Soc. 1976; 19: 53-56.

[22] Pillai RS, Chanda SK. Two new species of frogs (Ranidae) from Khasi Hills, India. J. Bombay nat Hist. Soc. 1977; 74 (1): 136-140.

[23] Pillai RS, Chanda SK. Amphibian fauna of Khasi Hills, Meghalaya. Rec. zool. Surv. India. 1979; 75: 383-395.

[24] Pillai RS, Chanda SK. Amphibian fauna of Garo Hills, Meghalaya with description of a new species of *Rana*. Rec. zool. Surv. India. 1981; 79: 159-168.

[25] Sahu AK, Khare MK. Field key of *Rana alticola* Annandale (Anura: Ranidae) tadpoles. The Journal of the Bombay Natural History Society. 1983; 80:144 –148.

[26] Hooroo RNK, Khongwir S, Gupta BBP. Record of *Kaloula pulchra* (Gray, 1831) (Anura: Microhylidae) from Cherrapunjee, East Khasi Hills District, Meghalaya, North eastern India. Hamadryad. 2002; 27(1): 146-148.

[27] Roonwal ML, Kripalani MB. A new frog, *Philautus cherrapunjiae* (Family: Rhacophoridae) from Assam, India with field observation on its behavior and metamorphosis. Rec. Indian Mus. 1961); 59 (1 & 2): 325-333.

[28] Das I, Rangad D, Tron RKL, Deuti K, Hooroo RNK. Rediscovery of the Endangered Khasi Hills Rock Toad, *Bufoides meghalayana* in Meghalaya, Northeastern India. Frog Log. 2009; 92: 1-4.

[29] Pillai RS, Chanda SK. *Philautus shillongensis*, a new frog (Ranidae) from Meghalaya, India. Proc. Indian Acad. Sci. 1973; 78B: 30-36.

[30] Chanda SK. *Rana mawlyndipi*, a new frog (Ranidae) from Khasi hills, Meghalaya, India. J. Bengal Nat. Hist. Soc. 1990; 9(1): 44 – 48.

[31] Pillai RS, Ravichandran MS. Gymnophiona (Amphibia) of India. A taxonomic study. Records of the Zoological Survey of India, Occasional Papers. 1999; 72, pp 117.

[32] List of amphibians [Internet]. 2021. Available from: https://amphibiaweb.org [Accessed on 12/07/2021]

[33] Frost DR. Amphibian Species of the World: an Online Reference. [Internet]. 2021. Available from: http://reserch, amnh.org/herpetology/amphibia/index.html. [Accessed on 12/07/2021]

[34] Sen N. Further notes on statewise distribution of the amphibian fauna of northeast India. Rec. Zool. Surv. India. 2004; 102(3-4): 105 – 112.

[35] Mahony S. Redescription and generic reallocation of *Rana mawphlangensis* Pillai & Chanda, 1977 (Amphibia: Ranidae). Hamadryad. 2008; 32 (2): 78 – 89.

[36] Mathew R and Sen N. Studies on Caecilians (Amphibia: Gymnophiona: Ichthyophiidae) of North East India with description of three new species of Ichthyophis from Garo Hills, Meghalaya and additional information on

Ichthyophis garoensis Pillai and Ravichandran, 1999. Rec. Zool. Surv. India. 2009; 309:1-56.

[37] Rangad D, Tron RKL, Hooroo RNK. New record of *Pterorana khare* from Meghalaya, East Khasi Hills District. Herpetological Review. 2007; 38 (1) pp. 99.

[38] Das I, Tron RKL, Rangad D, Hooroo RNK. A new species of *Leptolalax* (Anura: Megophryidae) from the sacred groves of Mawphlang, Meghalaya, north-eastern India. Zootaxa. 2010; 2339: 44-56.

[39] Mathew R and Sen N. Pictorial Guide to Amphibians of North East India. 2010; 1-144. (Published by the Director, Zool. Surv. India, Kolkata).

[40] Purkayastha J, Matsui M. A New Species of *Fejervarya* (Anura: Dicroglossidae) from Mawphlang, Northeastern India. Asian Herpetological Research. 2012; 3(1): 31-37.

[41] Kamei RG, Gower DJ, Wilkinson M, Biju SD. Systematics of the caecilian family Chikilidae (Amphibia: Gymnophiona), with description of three new species of Chikila from northeast India. Zootaxa. 2013; 3666 (4): 401-435.

[42] Mahony S, Sengupta S, Kamei RG, Biju SD. A new low altitude species of Megophrys Kuhl and van Hasselt (Amphibia: Megophryidae), from Assam, northeast India. Zootaxa. 2011; 3059: 36-46.

[43] Mahony S, Teeling EC and Biju SD. Three new species of horned frogs, Megophrys (Amphibia: Megophryidae), from Northeast India, with a resolution to the identity of *Megophrys boettgeri* populations reported from the region. Zootaxa. 2013; 3722 (2): 143-169.

[44] Mahony S, Kamei RG, Teeling EC, Biju SD. Cryptic diversity within the *Megophrys major* species group (Amphibia: Megophryidae) of the Asian Horned Frogs: Phylogenetic perspectives and a taxonomic revision of South Asian taxa, with descriptions of four new species. Zootaxa. 2018; 4523 (1): 1-96. https://doi.org/10.11646/zootaxa.4523.1.1

[45] Saikia B, Kharkongor IJ. Checklist of endemic amphibians of Northeast India. 2017; DOI: 10.26515/rzsi/v117/i1/2017/117283

[46] Sinsch U. Migration and orientation in anuran amphibians. Ethology, Ecology and Evolution. 1990; 2, 65-79.

[47] Gibbs JP. Amphibian movements in response to forest edges, roads, and streambeds in southern New England. Journal of Wildlife Management. 1998; 62, 584-589.

[48] Findlay CS, Houlahan J. Anthropogenic correlates of species richness in southeastern Ontario wetlands. Conservation Biology. 1997; 11, 1000-1009.

[49] Semlitsch RD. Principles for management of aquatic-breeding amphibians. Journal of Wildlife Management. 2000; 64, 615-631.

[50] Houlahan JE, Findlay CS. The effects of adjacent land use on wetland amphibian species richness and community composition. Canadian Journal of Fisheries and Aquatic Sciences. 2003; 60, 1078-1094.

[51] Baillie JEM, Hilton-Taylor C, Stuart SN. IUCN Red List of Threatened Species. A Global Species Assessment. IUCN Gland, Switzerland and Cambridge, UK. 2004.

[52] Carr LW, Fahrig L. Effect of road traffic on two amphibian species of different vagility. Conservation Biology. 2001; 15 (4),1071-1078.

[53] Bowne DR, Bowers MA. Interpatch movements in spatially structured populations: a literature review. Landscape Ecology. 2004; 19 (1), 1-20.

[54] Hasan MK, Akhtar S. Amphibians of Teknaf Wildlife Sanctuary. In: (Ed., Feeroz M. M.). Biodiversity of Protected Areas of Bangladesh. Vol. III: Teknaf Wildlife Sanctuary. Arannayk Foundation, Dhaka. 2013. pp 113-120.

[55] Tron RKL, Das I, Hooroo RNK, Rangad D. "Spring breeding and reproductive mode in *Leptolalax khasiorum* (Anura, Megophryidae) in North Eastern India," Russ. J. of Herp. 2015; 224-232.

[56] Khongwir S, Hooroo RNK, Dutta SK. Breeding and nesting behaviour of *Rhacophorus maximus* (Anura: Rhacophoridae) in Meghalaya, North East India. Curr. Sci. 2016; 110(6): 1102-1105.

[57] Shangpliang PW, Hooroo RNK, Dutta SK. Unique breeding activity and oviposition in Annandale's high-altitude tree frog, *Kurixalus naso* (Annandale, 1912) in Meghalaya, North East India. *Current Science*. 2016; 118 (3): 467-472.

[58] Herrmann T. Indigenous Knowledge and Management of *Araucaria Araucana* Forest in the Chilean Andes: Implications for Native Forest Conservation. Biodiv. Conserv. 2006; 15. 647-662. 10.1007/s10531-005-2092-6.

[59] Poffenberger M. Indigenous Forest stewards of North east India. Technical Report. Community Forestry International. Santa Barbara. 2007.

[60] Adhikari B, Falco SD, Lovett JC. Household Characteristics and Forest Dependency: Evidence from Common Property Forest Management in Nepal. Ecological Economics, 2004.48:245-257.

[61] Pokharel BK, Stadtmuller T, Pfund JL. From degradation to restoration: An assessment of the enabling conditions for community forestry in Nepal, Inter cooperation, Swiss Foundation for Development and International Cooperation, Kathmandu, Nepal. 2005.

[62] Poffenberger M., Communities and forest management in northeast India. In Background Paper No 12, Input to the study 'Development and Growth in Northeast India: The Natural Resources, Water, and Environment Nexus', Community Forestry International, USA, 2006.

[63] Anon., Articles 244(2) and 275(1), Constitution of India, Government of India, 1950.

[64] Tiwari B K, Tynsong H, Lynrah M M, Lapasam E, Deb S, Sharma D. Institutional arrangement and typology of community forests of Meghalaya, Mizoram and Nagaland of North East India. J. For. Res., 2013, 24, 179-186.

[65] Tynsong H, Tiwari BK. Diversity and population characteristics of woody species in natural forests and arecanut agroforests of south Meghalaya, Northeast India. Tropical Ecology. 2011a; 52(3), 243-252.

[66] Tynsong H, Tiwari BK. Contribution of Phrynium capitatum Willd. leaf a nontimber forest product to the livelihoods of rural poor of South Meghalaya, North-East India. 2011b.

[67] Meghalaya Forest Department [Internet]. 2021. Available from: http://www.megforest.gov.in/ [Accessed on 07/07/2021]

[68] Sawian JT, Jeeva S, Lyndem FG, Mishra BP, Laloo RC. "Wild edible Plants of Meghalaya, North-East India." N Prod Radiance, vol. 6, no. 5, 2007, pp. 410-426

[69] Tiwari BK, Barik SK, Tripathi RS. Biodiversity value, status, and strategies

for conservation of sacred groves of Meghalaya, India. Ecosystem Health. 1998; 4(1): 20-32.

[70] Malhotra KC. Village supply and safety forest in Mizoram: a Traditional Practice of protecting ecosystems. 1990.

[71] National Biodiversity Authority [Internet]. 2021. Available from: https://indiabiodiversity.org/ - [Accessed on 08/07/2021]

[72] Meghalaya Biodiversity Strategy and Action Plan. National Biodiversity Strategy and Action Plan India 2000-2002, Strategy and Action Plan for the state of Meghalaya 2004.Prepared by North Eastern Biodiversity Research Cell, North East Hill University, Shillong. 2004.

[73] Community Reserve Management Committee [Internet]. 2021. Available from: www.wiienvis.nic.in/ [Accessed on 08/07/2021]

[74] Community conservation: Govt taking measures to notify rich habitats of wildlife as community reserves [internet]. 2020. Available from: https://theshillongtimes.com/2020/06/07/community-conservation/ [Accessed on 08/07/2021]

[75] Secretariat of the Convention on Biological Diversity. Forest Biodiversity—Earth's Living Treasure. Montreal. 2010; pp 13.

[76] Ormsby A, Bhagwat S. Sacred Forests of India: A Strong Tradition Of Community-Based Natural Resource Management. Environmental Conservation. 2010; 37: 320-326. 10.1017/S0376892910000561.

[77] Khiewtam RS, Ramakrishnan PS. Socio-cultural Studies of the Sacred Groves at Cherrapunji and Adjoining Areas in North-Eastern India. Man in India. 1989; 69(1): 64-71.

[78] Lindenmayer D. Small patches make critical contributions to biodiversity conservation. PNAS. 2019; 116 (3): 717-719.

[79] Pathak MR, Abido MS. The role of Biotechnology in conservation of Biodiversity. Journal of Experimental Biology and Agricultural Sciences. 2014; 2(4): 352-363.

[80] Forest & Environment Department, Govt of Meghalaya (http://www.megforest.gov.in)

[81] Hayes T, Ostrom E. Conserving the World's Forests: Are Protected Areas the Only Way?. Ind. L. Rev. 2005; 38: 595-595.

[82] International Forest Policy- the instruments, agreements and processes that shape it [Internet]. 2007. Available from: https://www.un.org › forests › uploads › 2015/06 [Accessed on 12/07/2021]

[83] FAO Statement on Biotechnology [Internet 2021] http://www.fao.org/biotech/fao-statement-on-biotechnology/en/ [Accessed on 15/7/2021]

[84] Yanchuk AD. The role and implications of biotechnological tools in forestry. Unasylva. 2001;52: 53-61.

[85] Anticipated contribution to and scale of impact of biotechnology in forestry [Internet 2004]. http://www.fao.org/3/ae574e/AE574E09.htm [Accessed on 15/7/21]

[86] Kumar V, Rout S, Tak M & KR, Deepak. Application of Biotechnology in Forestry: Current Status and Future Perspective. Nature Environment and Pollution Technology 14; 645-653.

[87] Araújo, M.B., W. Thuiller, and R.G. Pearson. Climate warming and the decline of amphibians and reptiles in Europe. Journal of Biogeography, 2006; 33:1712-1728.

[88] Reading, C.J., L.M. Luiselli, G.C. Akani, X. Bonnet, G. Amori, J.M. Ballouard, E. Filippi, G. Naulleau, D. Pearson, and L. Rugiero. Are snake populations in widespread decline? Biology Letters. 2010; 6:777-780.

[89] Gibbons, J.W., D.E. Scott, T.J. Ryan, K.A. Buhlmann, T.D. Tuberville, B.S. Metts, J. Greene, T. Mills, Y. Leiden, S. Poppy, and C.T. Winne. The global decline of reptiles, deja vu amphibians. BioScience. 2000; 50:653-666.

[90] Rudel, T.K., O.T. Coomes, E. Moran, F. Achard, A. Angelsen, J. Xu, and E. Lambin. Forest transitions: towards a global understanding of land use change. Global Environmental Change. 2005. 15:23– 31.

[91] Wright, S.J. Tropical forests in a changing environment. Trends in Ecology & Evolution. 2005. 20:553-560.

Chapter 11

Approaches toward Community Participation Enhancement in Ecotourism

Thembinkosi Keith Gumede and Antonia Thandi Nzama

Abstract

A vast majority of scholarship share a similar view that collective participation of different stakeholders serves as a prerequisite for ecotourism sustainable development. Local community participation is considered to be an important pillar of ecotourism development as local communities are capable of influencing success or failure of ecotourism development projects. Socio-economic and socio-cultural well-being of local communities are crucial ingredients for maintaining rapport amongst stakeholders and sustaining ecotourism development. Despite being promulgated as a central pillar of ecotourism development, literature reveals that local communities have not been actively participating in planning and decision-making processes regarding ecotourism development. Adoption of Western-centric oriented participation frameworks by numerous state authorities coupled with lacking necessary skills have been identified as the main factors that hinder active participation of local communities in ecotourism development initiatives. It has therefore, been suggested that ecotourism destinations need to adopt and implement participatory approaches that suit their specific contexts and promote bottom-up ecotourism development procedures. Based on its potential for influencing review and amendment of existing tourism-related policies, a local community participation improvement model has been developed. The model is aimed at facilitating inclusive and active participation of all stakeholders in ecotourism development processes.

Keywords: Stakeholders, local community participation, neoliberalism, local community participation improvement model, ecotourism development

1. Introduction

Ecotourism as a concept and practical phenomenon became popular during 1980s resulting from its ecological and non-ecological benefits [1]. The phenomenon has been termed by Brechin, Wilshusen, Fortwangler and West ([2], p. 53) as 'pragmatic middle ground' resulting from its ability to concurrently stimulate ecological and non-ecological benefits to both the environment and local communities. By definition, ecological benefits are all the services rendered by natural environment within or adjacent to nature-based areas and they include: livestock fodder, fresh water, building material and herbs. While, non-ecological benefits encompass economic opportunities, employment opportunities, capacity building, multiplier effect, development and growth derived from ecotourism activities [1]. Inevitably, some tourism activities engender numerous antagonistic effects that are incurred by natural resources and local communities [3, 4]. As a consequence, ecotourism has been considered to be a more efficient sub-set of the tourism industry based on its ability to promote nature conservation and well-being of local communities [5, 6]. However, there has been a lack of consensus regarding a generally acceptable definition of ecotourism [7]. Consequently, a range of definitions that have been informed by scholars' and practitioners' specific focus has been used to define ecotourism [8].

Amongst popular definitions, is the one that was coined by a Mexican Ecologist, Hector Cebellos-Lascurain, who defined ecotourism as traveling responsibly to relatively fragile destinations for the purpose of studying, admiring and enjoying natural landscapes, fauna and flora, and cultural resources of the adjacent local communities [9]. Having taken into consideration a cascade of social, ecological, cultural as well as economic adverse impacts that can be triggered by tourism activities within the environment and local communities [3, 4], a need for a revised and local community-oriented definitions arose. Thus, subsequent definitions of ecotourism such as those coined by Moran-Cahusac [10]; Sangpikul [11] describe ecotourism as traveling responsibly to nature-based areas to experience and learn about nature and its habitat while taking into account cultural and socio-economic needs of the local communities. Accordingly, local communities and their participation have become the cardinal components of ecotourism development as most eco-destinations are located within the marginalized areas characterized by limited monetary resource, incapacitation and lack of mechanisms necessary for ecotourism development [12–14]. As a result, involvement of local communities in ecotourism development activities has been considered as a quantum leap by numerous tourism destinations [15]. It had been envisaged that involving local communities in ecotourism activities could assist in strengthening relationships between nature-based areas' authorities and locals while sustaining ecotourism development. Despite being promulgated as a lifeblood of ecotourism development and a means through which Sustainable Development Goals-2030 could be achieved [16], local community participation in development endeavors including ecotourism has been very limited, if not lacking, in many parts of the world [17]. Against this backdrop, the authors have developed a model by which local community participation in ecotourism could be enhanced.

2. Theoretical underpinning

As a sub-set of a broad modern-day international tourism industry, in many ways, the emergence and development of ecotourism is embedded within a globalized neoliberalism economic ideology [18]. The ideology is subsumed within a dominant logic of revenue accumulation and advocating market efficiency, promotion of material growth, minimal state interference and commodification of natural resources as a means for alleviating poverty and facilitating economic growth [18–20]. Accordingly, neoliberalism is defined by Fletcher [21] as a 'political ideology that advocates capitalist market system which is characterized by political and ideological antagonism towards state business regulation'. As a consequence, the emergence and implementation of neoliberalism in many parts of the world had triggered staging and commodification of natural resources for tourism and revenue generation purposes [22]. Staging and commodification of natural resources has been termed by McCarthy and Prudham ([22], p. 275); Duffy [23] as the 'neoliberalisation of nature'. This refers to a process whereby natural resources are increasingly subjected to market-oriented management and development systems. During this process, natural resources such as distinct environments, fauna and flora and their habitats are staged to entertain visitors [23]. In essence, natural resources have been used as a means by which neoliberalism, through ecotourism has been subtly promoted for the past four decades. In attempt to protect natural resources from potential extinction, numerous privately-owned nature-based areas adopted fortress conservation system. This system promotes intensive management of protected areas (PAs) with the aim of maintaining security and safety of natural resources by which potential tourists' desire to explore might be spurred [1]. However, fortress conservation instigated extensive tensions between local communities and PAs' management as former had been denied access to a land that at some stage had belonged to their forefathers [24]. In attempt to mitigate tensions between these parties, a vast majority of privately-owned corporations including nature-based areas redirected their focus from absolute fortress conservation toward promoting community development initiatives [23]. Notwithstanding, critics of neoliberalism maintain that the ideology remains a strategy by which state agencies, conservation organizations and private enterprises accumulate monetary gains through staging natural resources for tourism purposes while denying locals a voice in decision-making processes regarding ecotourism development [25].

3. Literature review

3.1 Stakeholders involved in ecotourism

Development, success and sustainability of ecotourism hinge mainly upon active participation of a variety of stakeholders who play different roles depending on their capacity, type of ecotourism and necessary level of participation [18, 26–28]. In the context of ecotourism, stakeholders can be understood as all those parties or actors representing similar or divergent interests but working collaboratively toward the success of ecotourism project [29, 30]. Based on their unique roles and level of commitment, stakeholders are capable of influencing the success of ecotourism development activities [31]. There is an array of stakeholders involved in ecotourism development, and they range from public sector, private sector, non-governmental organizations (NGOs), tourism operators, tourism sites' management, academic researchers and local communities [26, 32], and they can be categorized into primary and secondary segments [29]. The former refers to those who provide essential support without which ecotourism development could not take place, and they include: local communities, tourists, tourism operators, ecotourism sites' management, public and private sectors [29]. The latter comprises those who influence the operationalization of ecotourism development initiatives but do not qualify to be considered as interest groups, and these include: NGOs, community tourism committees/organizations as well as academic researchers [29]. As the preceding text highlights, each stakeholder plays a unique role in ecotourism development process. Government institutions (referred to here as public sector) are responsible for administering consultative processes through which development and enactment of policies and strategies reflecting aspirations and interests of the public concerning ecotourism development is ensured [33]. However, each state adopts and enact its unique policies and strategies meant to facilitate ecotourism development processes within the boundaries of their countries. This is normally done in accordance with the rules and regulations stipulated by the United Nations World Tourism Organization (UNWTO).

Established during 1975, the UNWTO is the international agency entrusted with the responsibility to ensure that responsible, sustainable and universally acceptable tourism activities are promoted [34]. The private sector plays a significant role as it provides financial assistance needed by most developing countries in facilitating ecotourism development projects [26]. NGOs have also been playing a crucial role in ecotourism development through their interventions in addressing institutional and financial constraints that may hinder sustainability of ecotourism development programmes. Subsequently, ecotourism sites where NGOs are actively involved have been better managed compared to those that are solely state-owned [35]. Ecotourism operators play a major role as they ensure that the demands and needs of tourists outside the boundaries of ecotourism sites are met. For example, they provide accommodation and hospitality services to ensure that tourists have access to proper food, beverage, entertainment and comfortable rest [36]. Ecotourism sites bear a responsibility for ensuring that tourists' quest for studying, experiencing, enjoying and admiring natural resources while considering the socio-economic and socio-cultural well-being of the local communities is satisfied [37]. Tourists form an integral component of ecotourism development as they purchase the services and consume the products offered within and outside the boundaries of ecotourism sites [38]. Academic researchers gather useful data in influencing decision-making regarding review of existing and enactment of new policies by which proper management and sustainability of ecotourism can be facilitated [26]. Local communities are considered to be one of the key stakeholders in ecotourism development process. Meng, Jun and Zhengzheng [39] uphold cogently that it could be a mirage to achieve ecotourism overall objectives if community members are excluded from participating in ecotourism development programmes. Accordingly, the concept 'community participation' has emerged and became popular as it has been considered by the United Nations Environmental Program (UNEP) as one of the essential elements and principles of ecotourism development ([40], p. 30; [21], p. 269). Likewise, numerous government policy documents regard community participation as an essential pillar with which Sustainable Development Goals-2030 can be achieved [41].

In the context of ecotourism, community participation refers to a process whereby local residents are voluntarily engaged or involved in ecotourism development initiatives undertaken within the vicinity of their communities [14, 42]. In the process of participation, local communities are expected to take full responsibility during planning and management processes regarding ecotourism development [43]. Garrod [44] describes participation of local communities in planning and management as a process whereby locals are provided with a platform for sharing their views during conceptualization and decision-making phases of ecotourism development project. Participation of local communities in ecotourism development processes may result into accessing opportunities such as self-governance and working collaboratively with other stakeholders in planning and management process, especially on issues affecting their well-being [14, 45]. Local community well-being is defined by Eshun, Adjei and Baah ([46], p. 4) as 'the totality of efforts towards social resilience of local residents inhabiting communities adjacent to ecotourism sites through minimal external control and provisioning of alternative livelihood strategies'. Consideration of local community well-being in ecotourism development has been triggered mainly by possible adversity from socio-economic and socio-cultural impacts of ecotourism activities [40]. Accordingly, Nkemngu [47] argues cogently that issues of community well-being are deeply embedded within the social exchange theory, which maintains that local communities tend to trade their support for projects in exchange for benefits that they stand to gain from ecotourism development activities. As identified by Garrod [44], there are at least five major indicators for successful community participation in ecotourism, and are illustrated in **Figure 1** and discussed in the subsequent writing.

Figure 1 illustrates that there is a need for a strong leadership during participatory planning process as different stakeholders may have different views, motives, preferences and objectives based on their expectations from ecotourism development project. For example, those who own accommodation and hospitality facilities may wish to host more visitors compared to ecotourism sites' managers who may want to regulate the number of visitors resulting from potential adverse impacts on natural resources due to ecotourism activities. In order to neutralize possible contrasting views and conflict of interest, a strong and assertive leadership is required for making resolute decisions that could amicably settle potential dispute while facilitating empowerment amongst stakeholders. Empowerment can be understood as the effort of ensuring that all stakeholders including local communities participate in planning and decision-making processes. This could assist in ensuring that stakeholders collectively set goals and objectives and take full responsibility for ensuring that they are timeously achievable [24]. This may in turn, strengthens local people's support and responsibility for sustaining ecotourism projects. Garrod [44] concurs that empowering local people could be instrumental in ensuring genuine and long-term support necessary for sustaining ecotourism projects. There are four types of community empowerment [48], and they are: economic empowerment, social empowerment, psychological empowerment and political empowerment. According to Scheyvens, economic empowerment is concerned with ensuring that local people are provided with an opportunity to fully engage in decision-making processes regarding non-ecological benefits of ecotourism. In essence, economic empowerment provides local people with an opportunity to access financial resources involved in ecotourism development activities. Social empowerment enables locals to determine the social impacts resulting from

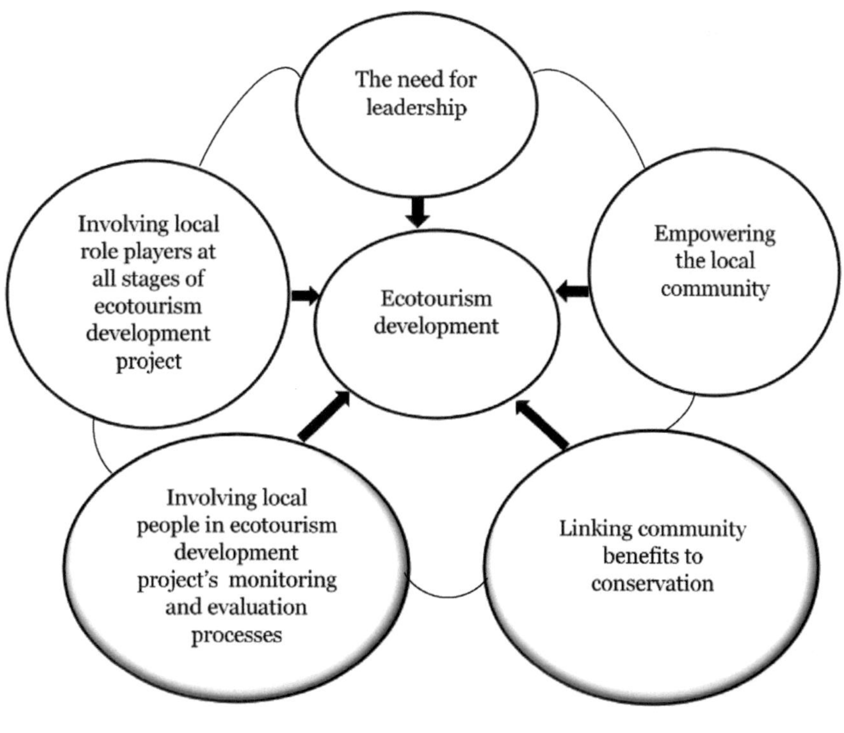

Figure 1.
Indicators for successful community participation in ecotourism. Source: Adapted from Garrod [44].

ecotourism activities. Thus, social cohesion and integrity of the locals form part of social empowerment. Psychological empowerment is concerned with shaping the attitudes that locals may develop over time toward ecotourism development activities. Whereas, political empowerment deals with the ability of the locals to express their concerns toward directing, formatting and accelerating ecotourism development activities.

There is general perception that ecotourism destinations' managers and state agencies responsible for ecotourism development pay minimal attention and allocate limited resources toward monitoring and evaluation of ecotourism development projects [44]. Involving local communities in monitoring and evaluation processes could therefore, increase efficiency while contributing positively toward sustainable ecotourism development. This could assist in enhancing the capacity of local stakeholders and other intended beneficiaries of ecotourism development projects [44]. For the fact that ecotourism development approaches adopted by many tourism destinations prioritize nature conservation over socio-economic and socio-cultural well-being of local communities [18, 21], revenues generated from ecotourism had been solely spent on protection and maintenance of natural resources. This has been done mainly to attract potential tourists and satisfy administrators' ulterior motives [44]. To the contrary, if the local communities are considered for beneficiation, they are more likely to bear a responsibility for custodianship and commit to sustain ecotourism development activity [41]. Butcher [49] echoes that participation of local communities and being considered as beneficiaries foster a sense of pride and ownership amongst them and create opportunities for establishing locally-based small, medium and micro enterprises (SMMEs) to benefit local entrepreneurs, residents and tourists. However, Stronza [28] argues that the relationship between economic incentives and community participation cannot be easily determined.

Despite being considered as a cardinal tenet of ecotourism development, there is paucity of literature focusing on the nexus between ecotourism and local community well-being [32, 46]. As the preceding writing alludes, despite the fact that most ecotourism activities take place within marginalized, distressed and impoverished communities [12], ecotourism sites managers are more concern with conservation and commodification of natural resources for monetary gains than well-being or welfare of local communities [18, 21]. As a consequence, numerous developing countries in most parts of the world have been declared as non-compliant to ensuring local community participation in ecotourism development processes [46]. Local communities' exclusion, apathy and lack of commitment to participate in ecotourism could result in numerous threats that may be detrimental to ecotourism development, and these include: vandalism, marauding and hostility of locals toward ecotourism sites' authorities and tourists [17]. Nevertheless, local communities in many parts of the world are still excluded from participating in decision-making processes concerning ecotourism development [17]. Linked to local communities' socio-cultural well-being, is the aspect of cultural sensitivity (CS). CS can be understood as the extent to which the adverse impacts related to natural and cultural environments are alleviated by the key stakeholders [50]. Donohoe suggests four ways by which CS can be observed:

- Ensuring that intercultural awareness and respect amongst stakeholders are encouraged;

- Adequate effort is made to sustain the existing heritage;

- Encouraging voluntary participation and ensuring empowerment of local communities; and
- Respecting socio-cultural values of the local communities.

In line with the above, West and Carrier [51] uphold that ecotourism incorporates socio-cultural element which can be manifested through aspirations to interact with local communities and commitment to observe their diverse customs, values and appearance in a respectful manner. Understanding, respecting and considering socio-cultural features of the local communities can play a crucial role in sustaining both ecotourism development endeavors and rapport amongst stakeholders. Ironically, there has been a paucity of empirical evidence on active participation of local communities in ecotourism activities [17]. Against this backdrop, some authors [17, 24] caution that prevalent exclusion of local communities from participating in ecotourism activities may jeopardize its development and sustainability, and result into considerable threats such as criminal offenses against tourists and fractured state between ecotourism sites' personnel and local communities. Tosun [14]; Nyaupane, Morais and Dowler [52]; Swemmer et al. [1] identify numerous factors that may limit participation of local communities in ecotourism, and these are discussed in the subsequent section.

3.2 Limitations to community participation in ecotourism

Participation of local communities in ecotourism may be hindered by numerous limitations, and these are: limitations at the operational level, structural limitations, cultural limitations and fortress conservation.

3.2.1 Limitations at the operational level

The likelihood of implementing a participatory development approach (PDA) has been hindered by factors that are related to operational procedures of the task. These hindrances include: centralization of public administration of ecotourism development, lack of co-ordination between involved parties, and poor dissemination of information to communities residing adjacent to ecotourism sites [14, 52]. Centralization of power to government agencies and privately-owned PAs' authorities restricts the influence of community-level groups during planning and implementation processes. It stifles local community participation thereby adopting a 'top-down-oriented' planning and implementation system. Lacking both co-ordination and co-operation between stakeholders may impact negatively on the quality of ecotourism product and impede effective implementation of participatory ecotourism development approach. It may also hinder potential opportunities for locals to participate in development processes affecting ecotourism sites adjacent to their communities. Inadequate ecotourism data resulting from inequitable dissemination of the available information could result into naivety amongst local communities regarding their role and responsibility in ecotourism development processes. It is therefore, important that all stakeholders including local communities are well informed about resources necessitating sustainable ecotourism development.

3.2.2 Structural limitations

Structural limitations are believed to have been stifling the emergence and implementation of participatory approach to ecotourism development. These limitations include: attitudes of professionals toward local communities, lack of expertise from politically deployed officials, elite dominance, lacking appropriate legal system, lacking trained human resources, relatively high costs of community participation, and lacking financial resources. Despite playing an important role in formulation of ecotourism policies especially in developing countries, it is apparent that negative attitude displayed by technocrats toward local communities may hinder the emergence and implementation of participatory approach to ecotourism development. Prioritization of local communities' involvement during policy-making processes remains essential for development and sustainability of ecotourism activities. In circumstances whereby responsible bodies lack expertise necessitating sustainable ecotourism development, it is fundamental that all stakeholders are guaranteed an opportunity to be trained on necessary skills and expertise. Tosun [14]; Nyaupane et al. [52] argue cogently that ecotourism development has been merely interpreted as ecotourism growth characterized by improved infrastructure. In essence, these elements do not suffice appropriate ecotourism development, and can therefore, be referred to as myopic ecotourism development approach. The authors suggest that these limitations can be addressed by considering a holistic approach rather than arriving at particular conclusions based on partial considerations which may disregard local community participation as an important contributor toward sustainable ecotourism development. In many parts of world, democratic benefits had been solely enjoyed by certain entrepreneurs and state elites. Equally, ecotourism development processes have been spearheaded by particular local elites in collaboration with international tourism operators. As a consequence, participatory ecotourism development approach has not been adhered to as numerous local communities had been deprived an opportunity to participate in local and national affairs [14, 52].

Local people, especially youths and entrepreneurs should benefit from ecotourism activities through capacity building and multiplier effect. Legal systems adopted by many developing countries do not allow local people a free role to take control of their own affairs. Instead, they exacerbate a participatory gap existing between the masses and authorities. A typical example is India, where community participatory attempts by the state agencies had become futile resulting from a legal system that has been bias toward authorities and against ordinary citizens [14, 52]. Lacking necessary qualifications by most human resources within the tourism sector had prompted importation of skills from foreign countries. This meant that the majority of local people had to occupy low status positions associated with hard labour and meager remunerations. As a result, local people are deprived of participating in planning and decision-making processes regarding ecotourism development as this prerogative has been enjoyed exclusively by foreign employees who occupy top positions. As highlighted in the preceding text, the sector should invest in local people by equipping them with skills necessary for ecotourism development through establishment of capacity building initiatives. In general, local community participation requires sufficient time and financial resources, and necessary skills to sustain. However, in most instances public sector bodies tend to be reluctant to spend colossal amounts of money on facilitating community participation initiatives whose benefits seem to take forever to be realized. Tosun [14]; Nyaupane et al. [52] maintain that high financial costs involved in participatory ecotourism development process had been a main impediment to launching participatory ecotourism development programmes in many parts of the world. Launching ecotourism activities within local communities may require colossal financial resources to be allocated to tourist infrastructural facilities. These facilities need to bear a resemblance to Western standards irrespective of a tourist destination's financial landscape. Financial resources required for ecotourism development are generally scarce, especially in developing countries. The scarcity of financial resources may hinder execution of participatory ecotourism development both in developing countries and relatively underdeveloped regions of developed countries [14, 52].

3.2.3 Cultural limitations

Culturally predisposed factors such as incapacity, apathy as well as limited awareness of local communities had been considered as limitations to the emergence and implementation of participatory ecotourism development processes [14, 52]. A vast majority of the inhabitants of developing countries encounter challenges with meeting their basic or felt needs as they perceive meeting these as a sole responsibility of the state. Apparently, meeting the needs of tourists take precedence over community development-related issues in many ecotourism destinations. Thus, for the fact that basic survival remains a challenge, participation in ecotourism development processes (as it may consume lots of time and energy) becomes a luxury that vast majority of local communities cannot afford. As a result of socio-cultural, economic as well as political constraints, the majority of the inhabitants of the communities adjacent to ecotourism sites had demonstrated apathy about participating in ecotourism development activities. Another issue of great concern is the fact that vast majority of local communities lack knowledge of ecotourism both as a concept and practical phenomenon. This suggests a great need for enhancing public dialogs and awareness about ecotourism as lack of awareness has been considered as a main barrier to effective local community participation in ecotourism development activities in many parts of the world. Fortress approach to conservation has also prompted numerous socio-economic challenges to local communities [24]. By definition, fortress conservation can be understood as creation and intensive management of protected areas (PAs) characterized by the exclusion of local residents from nature-based areas ([1], p. 5). The phenomenon agitates for eradication of human impact on natural environments as it lends itself well to the notion that nature-based areas should be protected against local communities either by force, coercion or any means necessary [1, 53].

The challenges engendered by fortress conservation include, but not confined to: land dispossession; extensive restrictions; barrier between humans and nature; increased poaching, vandalism and marauding incidents within PAs. Subsequently, there had been ongoing acrimonious relationship and extensive tension between conservation officials and local residents [24]. Historically, many PAs located in developing countries had been built on land from which local residents were forcefully and unlawfully removed. In this sense, fortress conservation denies local residents access to a land that had been expropriated from them through unlawful methods. It is for this reason that Hutton et al. [53] content cogently that local people should benefit from PAs by, amongst other things, participating in nature conservation activities and operating local enterprises that can either directly or indirectly benefit from ecotourism activities. This could strengthen relationships between locals and PAs' personnel, which may subsequently promote conservation of natural resources while sustaining livelihoods of the local residents. As Thondhlana and Cundill [24] echoed "promotion of local communities' inclination and participation in nature conservation activities could impact on transforming the manner in which local communities perceive nature-based areas". However, there are particular instances whereby fortress conservation had been perceived as an expedient intervention to overcome certain challenges that posed serious threat to conservation of particular prestigious animal species. For example, some prominent PAs such as Kruger National Park (KNP) in South Africa and Weza National Park (WNP) in Cameroon in collaboration with local communities had to reinstate and reinforce fortress conservation that had been dismissed resulting from rampant rhino poaching and frequent marauding incidents occurred in these nature-based areas [54].

Linked to the preceding background, Masberg and Morales [55] suggested five factors that need to be taken into consideration during ecotourism development endeavor, and these are: integrated approach to ecotourism development, proper planning and slow start, enforcing education and training, maximizing local benefits, as well as evaluation and feedback. First, the authors argue that all role players in ecotourism development including local communities should work collaboratively toward achieving a common goal. Second, the development of a business plan for the management of natural resources should be informed by the availability of adequate capacity for achieving anticipated outcomes. Third, all stakeholders including local communities should be trained on ecotourism-related skills. As in Garrod [44], facilitating capacity building amongst target groups serves as one of the advantages of participatory approach to ensure that ecotourism contributes to sustainable development of local communities. Fourth, economic gains from ecotourism activities should be equitably shared amongst all stakeholders including local communities. Finally, comparison between actual and anticipated results from ecotourism activities should be done in order to address and manage identified backlogs, issues and concerns. This could be achieved by ensuring that both formative and summative evaluations are undertaken timeously.

3.3 Approaches for enhancing community participation in ecotourism

In the context of ecotourism, there are various types of participatory approaches, some are internally initiated and driven, while others are driven externally [56]. These include: expert-assisted and expert-initiated approaches to ecotourism development. The former involves the participants that are most likely to reap the benefits derived from ecotourism activities that take place within their communities. The approach enables participants to make decisions and take actions that may influence or determine the sustainability of ecotourism activity. During this process, the participants play an essential role as they define problems, identify sustainability indicators, provide necessary information and generate final set of indicators. In effect, the participants collectively provide necessary information by which inadequate awareness regarding certain issues could be addressed and make appropriate judgments upon which the sustainability indicators are entirely dependent [57]. There are two types of stakeholder groups that are involved in the approach, and these are: [58] community-based stakeholders, and [15] system-based stakeholders. The former consists of local community members also referred to as end-beneficiaries alongside academic researchers who often provide assistance with facilitating discussions and allowing participants to define problems and suggest possible solutions [56]. Whereas, the latter comprises a variety of representatives from the governing, private and public sectors that are capable of influencing the operationalization of ecotourism activities. This stakeholder group engages in a joint information-sharing-system-dynamics modeling often characterized by collective undertakings and intensive level of commitment to time and resources. The participants take full responsibility for identifying indicators on the basis of established and modeled utility for monitoring ecotourism activity so as to fast-track the criteria as well as indicators analysis and consensus-building process [59]. The expert-initiated approach allows participants including external actors and non-local specialists to contribute toward developing conceptual framework, identifying a cluster of indicators that can be instrumental during execution of the ecotourism project and collectively participate in a risk assessment process [56].

The approach makes provision for both local-based and system-based participants as their role(s) in the ecotourism activity is determined mainly by an initially set list of objectives and goals, as well as the available resources. Local-based participants include members of the local communities and community representatives, whereas system-based participants involve external experts (i.e. tourism specialists, agents and academic researchers). External experts are responsible for defining problems and studying the activities to be undertaken as stipulated in the work-breakdown-structure (WBS) of the ecotourism development project. More so, they ensure that a mutual decision is made regarding indicators and management strategies that are adopted to sustain ecotourism development project [56]. It is therefore, important that external experts develop indicators that are informed by rigorous assessment of basic characteristics of ecotourism development project. Otherwise, the participants would not succeed in addressing the identified unique key issues and often fail to incorporate concerns and recommendations raised or made during participatory sessions or workshops [60]. Nonetheless, expert-initiated approach has been considered in many parts of the world as a vehicle through which sustainable, efficient and resource conservation-oriented ecotourism development could be achieved [56].

3.4 The need for the local community participation improvement model

Review of literature [14, 32, 39, 44, 45] reveals that local communities in collaboration with other stakeholders could play a significant role in ensuring development and sustainability of ecotourism development process. However, in many parts of world especially in developing countries, local communities have not been considered as important stakeholders who can make a significant contribution toward the development and sustainability of ecotourism endeavors. Consequently, they have not been considered for playing a role in planning and decision-making processes regarding ecotourism development [14, 39, 44, 46]. Moreover, local communities' socio-economic and socio-cultural well-being have not been considered as an important elements of ecotourism development by numerous state agencies and conservation authorities in many ecotourism destinations [12, 18, 21]. Although some authors [56, 57, 59] presented participatory approaches (i.e. expert-assisted and expert-initiated approaches) that may have been used in ecotourism development projects, there have been quite a number of limitations that are considered to have been impeding local community participation in such projects [1, 14, 52]. These impediments are: limitations at the operational level, structural limitations, cultural limitations and fortress conservation. In addition, Botes and van Rensburg [61] argue cogently that Western-centric oriented participation frameworks adopted by the vast majority of state administrators in many parts of the world neither suit the context within which they are implemented nor serve the intended purpose (i.e. enhancing local community participation in ecotourism development process). Against this backdrop, Gumede and Nzama [62] developed a model that sought to form conceptual basis for planning, formulation, implementation and management of policies related to participation of local communities in ecotourism-related initiatives. The model takes into account the main challenges that are reported to have been inhibiting a vast majority of local communities, especially in developing countries to actively participate in development initiatives undertaken within the boundaries of their residential setting. The model entitled: "Local community participation improvement model (LCPIM)" is presented (see **Figure 2**) and unpacked below.

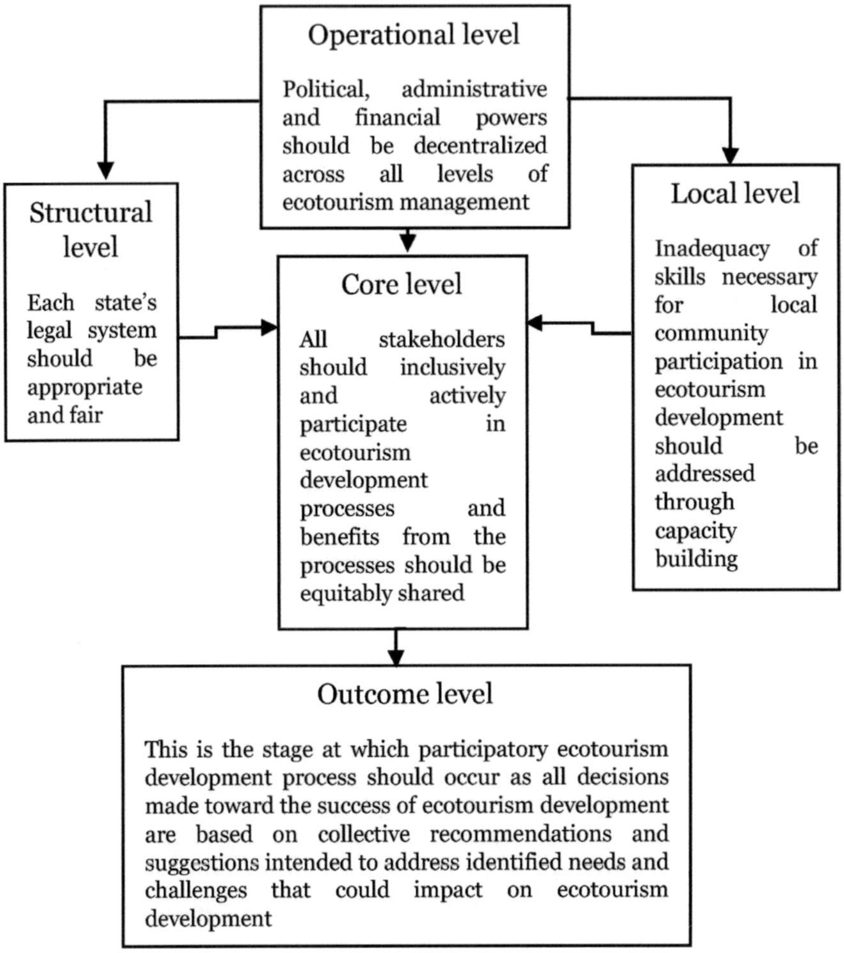

Figure 2.
A model for enhancing community participation in ecotourism activities. Source: Authors.

3.4.1 Components of the model

The model is comprised of five interconnected elements that are demarcated into different levels that have been considered as critical in enhancing participation of local communities in ecotourism development process. The levels (operational, structural, local, core and outcome levels) have been considered as fundamental for facilitating community participation in ecotourism activities based on two reasons: [58] flexibility to fit in a variety of contexts, and [15] capability for enhancing local community participation in ecotourism development processes resulting from their interconnectedness. The first three levels (i.e. operational, structural and local levels) relate to different institutional levels and their roles in ecotourism development process. Whereas, the other two levels (i.e. core and outcome levels) relate to the expected outcome resulting from the interconnectedness of the preceding levels.

3.4.1.1 The first level (operational level)

There are three elements that constitute the operational level. The first element focuses on the importance of decentralizing authority to administer ecotourism development process. The second element focuses on the importance of appropriate co-ordination of ecotourism resources, and the third element focuses on the importance of disseminating information across all governance spectrum.

3.4.1.1.1 Decentralizing authority to administer ecotourism development process

Decentralizing authority to administer ecotourism development process across all governance continuum remains a giant step toward facilitating community participation in development initiatives [14, 63, 64]. According to Ribot [65], decentralization of administrative authority could facilitate inclusive participation while promoting equitable and efficient management of development initiatives. In ecotourism perspective, decentralization of administrative authority could assist in reshaping the manner in which natural resources are managed by local institutions such that community participation is intensified to equitably benefit the authorities, tourists and all those who may affect or be affected by the usage of natural resources [65]. However, in most parts of the world, the overall authority for public administration has been vested in the central government under a direct management of political executives. Consequently, decentralization of power to manage natural resources has often been lacking local representation and downward accountability [14, 65]. It was for this reason that the World Bank [64] made a claim that 'decentralization has been haphazardly implemented'. As such, the influence of community-based groups regarding management of natural resources has been extensively restricted and widely characterized by vertical distance between those who are responsible for planning and a broad spectrum of masses [14].

In numerous developing countries; development, monitoring and administration of tourism policies have been solely undertaken by the central government agencies. As such, ecotourism- related policies have been used as a vehicle to fulfill predetermined national administrators' political and economic agendas [14]. Thus, the LCPIM advocates decentralization of administrative powers from national to provincial and local spheres of ecotourism governance. Although some government agencies such as the Ezemvelo KwaZulu-Natal Wildlife (EKZNW) based in KwaZulu-Natal province, South Africa, have been mandated by the country's laws and regulations such as the KwaZulu-Natal Nature Conservation Management Act, Act No. 9 of 1997, to administer conservation of natural resources while ensuring ecotourism development and promotion at provincial level, local authorities and communities barely participate in decision-making processes regarding ecotourism development. Against this backdrop, LCPIM seeks to assist in addressing 'top-down' geared public administration and serve as a linkage between different spheres of governance to create an enabling environment for local community participation in administration of their own affairs.

3.4.1.1.2 Appropriate co-ordination of ecotourism resources

Prevalent fragmentation and conflicts amongst state agencies, conservation authorities and local communities may engender poor co-ordination of ecotourism resources. Adversely, resentment and conflict amongst stakeholders could nullify proper implementation of participatory ecotourism development activities [66]. Generally, stakeholders hold diverse views and expectations which may result in poor planning and decision-making regarding ecotourism development process [44, 67]. A systematic approach is therefore, required to strengthen rapport and facilitate consensus amongst stakeholders. The LCPIM could be resourceful in addressing fragmentation and conflicts amongst different stakeholders involved in ecotourism development process as it advocates for collective problem-solving and decision-making initiatives. It seeks to provide a conducive platform for stakeholders, based on their capacity, to make optimum contribution necessary for facilitating participatory ecotourism development process.

3.4.1.1.3 Dissemination of information

Local communities in many parts of the world lack information on ecotourism resulting from insufficient availability of ecotourism-related data. The only available information has been disseminated to the public through incomprehensive means [68]. A study conducted by Bello et al. [66] indicates that local communities lack understanding of ecotourism resulting from concentration of tourism information within management agencies, government and NGOs. Consequently, most inhabitants of local communities do not understand their main role in ecotourism development process. In response, the LCPIM can be used to address poor dissemination of information as it advocates for equitable share of information amongst all stakeholders through comprehensive, efficient and effective means.

3.4.1.2 The second level (structural level)

At this level, the LCPIM makes suggestions on how legal systems can contribute toward facilitating participatory ecotourism development process.

3.4.1.2.1 Suitable legal system at all spheres of governance

Tosun [14] argues cogently that participatory ecotourism could contribute significantly toward creating platforms for legal structures to implement effective tourism-related legislation. Although existing legal frameworks (mostly adopted from the West as [61] upheld) may have not been impacting adversely on the aspirations of those who are actively participating in ecotourism development process, it may on the vast majority of defenseless local communities. Legal frameworks that are adopted in many parts of world, especially in developing countries do not make provisions for locals to influence decisions affecting their concerns [14]. Thus, a customized legal system that is aimed at protecting local communities' constitutional rights and promoting participatory development initiatives is necessary. The LCPIM can be used to sensitize policy-makers and legal specialist to disadvantages of foreign-centric legislation and acknowledging tailor-made and inclusive policy-making process.

3.4.1.3 The third level (local level)

This level of LCPIM focuses on how community development-oriented initiatives can facilitate impartial distribution of resources and benefits derived from ecotourism development activities.

3.4.1.3.1 Impartial distribution of ecotourism benefits

Impartial share of ecological and non-ecological benefits generated from ecotourism activities has remained amongst popular scholarly arguments for quite some time [69]. Numerous theories (i.e. stakeholder theory, social exchange theory and participatory theory) share a common view that human behavior including both inclination and apathy is determined by cost–benefit analysis. In a nutshell, people become inclined to participate in endeavors from which they reap benefits and vice versa [40, 66, 70, 71]. The main principles of ecotourism encompass enhancing socio-economic well-being of the local communities and facilitating inclusive ecotourism development participatory approach [6]. Accordingly, equitable distribution of benefits derived from ecotourism has been advocated by numerous agencies involved in ecotourism development [66]. Natural resources form part of ecological benefits derived from ecotourism and have been used by local communities as a source of life, food, shelter and clothing. Local people obtain medicinal herbs, firewood, fruits and vegetables as well as timber and grass for building purposes from animal and plant species found within nature-based areas. However, in many parts of the world local people have been either deprived of or under strict surveillance permitted by conservation sites' authorities to harvest as limited resources as possible [66]. In addition, some cultural beliefs prohibit local people to have direct contact with certain animal and plant species while others prohibit women from working in the same environment with male counterparts [72]. Against this background, the LCPIM seeks to facilitate consideration of local community well-being such that local people reap equitable share of ecological and non-ecological ecotourism benefits. Furthermore, it agitates for inclusive participation in ecotourism activities by all stakeholders irrespective of their race, gender and cultural beliefs.

3.4.1.3.2 Equitable distribution of resources

There has been a growing interest in local community development initiatives resulting from their capability for addressing socio-economic challenges [73]. Review of literature [74, 75] reveals that local community development initiatives differ in nature in terms size, uniqueness and complexity. Accordingly, resources that are necessary for achieving aims and objectives set for each development initiative are completely unique in nature. Westerveld [75] maintains that a specific set of resources that align with unique requirements, aims and objectives is required for ensuring sustainability of any local community development initiative. Equally, the model upholds distribution of resources in accordance with identified needs, aims and objectives as well as desired outcome of ecotourism development initiative. This can be achieved by ensuring collective engagements amongst stakeholders who work as a consortium in identifying and addressing challenges that may inhibit achievement of aims and objectives set for ecotourism development initiative. Expertise is one of the most essential resource toward successful ecotourism development. Ironically, expertise has been identified as a most sought-after resource in ecotourism development initiatives worldwide [14, 66]. According to the authors, there are two prevalent factors that contribute to lack of expertise necessary for ecotourism development amongst local communities, and these are: [58] the fact that the majority of local people is constituted by adults and youths who have not been trained on tourism-related skills, and [15] that governments in many developing countries have been reluctant to inject financial resources by which capacity development programmes could be initiated. In view of the above, the LCPIM suggests that governments and funding institutions should redirect their focus toward local communities as target beneficiaries for tourism-related capacity building programmes. This could assist in increasing a number of people who are capable of effecting positive change in ecotourism activities such that the activities are perceived as community treasure rather than a source of resentment and tension between stakeholders.

3.4.1.4 The fourth level (core level)

This level serves as the heart of the LCPIM upon which the existence and implementation of other components extensively depend. This is the level that places more emphasis on facilitation of inclusive participation and equal share of important status by all stakeholders.

3.4.1.4.1 Inclusive participation and equal share of important status by all stakeholders

While there are numerous ecological and non-ecological attributes of ecotourism, one can not dispute the fact that its success hinges extensively upon collective participation of different stakeholders [18, 26, 27]. Collective partnerships amongst stakeholders had been promulgated as an essential vehicle through which beneficial and sustainable ecotourism development can be attained [58, 76]. In her reiteration, the then Executive Director of the International Ecotourism Society (TIES), Martha Honey, had to emphasize:

> *Considering the importance of collective participation of all stakeholders in community-based initiatives is crucial for ecotourism development ([21], p. 269).*

Drawing from the above literary background, the LCPIM intends to ensure that inclusive participation and equal share of important status by all stakeholders is not only theoretically but also facilitated in a concrete manner.

3.4.1.5 The fifth level (outcome level)

This level serves to ensure that the ultimate goal (i.e. enhancing local community participation in ecotourism development process) for developing the LCPIM is achieved.

3.4.1.5.1 Participatory ecotourism development process

Participatory ecotourism development process is the expected outcome or an ultimate goal resulting from adoption and proper implementation of the LCPIM's interconnected components. Once the limitations to local community participation in ecotourism development process have been identified and addressed by applying LCPIM, it is anticipated that there could be a significant improvement in terms of local community participation in ecotourism activities worldwide.

4. Conclusion

The concept of community participation is regarded as an important tool to assist in ecotourism sustainable development and enhancing local socio-economies of numerous rural communities. While this initiative may be argued as an ideal toward achieving successful development objectives, there is always missing link in its implementation which results in a failure to achieve the expected ecotourism development overall objectives. Many literatures have reported that local communities have not been actively participating in ecotourism development process due to various factors such as socio-economic, lacking of skills, rural setting, misunderstanding of roles, etc. In response, a model that sought to facilitate active participation of local communities in ecotourism development initiatives had been developed. Although it may not be a panacea to all challenges faced by those involved and those who should be involved in ecotourism development initiatives, the model is aimed at ensuring that all stakeholders, based on their unique capacity, actively participate at all levels of ecotourism development project.

Acknowledgements

The authors wish to express heartfelt gratitude to the National Research Foundation (NRF) and editors of the book for ensuring that this piece of academic writing is shared with the public.

Conflict of interest

The authors declare no conflict of interest.

Declaration

The authors would like to declare that this chapter entitled: "Approaches toward local community participation enhancement in ecotourism" is their own work and has not been submitted elsewhere for publication purposes. All sources that are cited or quoted in the text had been duly acknowledged and included in the list of references.

Author details

Thembinkosi Keith Gumede*† and Antonia Thandi Nzama†
Department of Recreation and Tourism, Faculty of Arts, University of Zululand, KwaDlangezwa, South Africa

*Address all correspondence to: gumedetk@unizulu.ac.za; tkgumede8@gmail.com

† The authors contributed equally.

References

[1] Swemmer, L., Grant, R., Annecke, W., & Freitag-Ronaldson (2015). Toward more effective benefit sharing in South African National Parks. *Society & Resources*, 28(1), 4-20, DOI: 10. 1080/08941920.2014.945055

[2] Brechin, S.R., Wilshusen, P.R., Fortwangler, C.L., & West, P.C. (2002). Beyond the square wheel: Toward a more comprehensive understanding of biodiversity conservation as social and political process. *Society & Natural Resources*, 15(1), 41-64, DOI: 10. 1080/089419202317174011

[3] Carter, A., Durham, W.H., Driscoll, L., & Honey, M. (2015). Can ecotourism deliver real economic, social and environmental benefits? *Journal of Sustainable Tourism*, 23(3), 339-357, DOI: 10. 1080/09669582.2014.965176

[4] Rogerson, C.M., & Visser, G. (2004). Tourism and development issues in contemporary South Africa. *Africa Institute of South Africa*. Pretoria: South Africa. DOI: 126579068

[5] Kimbu, A.N., & Tichaawa, T. (2018). Sustainable Development Goals and socio-economic development through tourism in central Africa: Myth or reality. *Geojournal of Tourism and Geosites*, 23(3), 780-796, DOI: 10.30892gtg.23314-328

[6] Saurombe, H.A., Du Plessis, Y., & Swanepoel, S. (2017). An integrated managerial framework towards implementing an ecotourism culture in Zimbabwe. *Journal of Ecotourism*, 17(1), 1-16, DOI: 10. 1080/14724049.2017. 1293066

[7] Viljoen, J., & Tlabela, K. (2007). Rural tourism development in South Africa: Trends and challenges. *Human Science Research Council*, Cape Town, South Africa: HSRC Press. DOI: 169197676

[8] Chirozva, C. (2015). Community agency and entrepreneurship in ecotourism planning and development in the Great Limpopo Transfrontier Conservation Area. *Journal of Ecotourism*, 14(2-3), 1-19, DOI: 10. 1080/14724049.2015.1041967

[9] Fennel, D. (2008). Ecotourism and the myth of indigenous stewardship. Journal of Sustainable Tourism, 16(2), 129-149, DOI: 10.2167/jost736.0

[10] Moran-Cahusac, C. (2009). Ecotourism in Podocarpus National Park. *Journal of Sustainable Forestry*, 28(6-7), 777-798, DOI: 10. 1080/ 10549810902936615

[11] Sangpikul, A. (2010). Marketing ecotourism through the internet: A case of ecotourism business in Thailand. *International Journal of Hospitality & Tourism Administration*, 11(2), 107-137, DOI: 10. 1080/15256481003732782

[12] Coria, J., & Calfucura, E. (2012). Ecotourism and the development of indigenous communities: The good, the bad, and the ugly. *Ecological Economics*, 73, 47-55, DOI: 10. 1016/j. ecolecon. 2011.10.024

[13] Hidalgo, A.E.R., Kozak, R.A., Harshaw, H.W., & Tindall, D. (2017). Different effects of cognitive and structural social capital on empowerment in two community tourism projects in Ghana. *Society and Natural Resources*, 31(3), 1-17, DOI: 10. 1080/08941920.2017.1364454

[14] Tosun, C. (2000). Limits to community participation in tourism development process in developing countries. Tourism Management, 21(2), 613-633, DOI: 10.1016/S0261-5177 (00)00009-1

[15] Baksh, R., Soemarno., Hakim, L., & Nugroho, I. (2012). Community

participation in the development of ecotourism: A case study of Tambaksari Village, East Java Indonesia. *Journal of Basic and Applied Scientific Research*, 2(12), 12432-12437, DOI: 2090-4304

[16] UN (United Nations). (1981). Popular participation as a strategy for promoting community-level action and national development. *Department of International Economics and Social Affairs*, New York: United Nations.

[17] Myeza, J., Mason, R.B., & Peddemors, V.M. (2010). Socio-economic implications of the KwaZulu-Natal sardine run for local indigenous communities. African Journal of Marine Science, 32(2), 399-404. DOI: 10.2989/1814232X.2010.506991

[18] Kline, C., & Slocum, S. (2015). Neoliberalism in ecotourism? The new development paradigm of multinational projects in Africa. *Journal of Ecotourism*, 14(2-3), 1-14, DOI: 10. 1080/14724049. 2015.1023731

[19] Fitchett, J., Lindberg, F., & Martin, D., (2021). Accumulation by symbolic dispossession: Tourism development in advanced capitalism. Annals of Tourism Research, 86, 1-10, DOI: 10.1016/j. annals.2020.103072

[20] Fletcher, R. (2011). Sustaining tourism, sustaining capitalism? The tourism industry's role in global capitalist expansion. *Tourism Geographies*, 13(3), 443-461, DOI: 10. 1080/14616688.2011.570372

[21] Fletcher, R. (2009). Ecotourism discourse: Challenging the stakeholders theory. *Journal of Ecotourism*, 8(3), 269-285, DOI: 10. 1080/147240 40902767245

[22] McCarthy, J., & Prudham, S. (2004). Neoliberal nature and the nature of neoliberalism. *Geoforum*, 35(3), 275-283, DOI: 10. 1016/j. geoforum.2003.07.003

[23] Duffy, R. (2008). Neoliberalising nature: Global networks and ecotourism development in Madagascar. Journal of Sustainable Tourism, 16(3), 327-344, DOI: 10.2167/jost748.0

[24] Thondhlana, G., & Cundill, G. (2017). Local people and conservation officials' perceptions on relationships and conflicts in South African protected areas. *International Journal of Biodiversity Science, Ecosystem Services and Management*, 13(1), 204-215, DOI: 10. 1080/21513732.2017.1315742

[25] Büscher, B. (2010). Derivative nature: Interrogating the value of conservation in 'Boundless Southern Africa'. *Third World Quarterly*, 31(2), 259-276, DOI: 10. 1080/0143659 1003711983

[26] Snyman, S. (2017). The role of private sector ecotourism in local socio- economic development in southern Africa. *Journal of Ecotourism*, 16(3), 247-268, DOI: 10. 1080/ 14724049.2016.1226318

[27] Stone, M.T. (2015). Community-based ecotourism: A collaborative partnerships perspective. *Journal of Ecotourism*, 14(2-3), 166-184, DOI: 10. 1080/14724049.2015.1023309

[28] Stronza, A. (2007). The economic promise of ecotourism for conservation. Journal of ecotourism, 6(3), 210-230, DOI: 10.2167/joe177.0

[29] Garvare, R., & Johansson, P. (2010). Management for sustainability-a stakeholder theory. *Total Quality Management and Business Excellence*, 21(7), 737-744, DOI: 10. 1080/ 14783363.2010.483095

[30] Lacher, R.G., & Nepal, S. (2010). From leakages to linkages: Local-level strategies for capturing tourism revenue in northern Thailand. *Tourism Geographies*, 12(1), 77-99, DOI: 10. 1080/14616680903493654

[31] Foley, K. (2005). *Meta-management: A stakeholder/quality management approach to whole-of-enterprise management*. Sydney: SAI Global.

[32] Eshun, G., & Tichaawa, T. M. (2019). Reconsidering participation for local community well-being in ecotourism in Ghana. *Geojournal of Tourism and Geosites*, 27(4), 1184-1200, DOI: 10.30892/gtg.27406-425

[33] Wynberg, R. (2002). A decade of biodiversity conservation and use in South Africa: Tracking progress from the Rio Earth Summit to the Johannesburg World Summit on Sustainable Development. *South African Journal of Science*, 98(5), 233-243, DOI: 10. 10520/EJC97486

[34] UNDP (United Nations Development Programme). (2003). *South Africa Human Development Report 2003-the challenge of sustainable development in South Africa: Unlocking people's creativity*. Cape Town: Oxford University Press.

[35] De Santos, E.M. (2012). From paper parks to private conservation: The role of NGOs in adapting marine protected areas strategies to climate change. *Journal of International Wildlife Law and Policy*, 15(1), 25-40, DOI: 10. 1080/13880292.2011.650602

[36] Valentine, P.S. (1992). *Review: Nature-based tourism*. In: B. Weiler & C.M. Hall (eds.) Special interest tourism. London, Great Britain: Belhaven Press, 105-127.

[37] Cheng, M., Jin, X., & Wong, I.A. (2013). Ecotourism site in relation to tourist attitude and further behavioral changes. *Current Issues in Tourism*, 17(4), 1-9, DOI: 10. 1080/13683500. 2013.800030

[38] Chan, J.K.L., & Baum, T. (2007). Motivation factors of ecotourism in ecolodge accommodation: The push and pull factors. *Asia Pacific Journal of Tourism Research*, 12(4), 349-364, DOI: 10. 1080/10941660701761027

[39] Meng, Z., Jun, Y., & Zhengzheng, W. (2009). Reflections on the connotations of ecotourism. *Chinese Journal of Population Resources and Environment*, 7(2), 67-70, DOI: 10. 1080/10042857.2009.10684926

[40] Honey, M. (2008). *Ecotourism and sustainable development: Who owns paradise?* 2nd (eds.) Washington DC: Island Press, 29-31, DOI: 10. 1080/14724040802586414

[41] Mgonja, J.T., Sirima, A., Backman, K.F., & Backman, S. (2015). Cultural community-based tourism in Tanzania: Lessons learned and way forward. *Development Southern Africa*, 32(3), 377-391, DOI: 10. 1080/0376835X.2015.1010710

[42] Magi, L., & Nzama, T.A. (2009). Tourism strategies and local community responses around the world heritage tourism sites in KwaZulu-Natal. *South African Geographical Journal*, 19(2), 94-102, DOI: 10. 1080/03736245.2009. 9725336

[43] Wen, Y., & Ximing, X. (2008). The differences in ecotourism between China and the West. *Current Issues in Tourism*, 11(6), 567-586, DOI: 10. 1080/13683500802475927

[44] Garrod, B. (2003). Local participation in planning and management of ecotourism: A revised model approach. *Journal of Ecotourism*, 2(1), 33-53, DOI: 10. 1080/ 14724040308668132

[45] Li, W. (2006). Community decision-making and participation in development. *Annals of Tourism Research*, 33(1), 132-143, DOI: 10. 1016/j. annals.2005.07.003

[46] Eshun, G., Adjei, P., & Baah, A. (2015). Where is community quality

well-being in ecotourism research in Africa? A case from Ghana. African Journal of Hospitality, Tourism and Leisure, 4, 1-12.

[47] Nkemngu, A.-A.P. (2015). Quality of life and tourism impacts: A community perspective. African Journal of Hospitality, Tourism and Leisure, 4(1), 1-12.

[48] Scheyvens, R. (2010). Promoting women's empowerment through involvement in ecotourism: Experiences from the third world. *Journal of Sustainable Tourism*, 8(3), 232-249, DOI: 10. 1080/09669580008667360

[49] Butcher, J. (2005). The moral authority of ecotourism: A critique. *Current Issues in Tourism*, 8(2-3), 114-124, DOI: 10. 1080/1368350 0508668208

[50] Donohoe, H.M. (2011). Defining actual sensitive ecotourism: A Delphi consensus. *Current Issues in Tourism*, 14(1), 27-45, DOI: 10. 1080/1368350 0903440689

[51] West, P., & Carrier, J. (2004). Ecotourism and authenticity getting away from all? *Current Anthropology*, 45(4), 483-498, DOI: 10. 1086/422082

[52] Nyaupane, G.P., Morais, D.B., & Dowler, L. (2006). The role of community involvement and member/ type of visitors on tourism impacts: A controlled comparison of Annapurna, Nepal and Northwest Yunnan, China. *Tourism Management*, 27(6), 1373-1385, DOI: 10. 1016/j.tourman.2005.12.031

[53] Hutton, J., Adams, W.M., & Murombedzi, J.C. (2005). Back to the barriers? Changing narratives in biodiversity conservation. *Forum for Development Studies*, 32(2), 341-370, DOI: 10. 1080/08039410.2005. 9666319

[54] Büscher, B. (2016). Reassessing fortress conservation? New media and the politics of distribution in the Kruger National Park. *Annals of the American Association of Geographers*, 106(1), 114-129, DOI: 10. 1080/00045608. 2015.1095061

[55] Masberg, B.A., & Morales, N. (1999). A case analysis of strategies in ecotourism development. *Aquatic Ecosystem Health and Management*, 2(3), 289-300, DOI: 10. 1080/14634989 908656965

[56] Vaidya, A., & Mayer, A.L. (2014). Use of participatory approach to develop sustainability assessments for nature resource management. *International Journal of Sustainable Development and World Ecology*, 21(4), 369-379, DOI: 10. 1080/13504509.2013.868376

[57] Simon, C., & Etienne, M. (2010). A companion modelling approach applied to forest management planning with the Societe Civile des Terres du Larzac. *Environmental Modelling and Software*, 25, 1371-1384, DOI: 10. 1016/j. envsoft.2009.09.004

[58] Anderson, W. (2009). Promoting ecotourism through networks: Case studies in the Balearic Island. *Journal of Ecotourism*, 8(1), 51-69, DOI: 10. 1080/14724040802140584

[59] Marques, A.S., Ramos, T.B., Caeiro, S., & Costa, H.M. (2013). Adaptive-participative sustainability indicators in marine protected areas: Design and communication. *Ocean & Coastal Management*, 72, 36-45, DOI: 10. 1016/j. ocecoaman.2011.07.007

[60] Reed, M.S., & Dougill, A.J. (2002). Participatory selection process for indicators of rangeland condition in the Kalahari. *Geographical Journal*, 168(3), 224-234, DOI: 10. 1111/1475-4959.00050

[61] Botes, L., & Van Rensburg, D. (2000). Community participation in development: Nine plagues and twelve commandments. Community

Development Journal, 35(1), 41-58, DOI: 10.1093/cdj/35.1.41

[62] Gumede, T.K., & Nzama, A.T. (2020). Enhancing community participation in ecotourism through a local community participation improvement model. *African Journal of Hospitality, Tourism and Leisure*, 9(5), 1252-1272, DOI: 10.46222/ajhtl.19770720-82

[63] UNCDF (United Nations Capital Development Fund). (2000). Africa: Decentralisation and local governance. *Conference concept paper for decentralisation and local governance conference, March 2001*, United Nations Capital Fund, draft: Cape Town, South Africa.

[64] World Bank (2000). *Entering the 21st century: World Development Report 1999/2000*. Oxford University Press: Oxford.

[65] Ribot, J.C. (2003). Democratic decentralisation of natural resources: Institutional choice discretionary power transfers in sub-Saharan Africa. *Public Administration and Development*, 23, 53-65, DOI: 10. 1057/9781403981288

[66] Bello, F.G., Lovelock, B., & Carr, N. (2017). Constraints of community participation in protected area-based tourism planning: The case of Malawi. *Journal of Ecotourism*, 16(2), 131-151, DOI: 10 1080/14724049.2016.1251444

[67] Pattersen & Solbakken, 2002 [Pattersen, L.V., & Solbakken, H.C. (2002). Empowerment as a strategy for change for farm women in western industrialized countries. Sociogia Ruralis, 38(3), 318-330, DOI: 10.1111/1467-9523.00081]

[68] Marzuki, A., Hay, I., & James, J. (2012). Public participation shortcomings in tourism planning: The case of the Langkawi Islands, Malaysia. *Journal of Sustainable Tourism*, 20(4), 585-602, DOI: 10. 1080/09669582. 2011.638384

[69] Su, M.M., & Wall, G. (2014). Community participation in tourism at a world heritage site: Mutianyu Great Wall, Beijing, China. *International Journal of Tourism Research*, 16(2), 146-156, DOI: 10. 1002/jtr.1909

[70] Crapper, M.M. (1998). From hunters to guides: How some native communities have profited from ecotourism, Contact Peru, 3, 20-21.

[71] Homans, G.C. (1958). Social behavior as exchange. American Journal of Sociology, 63, 597- 606. DOI: 10.1080/222355

[72] Snyman, S. (2013). Household spending patterns and flow of ecotourism income into communities around Liwonde National Pack, Malawi. *Development Southern Africa*, 30(4-5), 640-658, DOI: 10. 1080/0376835X. 2013.832149

[73] Matarrita-Cascante, D., & Brennan, M.A. (2012). Conceptualising community development in the twenty-first century. *Community development*, 43(3), 293-305, DOI: 10. 1080/15575330.2011.593267

[74] Muller, R., & Turner, R. (2007). The influence of project managers on project success criteria and project success by type of project. Europe Management Journal, 25(4), 298-309, DOI: 10.1016/j. emj.2007.06.003

[75] Westerveld, E. (2003). The project excellence model: Linking success criteria and critical success factors. International Journal of Project Management, 21, 411-418, DOI: 10.1016/S0263-7863(02)00112-6

[76] Winson, A. (2006). Ecotourism and sustainability in Cuba: Does socialism makes a difference? *Journal of Sustainable Tourism*, 14(1), 6-23, DOI: 10. 1080/09669580608668588

Chapter 12

The Management of Ecotourism as a Way of Contributing to the Development of Protected Areas of Cuando Cubango-Angola

José Eduardo Ezaquias

Abstract

The management of ecotourism in local communities and protected areas faces the challenge of contributing to the achievement of sustainable development indicators, based on a system that integrates political institutional coordination, coherent local participation, endogenous potential, ecotourism and multidimensional interpretations. The objective of the research is to design a procedure for the management of ecotourism that contributes to the development of the protected areas of Cuando Cubango, in Angola. Thus, a procedure was designed for the management of ecotourism, with 5 stages: preparation; diagnosis; the definition of key factors; the objectives and actions and the evaluation and monitoring, which are fed back. Scientific and empirical methods were used, as well as research instruments for data collection and processing. To assess the procedure, the criteria of experts who classified it as high were used. That is, it is relevant for decision-making in Cuando Cubango.

Keywords: Management, ecotourism, development, protected areas

1. Introduction

The challenging planetary changes caused by Covid-19 exposed the sensitivity and vulnerability of today's world, where integration through globalization can pose a threat to its continuity, but also, it can be a fundamental factor to ensure its survival based on the interdependence and connectivity of countries and their peoples.

Covid-19, in 2020 and 2021, had an essential impact on wildlife conservation levels and assuaged the chances of animals being infected by the disease [1]. Although the pandemic lowered tourism rates from 4% in 2019 to 1% in 2020, due to the decrease of nearly 700 million visits in 2020, this segment may be the best adapted to the pandemic from new forms of management, which aim to redefine the demand for these areas.

Africa postulates an attraction that will be able to receive new tourists in time of covid, based on its diversity in geographical and natural riches. But Angola, based on the softer biosecurity measures it has been adopting to curb COVID, ecotourism in protected areas will be better inserted. For this, the Minister of Culture, Tourism and Environment proposed that it is necessary to define a concise, realistic and integrative management strategy of tourism, which contains micro-actions, which generate macro results. From the above, this research is derived, which assumes the need to propose a consistent, integrative management capable of boosting ecotourism. The province of Cuando Cubango, the field of our study, comprises 40% of protected areas, with abundant natural, historical and cultural resources.

In contradiction, it does not have a procedure with actions that guides the management of ecotourism in its protected areas, which conditions the existence of a theoretical-methodological tool for decision-making on the management of policies to contribute to sustainable development at the local level. In reality there is no defined management process, since it has been dismantled and does not have enough methodological body to maintain its existence, so this management system needs to be redefined and refined according to insufficient knowledge of local potential and strategies The situation exposed led to identify the following general objective: Design a procedure to improve the management of ecotourism that contributes to the development of the protected areas of Cuando Cubango.

2. Methodology

This is a descriptive research, inductive and deductive. To achieve the proposed objectives, scientific methods were used: analysis-synthesis; historical – logical; systemic – structural, and the other categories of empirical methods: observation; survey; documentary review and statistical-mathematical methods.

At first, the instruments were designed and tested to collect the primary information. Then the survey and the present observation were applied to communities of six municipalities of the study (Dírico, Kalai, Kuangar, Cuchi, Menongue and Cuito Cuanavale), which represents 66.6% of the nine municipalities of Cuando Cubango.

The research has an intentional sample, composted by community leaders who inhabit the protected areas of the province of Cuando Cubango (**Table 1**).

The article is divided into two parts: the first is the theoretical foundation on the management of ecotourism as a way of contributing to local development and in the second part is the design of the proposed procedure for the management of ecotourism for the development of the protected areas of Cuando Cubango. The relevance of the procedure shall be assessed through the expert judgment.

Stratum groups	Strata	Population	Sample	%
n1	Community leaders	50	27	54%
Total		50	27	54%

Source: own elaboration.

Table 1.
Sample size (n) by strata.

3. Theoretical rationale

The research assumes that ecotourism, because of a historical event, arose from the need to avoid threats of varying order to natural environments. Faced with this threat, the Mexican ecologist raised the need to establish a type of tourism that would curb the growing exploitation through mass tourism in the cultural and natural environment and that would mean a potential for local development [2–7].

One of the most notable examples was Africa. In 1960, Western began working to address the problem of natural resource conservation in the Kilimanjaro area of Tanzania, stating that it was time for a more proactive local role. Western said: "People should be the main beneficiaries and inspectors of natural resources".

Meanwhile, since its emergence in 1960, ecotourism has been known as an activity aimed at generating new sources of work, local structuring, zoning, community performance in rural and protected areas, through legal or other types of effective means in order to achieve the long-term conservation of nature and its ecosystem services and their associated cultural values [8].

It is inferred that ecotourism is a process of local interpretations and sensible scientific, technological, economic, psychological, socio-cultural and environmental relations between the local population, their homes and guests. Under this sense, the most recent concept was born: "responsible travel to natural spaces that conserve the environment, sustain the well-being of the local population and involve interpretation and education" [9]. In this definition, the political, coordinative aspects and the local self-management associated with nature conservation are absent, with multidimensional impacts [10–13].

However, we must continue to deepen the theory on the subject. For the author, the essence of ecotourism is concretized in three fundamental constructs, without one of them there is no ecotourism: leisure, nature and community: the practice of leisure, as a psychsociological implication - is the main and social function of ecotourism. That is, in the present, everyone has the right to rest, vacation and to protect the interests of present and future generations; nature conservation, culture and environmental education as an intermediate function of ecotourism, for which they must generate processes that promote environmental conservation and cultural preservation; e Impact on communities: the ultimate goal – must generate benefits to the community; be able to insert it into local and global society; raise the quality of life and control over their reality.

The links between Sustainable Development (SD) and ecotourism focus on the sensible and responsible outcomes that both demand. The term sustainability first appeared in the forest fields of Germany in 1713 [14]. Nevertheless, as a global concept, it was taken up by the Commission for Sustainable Development's document "Our Common Future", Brundtland Report, in 1987. In this, the concept was made known, as one that meets the needs of the present without involving those of future generations, that is, from a coherent and sensible current use of resources, well-being can be shared with others.

Conversely, it has to be approached from the local and from the vision of each individual that is part of this context. For this, Alonso [15] states that local development "are those who seek solutions with their own resources and are urged (...) local actors to become the protagonists of their own growth." In this line, Sachs [16] and the speeches of Fidel Castro, propose a foreign aid to the local, where the most developed become donors to the less developed and through the aid investments are made in 5 major areas: agriculture, health, education, infrastructure and water management. This approach is called the "top-down" development of the World Bank, which was adopted in 1975, but its materialization towards the local level is consequent to us due to a weak distribution mechanism, which makes it easier for these supports to actually reach the most deprived people.

Meanwhile, in turn, the term management dates from 1884 and comes from the etymological root gesto, which comes from the Latin gestos, defined as attitude or movement of the body, which in turn is derived from gerere, which means to execute and carry out. Management as an influence factor of sustainability at local level of protected areas is very recent, comes from the 80s, after the Brundtland report, and is today in a stage of theoretical-methodological redesign, assuming the name of local development management.

It turns out then to link the management from what is strategy, its operational component. In this way, the research led to the following definition: Strategy for the management of ecotourism: it is a logical plan based on the context and tourism resources (natural, cultural and historical), with coherent objectives and actions, to achieve sustainable local results; increase the tourist demand and manage spaces, in an average time.

For a better local coherence, the management has different techniques to assist in the making of strategic decisions to give rise to the characterization of the context, among them: the theory of the games, decision tree, Balanced Scorecard and the critical factors of event (FCS), etc. [17, 18]. But in the author's opinion, manipulating the FCS is an outstanding way to know the premises properly and guide the management, while other techniques have a more entrepreneurial vision.

4. Design of the procedure for the management of ecotourism

Procedures of international authors for the management of ecotourism were studied [10, 19–22]. These authors agree on the need for the design of a procedure for management and reinforce its systemic, scientific, dynamic, flexible and objective-oriented character.

In general, they do not have such a procedure for the management of ecotourism. Then there is no consensus. Most do not consider the conservation of environments as an imperative; some do not assume carrying capacity; the accuracy of local FCS in making decisions is insufficient; all do not incorporate the ideas of the community and there is little clarity in the strategies of commercialization of the local tourism potential, that is, it can be observed in the policies and actions, they are designed without the direct participation of those involved in the future destination.

The previous shortcomings led to the proposal of a procedure that is essentially based on the contributions of Rodríguez et al. in his study, incorporates community and management into a methodological design, suggesting that villagers have the need to propose their own ideas; exchange with tourists and integrate the management process, as shown below (**Table 2**):

4.1 I. First stage: pre-preparation

Step 1. Selection of actors who will participate in the process and establish the premises, scenarios and the management system.

They constitute tasks of the present stage, the selection of actors who will participate in the process and establish the premises and scenarios.

The proposed procedure comprises the following **foundations:** systematization; the philosophical; the sociological; the psychological; local self-management; training; participation and partnership.

The management of ecotourism as a way to bring development to protected areas must be formed by a local **hierarchical structure.** In the specific case of Cuando Cubango, the Provincial Governor is in office to coordinate and mobilize efforts to achieve the expected results.

Along with the governor, the team include other members of the local management: Intermediate: directors of the local office of Tourism, Environment and Culture; Directors of the Okavango Development Pole; private companies, middle managers, university professors and private sector officials, and at the grassroots: they include academics, members of non-governmental organizations and community leaders.

DOI: http://dx.doi.org/10.5772/intechopen.100333

Stage	Designation	Objective	Steps	Tools
I.	Pre-preparation	Establish the systematization of a group of participants and sectors of the ecotourism segment to comply with the established premises, taking into consideration the methodology of research-action-participation that incorporates the community.	1. Selection of actors who will participate in the process and establish the premises, scenarios and the management system	Checklist to check local conditions; constitution of the ecotourism management team (Consultation Equipment) and instrument to determine the stakeholders in the results; levels of hierarchy and functional system.
II.	Analyze the general environment	Perform a strategic analysis of the local diagnosis.	2. Diagnosis of the starting situation	Primary and secondary sources
III.	Defining key factors	Define, from the diagnosis, the SWOT matrix and the key factors to guide the objectives and actions of the strategy.	3. Determination of internal and external factors 4. SWOT Analysis 5. Definition of key factors	Documentary consultation, creativity, expert opinion, SWOT and MICMAC method.
IV.	Formulation of the strategy	Design the objectives and actions that can contribute to the development of ecotourism in Cuando Cubango.	6. Determination of strategic objectives 7. Definition of priority actions	Creativity, consultation group, expert consultation and documentary consultation.
V.	Evaluation and monitoring	Formulate evaluation indicators to assess the relevance and relevance of the proposed objectives and actions, responding to the dimensions of sustainability.	8. Assessment of the proposed procedure	Survey of experts to determine the relevance and relevance of the strategy formulated.

Source: own elaboration.

Table 2.
The procedure for the management of ecotourism.

The procedure, based on the previous studies and the objectives pursued, adopts the following **premises,** previously set to the participants: The local community; el potential that has the province of CuandoCubango; lto will and political participation, and a business sector.

For this, the following mission is proposed: to design a procedure for the management of ecotourism that contributes to the local development of Cuando Cubango and creates a theoretical-methodological basis for local policies.

In turn, the vision as the futuristic point of what is to be achieved consists of those assumed indicators: economic (one - tourist demand; two - new infrastructures); sociocultural (three - elevation of community quality of life) and ecological (four - conservation of flora and fauna).

4.1.1 Ecotourism system

For an approach of the relationships of these agents, below is a functional proposal of the structural system of ecotourism management, which evidences results and the functions of those involved. But you have to engage them and train tourism agents, to motivate them and have the sense of belonging to the need for development (**Figure 1**).

The proposed ecotourism system contains four zones, with endogenous (micro-destination) and exogenous (the source market) areas that interact creating an integrated circular system of scientific, technological, economic, psychological, sociocultural and environmental relations:

- Zone 1: the source market, where the tourist departs for the destination. This agent travels to the destination and in return gains a new experience;

- Zone 2: here resides the micro destination, where rests the fauna and flora, culture and historical heritage, accommodation and restoration, attractions and animation.

- Zone 3: note that the local community works (providing its time, effort, skill, hospitality, knowledge and agricultural products and chicken coop) in the micro destination and is considered the owner of the land that in turn also rents or cedes to entrepreneurs, under state control. From employment, renting and granting land, the community has income and income. With this income, the community can meet their needs and obtain benefits allocated by public companies.

- Zone 4: it is the transversal market that at the beginning is in an urban area where there is the coordination (local government) of everything and supports the micro eco-destination; provides assistance to tourists, creates public

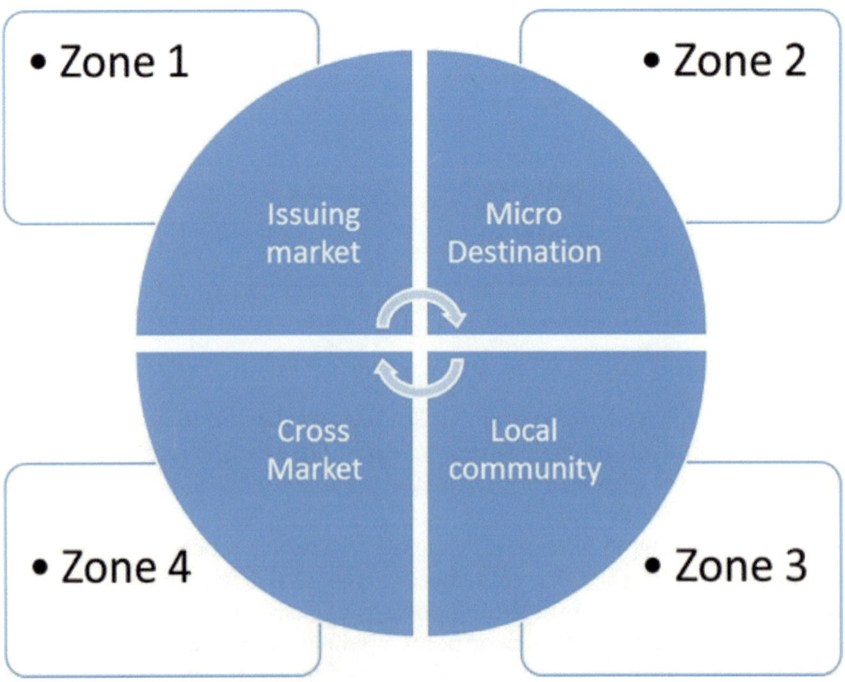

Figure 1.
Spatial circular model of ecotourism. Source: own elaboration.

and private benefits to the community, tourists and their agents. In his thesis, Escribano [23] states that in this area is: the effects with transport service, activities demanded by tourists in the places visited, car rental, accommodation, restaurants, travel agencies, after-sales service. There are also support activities such as infrastructure, health, school, road, human resources, logistics and technology that are decisive for the success of the system, in order to generate desired multidimensional impacts.

4.2 II. Stage: analysis of the diagnosis

Step 2: Diagnose the starting situation.

This step fulfills the objective of identifying variables of ecotourism management from the field of action. To that end, a brief report is made on the subject in Angola, then in Cuando Cubango, and then the methodology is applied to the assumed sample.

4.2.1 Brief summary of tourism in Angola

With an area of 1,246,700 km2, tourism in Angola is also facing the current consequences of the oil crisis. Although the weight of tourism in GDP went from 0.6% in 2014 to 3.5% in 2017, demand has been slowing down for years (**Table 3**).

Of this figure for 2017, only 13% came to the country on holiday, while 87% were work and business. This leads to the fact that Angola is not yet a known destination for tourism. According to MINTUR Luanda, Benguela, Huila and Huambo are the provinces that received the most tourists.

On employment, the information is modest. According to MINTUR, general tourism employed an average of 200,000 workers between 2015 and 2017 (**Table 4**).

- It is estimated that only about 10% of these employees work in hotel entities linked to conservation.

- Meanwhile, the urban and peri-urban hotel structure has been growing since 2010, the year of the African Football Championship which mobilized the national business community to invest in hotel units. In fact, Angola has renowned five-star hotels. But the weaknesses of the areas of tourist support and widespread social problems, make that the hotel quality does not have much effect on the attractiveness (**Table 5**).

While this information, the development of ecotourism management in protected areas of Angola is in a phase of redefinition, due to the absence of strategic

Years	Tourists	Europe	America	Asia	Africa	Austrália
2017	260.961	134.456	33.802	51.197	40.769	730
2016	397.485	213.051	61.731	68.756	52.686	1.261
2015	592.000	199.127	105.106	111.262	176.022	978
2014	594.998	325.970	83.605	77.204	107.269	950
2013	650.033	231.266	74.216	119.657	222.230	2.064

Fuonte: adapted from MINTUR [24].

Table 3.
Tourist issuers for Angola (2013 to 2017).

Years	2012	2013	2014	2015	2016	2017
Employees	157.954	173.478	202.766	219.349	221 847	223 965
Men			107.745		112.516	113.547
Women			95.021		109.331	110.418

Source: adapted from Institute of Foment of Tourism, INFOTUR [25].

Table 4.
Change in employment in the national tourism sector.

Years	2013	2014	2015	2016	2017	2018
Hotels, restaurants, travel agencies and the like	5.766	6.277	6.378	6.720	7.573	8.092
Number of beds	22.000	30.275	32.158	35.834	36.687	
Room number	20.788	22.115	21.201	21.718	27.148	28.462

Source: adapted from INFOTUR [25].

Table 5.
Information on the hotel and restaurant sector.

management mechanisms that link all actors, sectors and local resources and lack of investments. However, it was considered that 13% of the national territory are conservation areas, but any effort to ensure better control of these areas was left unchanged, by the government to reduce the ministerial of tourism and environment to secretaries, who could, from above, go articulating the management processes of the protected areas.

4.2.2 Ecotourism management in Cuando Cubango

In turn, Cuando Cubango the second largest province of Angola, after Moxico, with an area of 199,335 km2 and an estimated population of 535,838 inhabitants, which translates to 2.3 inhabitants per Km2 and to the total are about 59,537 families, an average of 9 people per dwelling, has 9 municipalities: Menongue (head), Cuchi, Cuito Cuanavale, Dírico, Rivungo, Mavinga, Kuangar, Calai and Nancova.

The province has a management structure that in parallel also coordinates the local ecotourism system. It functions with the power relationship of a governor appointed by the President of the Republic, at the top of the management; at the intermediate level it is helped by the offices (local administration, tourism, environment, culture, commerce, economic, migration, tax and fiscal) and at the base: there are the traditional leaders, social organizations, universities, churches and populations. Everyone is called upon to implement policies linked to tourism.

However, it is understood that management is not adequately structured; the relationships between the levels of hierarchy are not efficient and effective, a fact that disorients the objectives; it is evident the absence of a political attitude and knowledge of the dimension and opportunity that ecotourism reaches; structural political defects aggravate the social situation by failing to extract local problems; the width of the territory also makes it difficult to manage ecotourism and there is no knowledge exactly of the local critical factors, a situation that influences decision-making contrary to local needs.

As a result of this attitude, the province was in the last ten years the deforestation of thousands and thousands of trees that weakened the local ecological system.

Year	Tourists		Total
	National	Foreign	
2018	822	220	1.042
2017	2.512	146	2.658
2016			4.599
2015			10.557

Source: Provincial Directorate of Tourism [26].

Table 6.
Flow of tourists in Cuando Cubango.

Likewise, we are witnessing the halting of all actions that at this time are considered essential for the implementation of ecotourism. In 2017, there were 13 hotels and similar units, including hostels, which in total had 194 rooms and 248 beds, of which two reference hotels in conservation areas. It currently had an increase of 53.8% % in the hotel offer. In this regard, the structure of prices per room and A + B (food and beverages), is characterized by high prices in the local currency (Kwanza). But with currencies, consumption will be made more accessible by the devaluation of 85.7% of the local currency in USD.

On the other hand, in the last four years, the arrival of tourists suffered a sharp reduction as an effect of the weak application of policies oriented to ecotourism (**Table 6**).

The socio-economic situation of the protected areas where the fauna and flora rest is chaotic. There are problems related to education, with the absence of school institutions and teachers. In fact, there are children outside the system, while another part of these children, depending on the precarious situation of their families, are sent by their parents to work in the agricultural fields, while the school year is going on. As of 2019, an estimated 132,984 pupils at different levels were estimated, including 100,524 (75.5%) children and 6,127 (4.6%) adults in literacy (Annex 2.1.6). of them 159 graduates in Tourism Management for 3 years.

The management of these areas can be affected by lack of hospital structures, where the only one is located more than 600 km from the main ecotourism centers, where there are no roads, a factor that can constitute a problem in the management. That is, 63% of these areas do not have sanitary facilities. Thus, some areas were adapted inappropriate facilities, where hospital services are provided.

Electricity as a basic budget for management, until July 2021, reached about 17,378 families, meaning a coverage capacity of 29% of the territory (Only the villages of Menongue, Cuito Cuanavale, Kuangar, Calai and Dirico have access to this service).

However, all headquarters in the nine municipalities have mobile technology service that varies from 3G to 4G by zones, telephony services and mobile Internet. In the villages, there are shortages of services, so you can travel up to 200 kilometers without having a network to communicate.

4.2.3 Application of the methodology

4.2.3.1 Outcome of the instrument applied to host communities

It took into account the communities located in areas of interest for ecotourism, natural and cultural sites of Cuando Cubango, where a survey guide with 6 multidimensional points was applied in the last three years.

27 community leaders from 6 municipalities were surveyed. All appointed by paternal transition or imposition of the local administration. 100% of these leaders

Economic	Cultural	Environmental
• 60% of the population walks long distances (over 50 km) to access various services, including health, education and trade; • Community leaders mostly have developmental management problems and their problems are solved from the family decision – customary justice; • 100% of local communities do not have a public transport service; • communities live on outside supports; • They do not have mobile communication media.	• Most one-room dwellings are occupied by 5.6 to 11 people; • The houses are (80%) of precarious construction (mud and guano, covered with grasts) and 20% of permanent construction; • The quality of health service in communities is 20% weak and 80% non-existent; • poor quality of teaching and lack of teachers; • Most leaders and their members cannot read and write; • Only 14% of local communities have access to safe drinking water.	• 80% of the inhabitants, when they are sick, use traditional medicine, handling herbs and roots, which sometimes results in death; • Fauna and flora are sources of subsistence for the community

Source: prepared on the basis of the survey of the host communities.

Table 7.
Problems of communities.

are men, aged between 45 and 70 years and an average level of education of 5th grade, but most cannot read and write.

These communities have identical economic, socio-cultural and environmental features. Thus, the results were exhibited based on the methodology of Lira and Escudero [27] that propose two sets to describe them, nomadly: problems and potentialities. In this way, the difficulties arise first (**Table 7**).

- While these problems, these community have shown strength and persistence that has allowed them to subsist in conditions of precariousness, absence of social comfort, for several centuries, in a nomadic and resilient way, based on local capacity.

- They also have endogenous potentialities that guarantee living conditions and minimize any problems mentioned in advance (**Table 8**).

It can also be assumed as endogenous wealth the ancestral knowledge, hospitable and the instructive values of these communities, as an intangible heritage, which guides the inhabitants and is very influential in moments of interaction with visitors.

4.2.4 Presentation of local potentialities

The studies of the last 60 years have classified the region as possessing interesting and significant mineral resources, flora and fauna. Among them the well-known *Big Five* (lion, elephant, hippopotamus, buffalo and rhinoceros) made known to the world in 1961 by the renowned International Magazine of the National Geographic on Angola " um lugar*a descobrir*".

But when it was already believed that there was no natural heritage in the province because of the war, recent scientific studies proved otherwise. The

Economic	Cultural	Environmental
• The main activity: livestock breeding in the south, family farming, hunting, fishing, comércio and gathering wild fruits; • Typical food of the region (hard flour of Massambala, Massango, Kizaca, Mute, Mono and Kambambi (forest goat) and bull; • produce and drink drinks and meals at regular events (Tchissangua, Leche, Hidromel, Mutoho, Viku ndu, Kapuca, Kaporroto, katchipembe, Mundevele and Kapata); • Produce tubers (potato and cassava) Cereals (Corn, Massambala and Massango) Horticultural (Tomato, Onion and Cabbage) and Sugarcane; • A moderate number of animals were registered for sale and consumption (beef, pigs, chickens and chickens); • With local techniques and materials are produced machetes, knives, swords, spears, arrows and machados, as well as mud objects as panels, plates, glasses and water reservoirs.	• Strong cultural, linguistic, folkloric, religious and customary potential; • Ability to manufacture traditional musical instruments (drums, bows, malunga, riquembe, marimba) and organization of dances for the local celebration (Thianda, Kamandada, Maku, Katanga, Kandoa, Kuviala, Tungandzi, Kuvamba, Mivan dye, Massakuila, Tchileya, Makapo and Mbongo); • Circumcision rituals for boys (Vamba) and fiko for girls (Mukula); • The local community communicates fluently in the mother tongues in Ngangela (54%) Cókue (14%) and Umbundo (18%), also speaks Mbwela variants; Nyemba; Kuangar; Ngonjelo; Mpengo; Ngakala; Lucazi; Cisokola; Xambiu; Mbunza; Diric; Mbukuxu and Khoisan [28] • In the communities there are rituals of veneration and worship of the ancestors, however, the Christian tradition is already taking hold, but obscure rituals are still frequent.	• Beauty that develops on vegetation • Almost untouched ecotourism sites (not yet commercially explored); • High mastery of the behavior of fauna; • Knowledge of flora and water resources; • All local communities are close to tourist resources, where different ecotourism activities can be developed; – It lives on rudimentary hunting, not only for food, but for the sale of its derivatives.

Source: prepared on the basis of the survey of the host communities.

Table 8.
Endogenous potentialities.

Researches of [28–30] revealed that the end of the armed conflict and consequent Declaration of Peace and National Reconciliation, on April 4, 2002, evidenced an intact, guarded, true and genuine rest of the national flora and wildlife, with a unique biodiversity and powerful rivers.

4.2.5 The existing ecosystem in Cuando Cubango

To know the supply of ecosystems, we substantially reviewed the data of the NGOWP that since 2015 makes a scientific task to study the biomes of the Rivers Cuando and Cubango, in order to make known to the world.

Table 9 shows the results of the first stage of the research.

The amount of fauna and flora resources recorded evidences the current situation, but they are not the totality of the existing natural heritage, as will be noted below.

Species of flora and fauna	Quantity recorded
Plants	1.050
Fish	3.000
Reptiles and amphibians	99
Poultry	443
Mammals	43
Flora and fauna: new discoveries for science	
Typicality	Quantity
Plants	40
Poultry	4
Fish	5
Reptiles	3
Source: NGOWP.	

Table 9.
Faunal and floristic resources.

4.2.6 Description of major wildlife resources

For the present study, the author presented mainly the animals that have been seen and studied in the last three years of empirical research within the protected area. Among them, you can find:

Fish: clams, catfish, sardines, freshwater crab, mullet, trout and limbombo.

Reptiles: alligator (order crocodile), lizards, chameleons and snakes.

Mammals and wild: black elephants (one of the highest populations in Africa), cheetah, hyena, leopard, wild dog, zebra, hippopotamus, lion, hyena, royal lever, nguelengue, buffalo, rabbit, gazelle (bambi, local name), jaguar, badger, inhala, puku, chango, topi, impala, oribi, steenbok, lechwe, olongo and boar that need natural and ample spaces to move.

Birds: ostrich, bico-de-serra, peru-do-mato, guinea fowl, Angolan hen, red throat grace, carunculadosted guru and secretary.

Livestock and poultry: cattle, goats, pigs and horses. The populations also raise ducks, chickens and pigeons.

Flora, mineral and river resources.

In this region of Africa there is an ecosystem characterized by savannah, where small shrubs and trees are found [31]. Among the main resources of the flora in Cuando Cubango, is:

Flora: muxumba, mucussi, muvambo, mumwe, mucoxo, mucuvi, muxexe, mungolo, mussivi, muvambe, girassonde or mucula and palmeiras. The flora is made up of forests of zambezia miombo and xerophilous forests and savannas. Areas with a water table near the surface are occupied by grasslands (chanas). In the riverbed there are palustrial and aquatic communities with a preponderance of grasses existing at the northeast end of the Okavango sub-basin and most of this unit is occupied by rooted vegetation that corresponds to palustr reeds and thorny trees (Manja) and there are also several extensions. However, part of the territory is almost a desert. You can also see swamps, virgin areas and observation.

Agriculture: maize; beans, macunde beans, green beans; cassava; massango, massambala and nuts and wild fruits.

Rocks: in the municipalities of Cuchi and Menongue it is possible with techniques to findethyst, tourmaline, seawater, gunza iron, diamond, varied clays and other ornamental minerals.

Water resources: its two important rivers are Cuito and Cubango.

4.3 III. Stage: definition of key factors

It is reasoned the need to know exactly the internal and external factors that affect the normal local development. Of all the diagnostic analysis process, 58 variables were accurately identified: 28 (51.8%) economic; 19 (35.1%) sociocultural and 11 (20.3%) environmental. In this stage, the action is to identify which of these variables have the greatest influence on others. To carry out this analysis, the research was based on the opinion of experts.

13 experts were chosen, under the following characteristics:

- Experience in topics on ecotourism management (five years minimum);

- The high level of preparation, knowledge and expertise on ecotourism management in support of sustainable development processes at the local level;

- Those who occupy functions in the different sectors of the tourist plant of Cuando Cubango keys to the takeoff of local tourism.

Of these, 7 (53.8%) are Doctors of Science, 5 (38.4%) Master and 1 (7.6%) Graduate. The selected experts present a Kendal Coefficient of 0.79 very close to high, which is considered valid for the present analytical stage.

An instrument was applied to qualify each variable. For this step, the methodology containing three phases was fulfilled (1 – estimation of the variables, assuming the frequencies, 2 – SWOT matrix and 3 – obtaining the key factors). All inherent mathematical-statistical process was processed by Excel 2013 and the MicMac method.

Step 3. Determination of internal and external factors.

Identifying internal and external factors is a tool that makes it possible to know and evaluate the real operating conditions of the management, in order to propose actions for its benefit [19].

In this sense, 58 variables were identified from the environment. Of these, 24 (44.4%) internal and 34 (55.5%) external. Among the 24 internal variables: 12 (50%) are economic; 9 (37.5%) are sociocultural and 3 (12.5%) are environmental and among the 34 external variables: 16 (47%) are economic; 10 (29.4%) are sociocultural and 8 (23.5%) environmental.

To evaluate each variable, the Likert method was adapted to (1 – Low, 2 – Medium and 3 – High). In the expert consensus, no variable averaged 3 values. The closest, with 2.8 was: theprovince has a gran ecotourism potential. Factor que puede se asumido com una base fundamental para los objectivos de la investigação.

4.3.1 Internal

From a total of 24 internal factors put to the assessment of the experts (Delphi Criterion), they chose 10 factors: 5 Strengths and 5 Weaknesses (**Table 10**).

4.3.2 External

It is related to the outer part of the territory, over which you cannot have direct control. Of the 34 variables, experts selected 10 (**Table 11**).

Strengths
Territory preserved and still little altered by the human hand;
Endogenous knowledge and hospitality;
The tourist potential of Cuando Cubango is little explored;
Rivers sailable in long extension;
The province has a great tourist potential, comprised of natural and cultural resources;
Weaknesses
Lack of accompaniment and monitoring of the When Cubango Tourism Master Plan (2012–2020);
Weak capacity of the hotel industry for visitor accommodation;
Insufficient integration of university staff;
Accessibility and infrastructure problems in the potential area of ecotourism (roads, bridges, schools, hospitals and private companies);
Weak investment in ecotourism;
Source: based on expert assessment.

Table 10.
Internal factors.

Opportunities
Growing interest in wildlife destinations;
Good weather conditions;
Angola has immense natural and cultural potential;
Political commitment in the National Development Plan (2017–2022);
Free movement in the African region and proximity to the main ecotourism areas of Africa (Okavango Delta, Victoria Folls, Livingston, Ethosa and Johannesburg
Threat
Growing dynamics of hunting movements and environmental deforestation
Difficult access and long distances between tourist spots;
There is no official information on tourist resources;
Poor integration of the province of Cuando Cubango in the tourist route in the southern African region;
Lack of qualified personnel for the hotel and tourism industry;
Source: drawn up fromv

Table 11.
External factors.

Step 4. SWOT Matrix Analysis (Weaknesses, Threats, Strengths and Opportunities).

The experts evaluated the influence between the internal and external variables. This process was completed with the search for consensus for the evaluation of each relationship of the SWOT matrix (Weaknesses, Threats, Strengths and Opportunities). As a result, it was obtained that the largest number of greatest impact was located in the upper right quadrant of the matrix (2.5228).

The SWOT matrix presents the highest result of 2.5228 in the Maxi-Mini area, which translates into a defensive strategy, which must minimize threats against strengths, or take advantage of strengths to reduce those interferences that still slow down management processes.

Step 5. Determination of key factors.

Based on the 20 factors (internal and external) that most influence decision-making about the context, the MICMAC method was used to analyze the

No.	Variable	Total number of rows	Total number of columns
1	Angola has immense natural and cultural potential;	39	35
2	Absence of integration da comunidade local no Plan Maestro de Turismo De Cuando Cubango (2012–2020)	33	37
3	Weak capacity of the hotel industry to accommodate visitors;	28	40
4	Public and private disarticulation on ecotourism development trends;	38	32
5	The tourist potential of Cuando Cubango is explored spontaneously and isolatedly;	38	39
6	Tourism is a strategic axis, but there are still no priority actions;	38	36
7	Lack of qualified personnel for the hotel and tourism industry;	41	37
8	Weak economic dynamism in the province of Cuando Cubango;	37	34
9	Relaxation of entry visas (political commitment);	36	38
10	Hospitality and the endogenous knowledge of its inhabitants.	38	40
11	Insufficient environmental education in communities;	34	40
12	The lack of free movement in the African region.	34	37
13	The province has a great tourist potential	37	40
14	Poor integration of the Province of Cuando Cubango in the tourist route in the region of southern Africa;	34	37
15	Accessibility and infrastructure problems in the potential area of ecotourism (Roads, Bridges, Schools, Hospitals and Private Companies);	43	39
16	Proximity to the main ecotourism areas of southern Africa (Delta do Okavango, Victoria Fols, Livingston, Rundo and Johannesburg.	40	36
17	Rios navegáveis em larga extensão;	38	45
18	Growing interest in wildlife destinations;	44	39
19	Road and rail access routes between provinces;	40	44
20	Territory preserved and still little altered by the human hand;	54	39
	Totals	764	764

Source: MicMac.

Table 12.
Direct influence matrix.

interdependence and influence of each variable, to identify the most influential among them. This method allows you to verify the dependence of each of the variables with the others in the system (**Table 12**).

The use of the MICMAC method allowed to identify three variables that have a strong influence on the design of the current ecotourism management strategy. These variables or key factors are:

1. Increased interest in wildlife destinations;

2. Accessibility and infrastructure in the potential area of ecotourism (roads, bridges, schools, hospitals and private companies);

3. Territory preserved and still little changed by man and industry.

From these traits derived from the internal processes of the investigation, the critical factors of success of the strategy were determined, which are: (1) interests in nature; (2) accessibility and infrastructure; and (3) preserved territory.

In this way, it can be inferred that the management of ecotourism is the dependent variable (effect) and the current independent variables (cause) are: the growing interest, accessibility and potentialities, depending on the transversality of ecotourism.

4.4 IV. Stage: formulation of the strategy

Step 6. Identification of strategic objectives.

Based on the approach pursued, the research seeks to operate the following general strategic objective: to design a procedure for the management of ecotourism that contributes to the local development of Cuando Cubango and creates a theoretical-methodological basis for local policies.

To follow up on the proposed general objective, three strategic objectives were raised, the critical factors of success of the strategy:

OE1. Take advantage of the growing interest in wildlife destinations, achieving the participation of communities to enhance the benefit of the natural and cultural potential of the province.

OE2. Promote accessibility and infrastructure in the potential area of ecotourism, motivating the competent bodies and private companies to improve roads, bridges, schools and hospitals.

OE3. Promote the care of existing natural, historical and cultural resources, in order to guarantee the conservation of nature and ethnic identity of its inhabitants towards sustainable development.

Step 7. Definition of priority actions.

The actions carried out serve to use the strategic objectives set. In this regard, an action plan was developed that connects local aspirations, which had contributions from members of the hierarchical management structure, experts and the creativity of the researcher.

Methodologically, each strategic objective proposed has a set of actions in the plan, which in turn, integrate: start date (2021–2031); the resources required (local activity, programme or project); the person responsible (level of management hierarchies) and the specific place to develop (sectors or premises).

4.4.1 Priority actions by strategic objectives (APOE)

In the practical context, the priority actions were designed to achieve the following multidimensional goals within ten years:

Strategic objejctivosos	Priority actions	Date	Resources	Responsable	Place
Strategic objective 1. Take advantage of the growing interest in wildlife destinations, engaging communities to drive the use of the province's natural and cultural potential.	1. Develop small conservation tourism projects linked to ecotourists that raise awareness and make these agents responsible for nature and the community;	February 2021	Project	Community leader	Communities of Leapeka
	2. Determine areas for natural wildlife sanctuaries such as elephants, hippos, buffaloes, lions with spaces for tourists;	January 2022	Project	Government, Universida and researchers	Luengue-Luyana and Mavinga
	3. Establish public incentives for the private sector to create small hotel conservation units;	Regular	Employer support programmes	Government	Ecotourism promotion zones
	4. Create financing lines for community projects with favorable interest rates and medium-term repayment;	Regular	Financial products	Government and Banks	Ecotourism promotion zones
	5. Promote internal, external and main interconnected commercial production chains, supporting the sector and generating income for the inhabitants;	All year round	Tax incentive projects	Government and business	Ecotourism promotion zones
	6. Disseminate the natural and cultural potential of Cuando Cubango at the local, national and main source markets for the modality of ecotourism;	All in year	Marketing and promotion campaigns	Government and private	Local, National and international

Strategic obejctivosos	Priority actions	Date	Resources	Responsable	Place
Strategic objective 2. Promote accessibility and infrastructure in the potential area of ecotourism, motivating competent organizations and private companies to improve roads, bridges, schools, hospitals and other support structures.	1. Develop concrete and gradual projects to improve accessibility to the main tourist attractions of Cuando Cubango;	May 2021	Public Investment Project	Central government	International road with Nkurenkure
	2. Create priority and essential infrastructures in Dírico - Angolan headquarters of the Okavango-Zambezi conservation area;	January 2022	Proyecto do Polo Okavango	Official Government and Organization	Municipality of Dirico
	3. Support projects of the micro-agriculture, livestock and river fishing communities in remote areas;	Agricultural times	Agricultural development programme	GovernmentCooperativesCountry folk	The whole province
	4. Providing accommodation with a low environmental impact that does not deplete local resources and destroys the environment, but provides ample opportunities for a better understanding of the environment and sensible exchange with local communities;	January – 2022	Conservation projects	Companies and families	Ecotourism development areas
	5. Create an ecotourism develop-ment agency for the coordination, implementation, supervision and monitoring of programs, actions and projects, guaranteeing the impact to its community agents, in the territory and promoting ecotourism;	Sept 2022	Creating a Structure Project	Government	Menongue

DOI: http://dx.doi.org/10.5772/intechopen.100333

Strategic obejctivosos	Priority actions	Date	Resources	Responsable	Place
Strategic objective 3. Promote the care of natural, historical and cultural resources, preserving nature and the ethnic identity of its inhabitants for a sustainable development of ecotourism.	1. Create training, literacy projects; training and preparation for community agents to manage ecotourism processes, without losing their originality and ancestral charm;	Sept 2020	Community training projects	University and Social Organizations	Communities studied
	2. Carry out a detailed survey of the tourist resources of Cuando Cubango;	Sept 2017	Activity	GovernmentCoordinationResearchersUniversity	When Cubango
	3. Develop a program to identify, describe and record the main tourist attractions of Cuando Cubango, with fixed and durable signs along the main roads, where additional information about the resource must also be placed;	Sept 2017	Program	GovernmentResearchersUniversity	When Cubango
	4. Create, with the support of the Government, the University and Non-Governmental Organizations, zones for the promotion of ecotourism;	January 2022	Project	GovernmentUniversity Social organizations	Ecotourism Promotion Zones
	5. Define standards and methodologies for the control and protection of fauna, wild flora and all the natural heritage of the province of Cuando Cubango;	January 2022	Project	Environmental Prosecutor	Potential ecological zones

Source: own elaboration.

4.4.2 Economic

1. Increase the number of "international" tourists by 84%.

2. Improve infrastructure to support ecotourism by 27%.

4.4.3 Cultural

Contribute to the improvement of the quality of life of 20 ecotourism communities.

4.4.4 Environmental

Raise the level of natural conservation and cultural preservation from 50–80%.

All the above fact, are challenges of local management that will be articulated from the dynamism of the agents of the ecotourism market.

4.5 V. Stage: evaluation and control

Step 8. Assessment of the proposed procedure.

The purpose is to formulate evaluation indicators to assess the relevance and relevance of the proposed objectives and actions. To validate five aspects of the procedure, the expert approach was used, through a Guide to the assessment of the proposal.

It was applied to a group of experts made up of Cubans, Angolans and Ecuadorians linked to the subject and the field of action. 40 experts were consulted, where 7 did not express interest in participating. Of the 33, 12 with a lower competition coefficient of 0.60 were ruled out, leaving 21, representing 63.6% of the total. However, in the territory of Cuando Cubango few experts working on the subject of ecotourism management were identified, for this reason, the Delphi Method was manipulated, which in turn allowed to investigate assessments of experts located in different geographical latitudes.

Of the 21 experts, 11 (52%) had a high level of competence and 10 (47.6%) had a medium level. Among them 8 (38%) Doctors of Science, 8 (38%) Master and 6 (28.5%) with bachelor's level. An average level of competence among experts of 0.80 (High) was reached. The average service time is 20 years. These requirements meet the requirements for expert judgment [32].

The first round by the Delphi Method achieved high consensus in the five aspects consulted. The ecotourism management procedure obtained a classification of 89% translated into high relevance for the province of protected areas of Cuando Cubango.

5. Conclusions

It has been possible to design 15 priority actions that are part of the initial stage of ecotourism management in Cuando Cubango taking advantage, for this, of the interest that I have expressed for wildlife and the constant possibilities of socio-economic and environmental development of the protected areas.

However, the proposed actions were designed within the new normal created in the world by Covid-19. Together they are a way for When Cubango to respond to the challenges of the following documents:

- The 2030 Sustainable Development Goals (goals, 1, 8 and 15);

- The African Agenda 2063 (aspirations number 7);

- Okavango Zambezi Transboundary Conservation Area Treaty of 2011 (task 4 and objective 4); and.

- The Government Programme of Angola 2022 (p.38) (section IV. 2.2.4. A – f).

Through the expert method, an 89% validity of the five internal aspects of the procedure was achieved, as an instrument with theoretical-methodological and practical contributions, which usually perfects the variables-keys for the management of ecotourism, in order to contribute to the development of protected areas of Cuando Cubango.

However, it is necessary to continue to deepen studies on this subject, in order to apply the actions proposed, as a way to improve the management of protected areas, with the linking of all internal actors, allowing effective conservation, based on the increase in the animal population and forest repopulation, assigning forms of sustent from the dynamism of ecotourism.

Author details

José Eduardo Ezaquias[1,2]

1 Cuito Cuanavale University, Angola

2 University of Havana, Angola

*Address all correspondence to: drezequias@gmail.com

References

[1] Zandonai, R. (2020). Impacto do Covid-19 ao ecotourism atinge vida selvagem e comunidades. Uneo.org.

[2] Castro, H. A. (2009). Cultural tourism and cultural consumption in tourist trips. Cuba and the Historic Center of Havana. Retrieved from. http://www.econdesarrollo.uh.cu/index.php/RED/article/view/354 on November 28, 2020.

[3] Ceballos-Lascuráin H. Ecotourism. Mexico: Nature and Sustainable Development. Editorial Diana; 1998

[4] Fernández, R.M. (2009). Tourism Models and the main determinants of international tourism flows. University of Havana. Faculty of Tourism. Retrieved from file: /// C: /Users/hp/Downloads/Estudiobasefordeterritorialvocation of thejama-Perdernales dedistrito.pdf on July 10, 2020.

[5] Oliveira SGE. Theoretical models applied to tourism. Studies and perspectives in tourism. Buenos Aires, Argentina: Centro de Investigaciones y Estudios Turísticos; 2017

[6] Vera, R. P. (2013). Procedure as a contribution to the perfection of the management of the natural tourism product. PhD thesis. University of the East. Santiago de cuba.

[7] Witt, L. (2011). An ecotourism model for South African National parks. The School for Business Management. Africa do Sul.

[8] Huamanchumo, R.; Flores, C. & Barrantes, D. (2020). Ecotourism as a reactivator of local enterprises in protected natural areas. Recovered in http://scielo.sid.cu/scielo.php?script=sci_arttex&pid=S2218-36202020000400436 recovered on June 12, 2021.

[9] International Ecotourism Society, TIES (2015). What Is Ecotourism? Retrieved from https://ecotourism.org/what-is-ecotourism/ on March 8, 2021.

[10] Almeida ALM. Perception of Ecotourism: the case of IPP Tourism students. Porto: Dissertation. Superior School of Hospitality and Tourism of Porto; 2019

[11] Baez AL. Systematization of ecotourism / sustainable tourism experiences and their contribution to the conservation of Costa Rica's biodiversity. San Jose: Ministry of Environment and Energy; 2017

[12] Pilar M. Ecological and sustainable tourism: profiles and trends. Barcelona: OSTELEA; 2017

[13] Wardana, I.M., Sukaatmadja, I. P. G., Ekawati, N. W., Yasa, N. N. K., Astawa, I. P., & Setine, M. (2020). Policy models for improving performance to build quality tourism experience and sustainable tourism. Retrieved from growingscience.com/msl/vol11/msl_2020_320.pdf on February 12, 2021.

[14] Kuhlman, T. & Farrington, J. (2010). What is sustainability? Retrieved from https://www.researchgate.net/publication/47697344_What_is_Sustainability on March 22, 2020.

[15] Alonso, G. J. (2013). Review of the concept of local development from a territorial perspective. Leader Vol. 23. pp 9-28.

[16] Sachs J. Economy for a crowded planet. Argentina: Debate. Translated by Ricardo García Pérez. Buenos Aires; 2008

[17] Moreno, R.M. A. (2013). Strategic planning. Concept consultant

[18] Rodrigues J. Gestão estrategica de instituções financeiras. Lisbon: School Editor; 2012

[19] Enamorado GP. Procedure for environmental management in charming hotels. Partial application in the Hotel Caballeriza. Dissertation: University of Holguín; 2020

[20] Farías, M. F., Sonaglio, K. E. & Ferreira, L. V. F. (2019). Science, tourism and complexity: theorizing about transdisciplinarity for the interpretation of the tourist phenomenon. Retrieved from https://www.eumed.net/rev/turydes/26/turismo.htmlhttp://hdl.handle.net/20.500.11763/turydes26turismo on September 15, 2019.

[21] Klimenko Y. Modeling of factors of the development of environmental tourism in the innovation economy. Kazan National Research Technological University. Russia: Kazan; 2018

[22] Paez, C.C. T. (2015). Management Model of Territorial Policies of Local Development the Municipal Scale in Cuba. PhD thesis. Pinar Del Río. University of Pinar Del Río "Hermanos Saíz Montes De Oca".

[23] Escribano C. The articulation of the productive chain of nature tourism in function of local development. Thesis in option to the scientific degree of doctor in economic sciences: University of Camagüey; 2018

[24] Ministério do Turismo, MINTUR. Anuário de Estatística do Turismo – 10ª Edição. Luanda: MINTUR; 2015

[25] Institute of Tourism Promotion of the Ministry of Tourism, INFOTUR (2019). Tourism in Angola. World Tourism Forum. Luanda. National Institute of Statistics retrieved from https://www.ine.gov.ao retrieved 3 September 2019.

[26] Provincial Directorate of Tourism. Report on tourism in Cuando Cubango. Menongue: Provincial Governor; 2019

[27] Lira, I. S. and Escudero, C. S. (2012) Methodology for local development strategies. It would be Manuel. ILPES 50 years, Santiago de Chile.

[28] Correia, M. A & Ornelas, E. (Coords) (2016). Rumo ao Cuando Cubango: terras do pogresso. Dilazo. Lisbon

[29] Mendelshon, J. & Obeid, S. (2004) Rio Okavango, to Fonte da Vidas, editora struik publishers. Cape Town.

[30] Cassela D. Economy and management of natural resources in Angola. Luanda: Cicero; 2015

[31] Amabis J, Martho G. Fundamentals of modern biology. Fourth ed. St. Paul: Modern; 2006

[32] Ferrer, Y., Cruz, M., Pravia, M. and Cárdenas. (2020). The Method of Expert Opinion in Educational Research: A View from a Sample of Doctoral Theses. Retrieved from

Chapter 13

The Importance of Partnerships for Effective Protected Area Management

Mirjam de Koning and Oliver Avramoski

Abstract

Protected areas work in complex environments in which they have to liaise with governments, scientific and civil society organizations, volunteers, local stakeholders, visitors, and funders. This requires next to thematic expertise on conservation, among others legal, management, financial, administrative and communications skills and capacities. Especially the smaller protected areas struggle to efficiently operate in all these specialized fields and often lack enough in-house capacity and resources. This chapter highlights the lessons learned and evolvement of various forms of partnerships in different countries on different continents (collaborative arrangement in Laos and different formal and informal arrangements in the Western Balkans). Core to the success is to build sufficient capacity within the protected area management authorities so they understand the priorities and the resources needed to fund, manage and implement these priorities. Specialized skills and capacities needed for effective protected area management are limited in most countries and it is inefficient and too expensive to build this capacity in-house. Having a clear vision on what needs to be done and building a strong cooperation between partners through effective communication is the key to success to come to more effective protected area management either on a national, regional or transboundary level.

Keywords: protected area management effectiveness, partnerships, collaborative arrangements, Laos, Western Balkans

1. Introduction

Historically, protected areas controlled by governments have been a primary mechanism for conserving the world's biodiversity. Since the beginning of the new millennium, the terms 'management' and 'governance' are often used concurrently to denote both technical and power-related aspects of nature conservation, respectively [1]. Over the past decades protected area governance and management have diversified, with significant growth in private and community-based management, as well as a variety of partnership-based models [2]. This diversification has been driven by both ethical and pragmatic needs to take into account local community dependence on ecosystem goods and services, respect the rights of indigenous peoples, and address failures of top-down governance to deliver expected outcomes [3–8]. Under these influences, power has been redistributed

IntechOpen

across multiple public, private and civil society organizations, and collaborative arrangements are now widespread [9, 10].

Collaborative governance and management of protected areas should be beneficial to stakeholders involved in the partnership to be sustainable. Biodiversity benefits for governments, the scientific community and non-governmental organizations go hand in hand with socio-economic benefits for the private sector and local communities. Bringing different skills and resources to the table and reaching consensus can lead to so-called win-win situations [11]. Such arrangements often form cost-efficient solutions for effective protected area management. In addition, the increased knowledge, capacity, trust and learning by doing can result in less conflicts between the partners through an improved understanding [12, 13].

Protected area management authorities work in complex environments in which they have to liaise with national and local governments, scientific and civil society organizations, volunteers and local stakeholders, visitors and potential funders. This requires next to thematic expertise on conservation, among others legal, management, financial, administrative and communications skills and capacities. Especially the smaller protected areas struggle to efficiently operate in all these specialized fields and often lack enough in-house capacity and resources. Therefore, it is important for protected area management authorities to build effective partnerships to ensure certain resources through third parties instead of trying to do everything themselves. This can be in the form of collaborative arrangements, partnerships regulated through a Memorandum of Understanding (MoU) or more informal partnerships. This chapter will highlight some of the lessons learned and the evolvement of various partnerships in different countries on different continents. The first case study highlights a formal collaborative arrangement in Laos. The second case study describes various partnerships regulated through different MoUs or informal arrangements in the Prespa-Ohrid Ecoregion in North Macedonia, Albania and Greece.

2. Methodology

Two different case studies are described in this chapter using data from 2013 to 2016 in the Hin Nam No National Park in Laos in South-East Asia [14–17] and from 2017 to 2021 in the Prespa-Ohrid Ecoregion in North Macedonia, Albania and Greece in the Western Balkans [18, 19].

For both case studies the main lessons learned are derived from identified building blocks using the 'solutioning approach'. The PANORAMA - Solutions for a Healthy Planet Partnership is a global partnership that supports both the long-term strategic framework for capacity development and the knowledge management component of the draft post-2020 Global Biodiversity Framework [20]. Based on theories of knowledge transfer, peer learning, and social-ecological resilience, drawn from psychology, education, ecology, and conservation biology, PANORAMA documents and promotes verified examples of inspiring, replicable solutions across a range of conservation and sustainable development topics, enabling cross-sectoral learning and inspiration [21]. It allows for communication among solution providers and users through a virtual online platform (www.panorama.solutions) and further face-to-face and virtual formats. Developed by International Union for Conservation of Nature (IUCN) and Deutsche Gesellschaft für Internationale Zusammenarbeit (GIZ), the PANORAMA 'solutioning approach' was launched at the IUCN World Parks Congress in Sydney in 2014. PANORAMA enables easy communication among practitioners, often being a source of inspiration and supports mutual learning in and for protected areas. The idea is that practitioners replicate workable solutions instead of re-inventing the wheel.

Each peer-reviewed and published solution is analyzed to identify the factors or building blocks that contribute to its successful implementation, and the online platform allows users or solution seekers to discover and access this knowledge, the solution providers, the relevant communities of practice, and also to compare and contrast solutions across geographies and sectors. PANORAMA has grown both in size and scope over several years. By April 2021, it included 868 solutions from 614 solution providers from 117 countries. Out of the 868 solutions, 431 are protected area solutions. From its inception, PANORAMA's relevance and contribution to the implementation of the Strategic Plan for Biodiversity 2011–2020, progress towards the Aichi Targets, the Sustainable Development Goals and the draft post-2020 Global Biodiversity Framework has been recognized specifically [20].

3. Case study I: Hin Nam No National Park, Laos

Functional and sustainable collaborative partnership arrangements in Laos and in Southeast Asia are not yet very common. Lack of communication and participatory decision making often leads to centralized efforts by the government resulting in lack of understanding and conflicts with local people living in or adjacent to protected areas. The definition of 'participation' is understood differently by the main stakeholders involved mixing up terms such as information sharing, consultation or real involvement in decision making. For effective and sustainable collaborative governance and management of protected areas to achieve biodiversity conservation and natural resource management objectives an equitable approach is needed [22].

Since the early 1990's, Lao Government policy for protected areas has focused on developing a partnership approach, which advocates peoples' involvement in conservation, especially with the locals who depend on the natural resources for their daily livelihoods [23]. To transform from a paper park approach to effective protected are management there is a need for the Lao Government to allocate sufficient resources for the management of each protected area and/or to establish functioning collaborative arrangements. The latter requires a clear division of roles between co-managers; ensuring that the transfer of responsibilities goes to the locals with customary rights; and promoting good governance and capacity development at all levels (especially if the poor are to benefit) [24].

This case study describes the lessons learned from a multi-level collaborative governance system in Hin Nam No National Park in central Laos following the 'PANORAMA solutioning approach' [25]. Five so-called 'building blocks' of the experimental collaborative governance model in Hin Nam No were identified [14].

Hin Nam No National Protected Area, in brief Hin Nam No, has been recently enlarged and declared as a national park (January 2020). Hin Nam No is located in Boualapha District, Khammouane Province. Containing 94,000 ha, the area is one of the largest karst landscapes in Southeast Asia, being contiguous with Phong Nha–Ke Bang National Park in Central Vietnam (see **Figure 1**). A total of 18 villages lie in immediate proximity to Hin Nam No, with a total population of about 8,000 people, many of whom are ethnic minorities. Like other national protected areas in Laos, Hin Nam No had for a long time insufficient resources with only a part-time director and no full time staff on site. The lack of limited human and financial resources allocated by the government resulted in a lack of capacity, skills, information, and law enforcement to effectively manage and monitor the protected area.

Technical and financial support by the German Government has facilitated high levels of external support, both at the management level and in the different

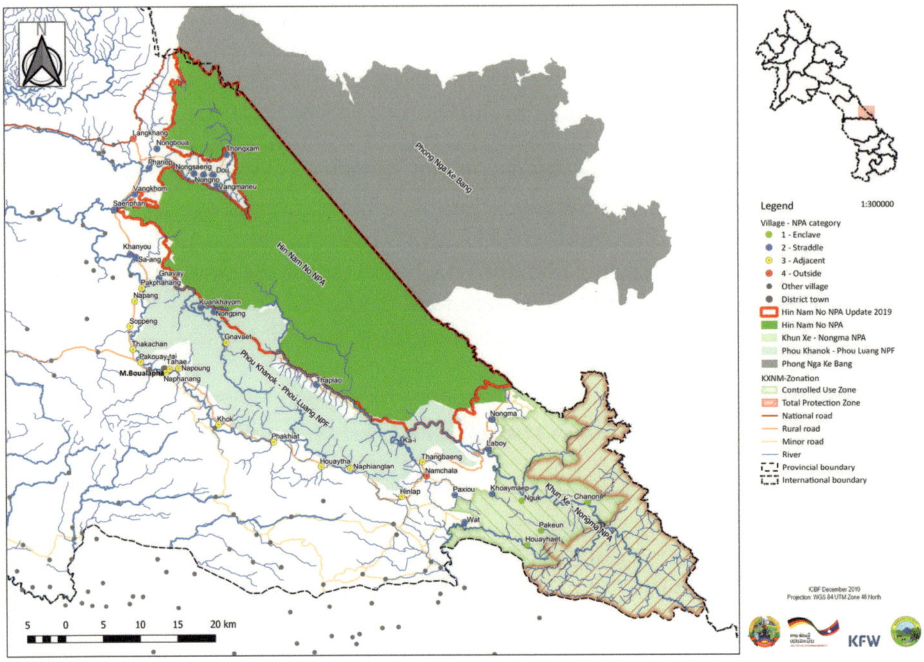

Figure 1.
Location of Hin Nam No National Park in Khammouane Province in Laos (map prepared by Ronny Dobbelsteijn).

specialized fields. Experts have provided on-going support to the establishment and maintenance of the collaborative governance and management system.

The five identified building blocks of the PANORAMA Solution are:

1. Governance assessment through participatory consultation

2. Setting-up a multi-level collaborative management and governance structure

3. Participatory zonation based on traditional knowledge and customary rights

4. Collaborative governance agreements

5. Local people as additional protected area management manpower.

3.1 Governance assessment through participatory consultation

To better understand the governance and management status of the Hin Nam No, a governance assessment was implemented in February 2014 at various levels: village, village cluster, district and province. The collected data led to a set of proposed interventions implemented over a period of two years. The results are presented in **Table 1**.

The participatory assessment was a good starting point for improved communication and understanding between the co-managers. It led to the creation of a joint vision and a proposed division of roles. As part of the assessment a Management Effectiveness Tracking Tool (METT) session was included. The METT developed by the ASEAN Centre for Biodiversity [26] is similar to the conventional used METT but has an additional focus on governance. In addition, a more detailed

Outcome governance assessment (February 2014)	Proposed intervention, progress (February 2016)
No clear delegation of decision making or implementation authority to guardian villages (building block 2 and 3)	Hin Nam No Management Authority identified tasks to be delegated to villagers
Governance system is ad hoc and top-down, with lack of systematic benefit sharing (building block 2 and 4)	Participatory reporting/planning system was developed at village (18), village cluster (5) and protected area level. Participatory co-management agreement, including benefit sharing mechanism, was developed and approved.
Lack of skills and capacity; lack of involvement by women (building block 2)	Capacity development plan has been elaborated; recruitment of five female Lao Government volunteers (trainees)
Unclear zonation of Hin Nam No into manageable units per guardian village. A guardian village is actively involved in the protection of the protected area based on their customary rights (building block 3)	Participatory zonation and trail mapping carried out in 18 priority guardian villages
Local rules exist but are unknown or not implemented by outsiders (building block 4)	Establish general rules for the different zones in each guardian village and disseminate the information broadly
Willingness of guardian villages/village rangers to be involved in Hin Nam No management (building blocks 4 and 5)	Monthly participatory biodiversity monitoring and patrolling system established using motivated village rangers who are compensated based on performance
Law enforcement system is unclear, slow and ineffective (building blocks 4 and 5)	Some delegation of law enforcement to villagers ensures a more rapid and effective response

Table 1.
Governance assessment results and subsequent interventions.

questionnaire adapted from annex 3 of the IUCN publication 'Governance of Protected Areas' was developed and used to assess good governance criteria [2].

3.2 Setting-up a multi-level collaborative management and governance structure

To have a better understanding of the tasks and to achieve more effective protected management a new management structure for Hin Nam No was established in 2013 and early 2014 including six technical units. This process was supported by GIZ and the National University of Laos. An overview of the main conservation actions was developed for each technical unit and tasks to be delegated to the villagers were identified.

The new Hin Nam No management structure and its six technical units had a total of 27 staff (out of which 19 volunteers) to manage the protected area (August 2016). None of the staff had sufficient capacity to lead one of the technical units in any of the specialized fields of management. **Figure 2** shows the institutional arrangements of the management authority of Hin Nam No in 2016. The implementation of the protected area management tasks was decentralized to the district level.

Stakeholders bringing different skills to the table need to be involved to ensure effective collaborative governance and management. Primary stake- and rights-holders are the villagers and protected area management authorities. The participation of secondary stakeholders is needed for effective strategic and operational steering in topics such as coordination, patrolling and law enforcement. This will help to mitigate

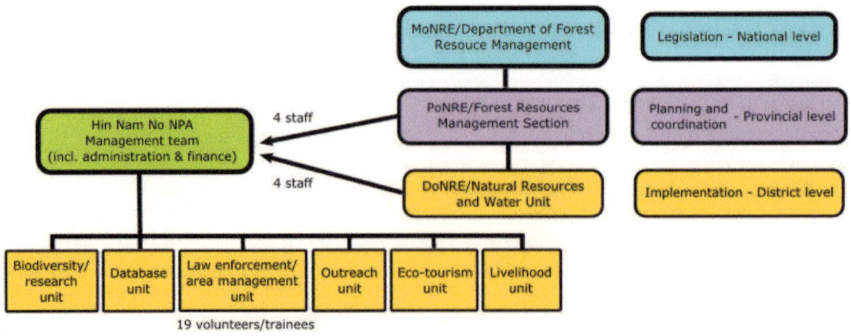

Figure 2.
Institutional arrangements of the Hin Nam No management authority and its six technical units.

against threats such as illegal logging and poaching often initiated by outsiders. Strategic alliances with assisting partners for institutional support, capacity development and funding can strengthen the collaborative arrangement to make it more effective and enabling collaboration among the stakeholders towards a common goal.

A District Co-Management Committee was established bringing together 13 appointed government officials from district level as well as village representatives from village cluster level [25]. Villages report to village cluster level, which thereon report to the higher levels. The functioning of this bottom-up process is monitored via the annual management effectiveness and good governance self-assessment in which villagers participate. Higher levels take the inputs and needs of the village levels into account and strategic decisions are communicated back to the operational village levels.

This institutional set-up ensures that all stake- and rights-holders can participate in decision-making processes. Transparent sharing of information, experience, and knowledge enhances the capacity for natural resource management among all parties to achieve the common goal of biodiversity conservation and poverty alleviation in and around Hin Nam No. A balance needs to be found between the need to involve people in the management i.e. doing the work in the forest (village rangers) and the need to involve people in the governance who can validate decisions (village authorities and high level officials).

3.3 Participatory zonation based on traditional knowledge, customary rights and biodiversity values

Participatory zonation is an essential tool for local communities to engage in collaborative governance and management – especially when the process takes into account local knowledge and respects existing customary rights. The participatory zonation process started in 2014, based on the agreed interventions of the governance assessment (see **Table 1**). In order to divide the work between the villages surrounding Hin Nam No, it was necessary to clarify areas and boundaries, based on used trails and customary rights of villages. Village rangers mapped the trails and collected data on important features, biodiversity and threats. Villagers were asked to define areas they need for collecting natural resources, areas that are inaccessible due to the rugged terrain, and areas that should be left alone to protect wildlife for breeding purposes.

Based on the proposals by the villagers, the Hin Nam No was geographically divided into 18 areas to be managed by the villages. The zonation process identified the Controlled Use Zones (CUZ) prescribing the traditional village lands of the 18 villages. In a second step, management rules for the CUZs were formulated, based on the customary rights of the villagers. The Total Protected Zones (TPZ) comprise all parts of Hin Nam No beyond the CUZ. They can be divided into inaccessible parts, and areas considered of high biodiversity value [25]. The process of participatory mapping of trails and the subsequent selection of key trails for regular monitoring led to a clear agreement on which area should be monitored by which village. This led to a 'de-facto' delineation of village areas of responsibility within Hin Nam No. In total, 86 per cent were proposed by the villagers as TPZ and 14 per cent as CUZ [15].

The basic rules and regulations governing the access and use of the proposed TPZ and CUZ are stipulated in the Forestry Law of 2007 and in the collaborative agreements that have been approved by the District Governor of Boualapha. The District Co-Management Committee agreed that further meetings with the villagers were required to discuss and agree upon more detailed resource use rules for the CUZ to prevent unsustainable use by villagers and outsiders with the final zonation system to be approved by the District Co-management Committee.

3.4 Collaborative agreements

Collaborative agreements were drafted with the help of a neutral facilitator and taking the inputs of the villagers into account. Based on the results of this process the local authorities decided to generate one uniform collaborative agreement in the form of a district by-law, including benefit-sharing arrangements based on customary rights. The district by-law went through several meetings and due diligence processes involving legal government offices before it was officially approved by the Boualapha District Governor. The final version was disseminated to all 18 villages and also in the adjacent Phong Nha-Ke Bang National park in Vietnam.

3.5 Local people as additional protected area management manpower

The Hin Nam No collaborative arrangement involves local villagers actively in the management of the protected area. First of all the villagers were willing to participate and secondly their knowledge about the area is invaluable. This formed a cost-efficient addition to the limited resources provided by the government. In total there were 87 democratically elected co-management committee members, spread over 18 villages and five village clusters, involved in participatory planning and reporting. Village rangers coming from the 18 villages were compensated for making regular trips into the protected area to record wildlife sightings and threats and to be involved in patrolling for law enforcement. Fees for the village rangers were agreed through negotiations and based upon fair compensation for the hard and dangerous work of climbing in the mountains.

A total of 110 villager rangers were trained in the use of GPS equipment and in recording sightings in coded booklets. Data and information collected by the village rangers were inserted into the Spatial Monitoring and Reporting Tool (SMART) system on a quarterly basis. The database unit analyzed the data and presented the main wildlife sightings and threats to the District Co-Management Committee and the Hin Nam No Director in quarterly reporting and planning meetings by using maps.

At the end of 2016 about 35 households in four villages were involved in the provision of eco-tourism services such as guiding, boating services as well as guesthouse and home-stays. The local service providers were trained to ensure a certain standard of services. The collaborative arrangement between the Hin Nam No management authority and the local service providers was captured in a conservation agreement to ensure the protection of the environment and benefiting the local people at the same time.

3.6 The way forward: Hin Nam No National Park

After the establishment of Hin Nam No as a national park in 2020 the resources for effective management have increased. However, in August 2016, Hin Nam No still had very low human and financial resources and therefore effective management needed to be improved. To address the challenges, the Hin Nam No management authorities and GIZ developed an innovative collaborative system in which technical and administrative agendas were mixed (socializing protected areas), in line with relevant legislation on decentralization and based on customary rights. This increased the political and local support for collaborative governance and management and was different from previously tested approaches in Laos.

The description of the building blocks and their interlinkages enabled a relatively simple and structured write-up and subsequent communication of the three-year process that was followed to set up the multi-level collaborative system. The collaborative model brought positive results (increase in management effectiveness) with opportunities to the entire protected area system in Laos, up to now often referred to as a 'paper park' system [27]. More work on 'sustainable financing' and 'adaptive management' through actual implementation is required to sustain this model.

4. Case study II: Prespa-Ohrid Ecoregion, North Macedonia, Albania and Greece

Spanning the borders of Albania, Greece and North Macedonia in the Western Balkans, the Prespa-Ohrid Ecoregion has been identified as one of Europe's biodiversity hotspots (see **Figure 3**). At the heart of the region are Lake Ohrid and the Prespa Lakes. Lake Ohrid, shared between North Macedonia and Albania, is possibly the oldest lake in continuous existence in Europe with an estimated age of 1.4 million years; it is also one of the most voluminous freshwater bodies in Europe. Due to the karstic bedrock, water from the Lake Prespa basin contributes significantly to the water inflow of Lake Ohrid. The Prespa basin includes the Greater Prespa Lake (shared between the three countries) and Lesser Prespa Lake (shared between Greece and Albania).

There are many protected areas in the Prespa-Ohrid Ecoregion established to protect its extraordinary biodiversity. International designations include the transboundary Prespa Park, Natural and Cultural Heritage of the Ohrid Region, a transboundary mixed (natural and cultural) World Heritage Site, the Ohrid-Prespa Transboundary Biosphere Reserve, and several Ramsar sites. There are two Natura 2000 sites in the Greek part of the region and several Emerald sites in the Albanian and Macedonian parts. Following the IUCN typology, the governance of protected areas in the Prespa-Ohrid Ecoregion falls in the governance by government model [28]. Nonetheless, conservation is not a priority for the three national governments, and protected area management authorities are both understaffed and underfunded or absent altogether.

So far transboundary cooperation functions informally. The most important conservation challenges in the Prespa basin are related to water quality and eutrophication, exacerbated by the recent significant water level decrease and climate change. Habitat degradation and urbanization along the lake shores top the long list of threats to the Lake Ohrid ecosystem.

On a socio-economic level there are high unemployment rates resulting in young people leaving the area. The ethnically diverse mix of people are living under poor local economic conditions with difficulties in trading local products, and a lack of basic

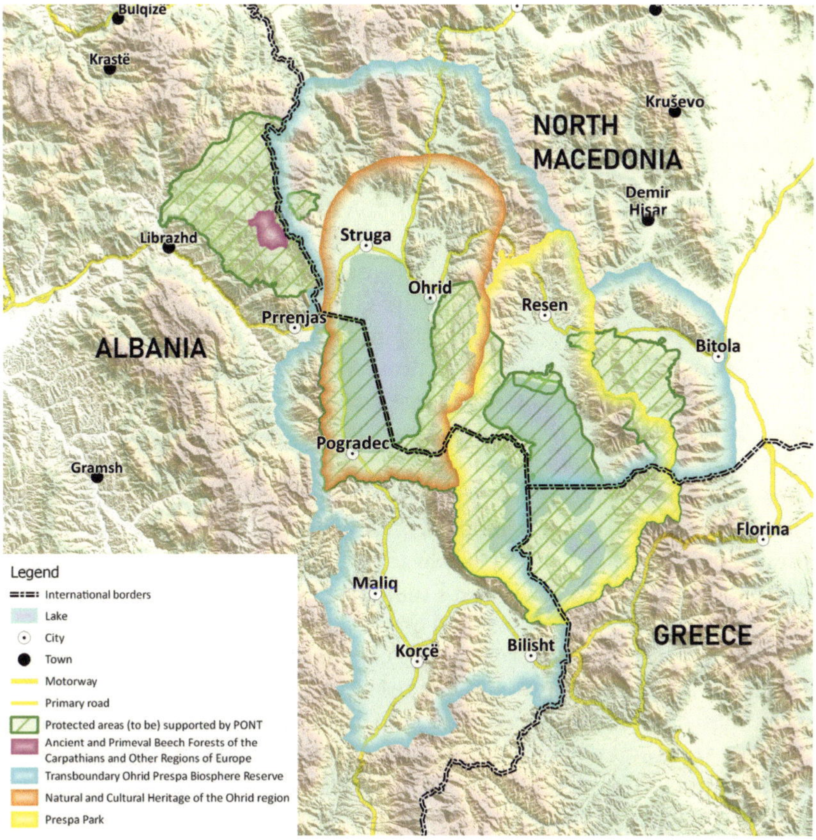

Figure 3.
Location of the Prespa Ohrid ecoregion in the Western Balkans (map prepared by Ronny Dobbelsteijn).

infrastructure. Civil society in the Prespa area is weak especially in Albania and North Macedonia. The rural area is dominated by agriculture with some income from stock-breeding, fisheries, forestry and tourism [29]. The secondary and tertiary sectors, in particular tourism, have a dominant role in the economy of the Ohrid region [30, 31].

The transboundary cooperation in the Prespa basin functions informally, despite the decades-long efforts to establish formal institutions. To address the lack of formal functioning transboundary institutions the three main conservation Non-Governmental organizations (NGOs), the Macedonian Ecological Society (MES), the Protection and Preservation of Natural Environment in Albania (PPNEA) and the Society for the Protection of Prespa in Greece (SPP), formed a network in 2013 called PrespaNet. The three partners work together to protect the transboundary Prespa lakes basin for the sustainable benefit of both people and wildlife. The joint Lake Ohrid Watershed Management Committee, established by the Albanian and Macedonian governments in 2004 meets irregularly and its Secretariat has been mostly inoperative [32].

4.1 Prespa Ohrid Nature Trust

Cooperation within and across borders is crucial to ensure sustainable conservation and effective management of protected areas. Prespa Ohrid Nature Trust (PONT) established in 2015 is a transboundary conservation trust fund providing long-term financing (~€1.5-2million/year drawdown until at least 2030), which is additionally used to leverage the co-financing of activities.

PONT enables protected areas in the Prespa-Ohrid Ecoregion to develop and implement their management plans to conserve nature through sustainable co-financing of operational costs. PONT supports the protected area staff with the development and use of standard operational planning and reporting systems, in line with the management plans, for the implementation of protected area programmes. Biodiversity monitoring systems are developed in which data collection, data analysis and habitat management are implemented, with an efficient division of what can be done by the protected area staff themselves, local people and what to outsource to third parties. Enabling the protected area staff to increasingly use scientific data in managing the area is included in the third-party contracts. The inclusion of minimum Natura 2000 requirements will gain importance in the coming years. **Figure 4** illustrates the PONT protected area grant programme.

Qualified NGOs, municipalities and research institutes with a local presence in the Prespa-Ohrid Ecoregion and with a strategy in conservation, developing society, improving communities, and promoting citizen participation in conservation are eligible for PONT grants to environmental actors. PONT funding priorities for environmental actors are mainly focused on transboundary conservation activities, some important specialized conservation activities and work related to nature-based tourism and Non Timber Forest Products. In addition, environmental actors are supported with their operations and organizational development with the aim to make the organizations more robust and capable of sourcing third party funding.

PONT financial support to NGOs and research institutes is directed towards the implementation of activities identified in the management plans for the respective protected areas and where there is a lack of capacity within the protected area management authorities. Applied research, with the involvement of protected area staff, directly focused on species or habitat management is also supported, but fundamental research is not. **Figure 5** illustrates PONT's environmental actor grant programme.

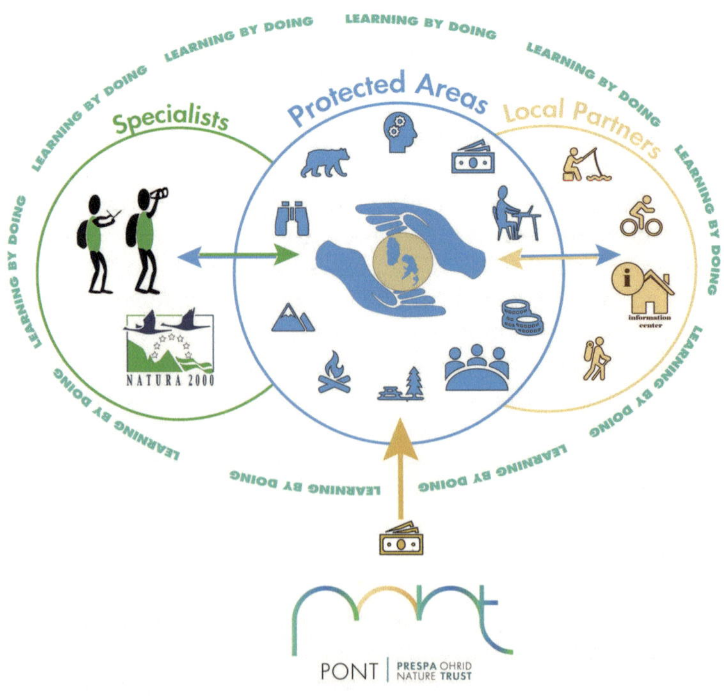

Figure 4.
PONT's protected area grant programme.

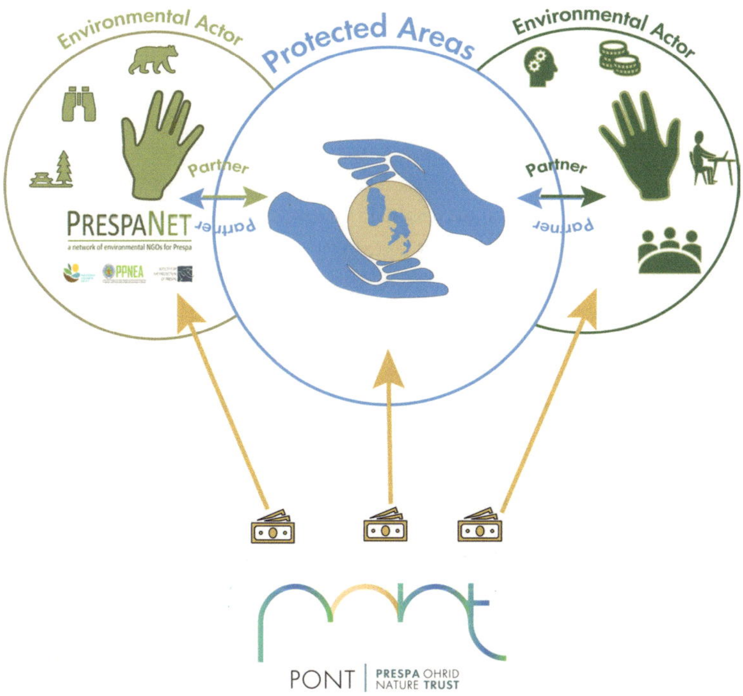

Figure 5.
PONT's environmental actors grant programme.

Relevant building blocks were distilled from two PANORAMA solutions to showcase the importance of partnerships for effective protected area management [18, 19]. Achievements and lessons learned per building block are described in more detail in the following sections.

1. PONT strategy promotes transboundary cooperation between government and non-government partners

2. Taking operational planning seriously

3. Core funding secured

4.2 PONT strategy promotes transboundary cooperation between government and non-government partners

The PONT ten year strategy for Prespa is based on the experiences of the PrespaNet partners who have worked in the area for a long time. By working directly with the protected area management authorities and the main NGOs the conservation and capacity development objectives were determined. Instead of re-inventing the wheel the priority gaps for financing were gathered by PrespaNet. This was done under coordination of World Wildlife Fund for Nature (WWF) Greece who knows the area very well and could verify the results. The recommendations for the conservation objectives were verified by the protected area management authorities and this formed the basis for the conservation objectives for the PONT ten year strategy for Prespa. With the help of the more social science oriented expertise by PONT the objectives for the inclusion and beneficiation of local stakeholders, organizational development and capacity development were identified and added. This resulted in a participatory developed strategy and conflict assessment study that were accepted by both government and non-government stakeholders within and across state borders.

With a focus of financing of the identified gaps the selection of grantees was made based on their mandate, vision, proven track record and expertise to work in the area [18]. This enabled the rapid granting of first grants to NGOs and protected area management authorities focusing on action-oriented implementation. Remaining identified gaps were addressed by several open calls for proposals. After four years of operation PONT has one five-year grant and ten three-year grants for their long-term partners. Through this process based on previous learning a strategy was established focusing on the financing of the gaps and building of strong partnerships with stakeholders that have a mandate, vision and the expertise to achieve the conservation results in the Prespa area. The PONT strategy also promotes transboundary cooperation between government and non-government partners based on previous lessons learned.

4.3 Taking operational planning seriously

Protected area managements plans have been gradually integrated into the long-term management cycles for the protected areas in the Prespa-Ohrid Ecoregion. However, there is still a wide gap with annual or operational planning. The gap exists due to a number of reasons, including unrealistic and non-operational management plans, lack of knowledge and skills, inadequate work procedures, missing or ineffective decision-support systems, as well as insecure funding. The heavy dependence over the past two decades on short-term international project funding and external consultants have often perpetuated these weaknesses. The resulting ad-hoc and inconsistent management hampers the effective implementation of the management plans.

Protected area management authorities in the Prespa-Ohrid Ecoregion use a template developed by PONT in Microsoft Excel to prepare annual operational plans and budgets that are part of their grant applications to PONT. These operational plans state the actions to be implemented each year to achieve the objectives set out in the management plan for the protected area concerned. While most of the actions are selected from among those identified in the management plans, additional actions arise from the (annual) METT assessments, by learning from experience, or in response to uncertainty and change. The operational plans integrate both recurrent (routine) activities and non-recurrent activities (investments, projects) to ensure resources are adequately distributed across the different functional areas. The operational plans only include activities that are currently achievable with existing staffing, technical and financial resources, including the co-financing from PONT.

Using pre-defined templates developed by the national authorities on protected areas in both Albania and North Macedonia, the protected areas in the Prespa-Ohrid Ecoregion prepare annual (operational) plans that are subject of formal approval by the national authorities. Aside from the budget that is more detailed, the template developed by PONT is similar in content to those used under national legislation. The operational plans and budget are prepared at the end of each calendar year for the subsequent one and constitute the key element of the grant applications submitted to PONT; the grant application process of PONT is aligned with the national system planning and reporting cycles to avoid duplication of work.

Although operational plans have been in use for about a decade in North Macedonia and for several years in Albania, management and on-ground work continued to be largely ad-hoc and inconsistent. The PONT template and the input from the regular METT assessments enable protected area managers develop more realistic annual operational plans and budget. The PONT template prompts the managers to plan in more detail the deployment of human, financial and technical resources related to the basic functional areas, such as biodiversity monitoring, patrolling, habitat restoration, environmental education or visitor management that were often neglected in the past. This proved to be quite a challenging task due to the lack of adequate procedures and systems in place, in particular for functions and activities where no prior experience exists.

4.4 Core funding secured

Improved operational planning enables the protected area managers clearly define their capacity gaps and most critical resource requirements for effective implementation of the activities. PONT's long-term co-financing enables the protected area managers in the Prespa-Ohrid Ecoregion to recruit new staff and deploy resources to sustain their core management functions. Using the budget template developed by PONT, protected area managers develop a detailed budget for each action that is broken down into five cost categories: staff costs; consultants; equipment and infrastructure; travel, meeting, and training costs; and consumables, operating and other costs. PONT co-financing amounts up to 50% of the total annual budget and is used for covering both recurrent and non-recurrent cost related to the core management operations, except for procurement of equipment and construction of new infrastructure exceeding € 20,000.

The PONT budget template helps protected area managers combine effectively PONT's co-financing with funding from the government or the revenue they generate, as well as projects implemented by conservation NGOs or international donors and agencies.

Lack of detailed data on protected area management costs hamper effective conservation planning and management. Protected areas in the Prespa-Ohrid Ecoregion lack systems in place that connect financial data with the on-ground conservation actions. Financial information is commonly managed for the purposes of meeting national financial reporting requirements, that are general in nature, rather than management. The annual budget using PONT's template is organized in a way that permits costs aggregation and analysis by results that are in turn linked to management objectives. This also informs the operational planning in the subsequent management cycle and helps identify opportunities for improved productivity and effectiveness. Further progress in operational planning would depend on the capacity to improve the estimates of the required costs of different functional areas of work and also of the levels of management performance.

4.5 The way forward: Prespa-Ohrid Ecoregion

Having a secured total budget allocation for the year in the long-term enables the protected area managers in the Prespa-Ohrid Ecoregion to develop and maintain the key functional areas and programmes, based on the management plan and thereby increase the management effectiveness. Especially recurrent activities such as regular monitoring of biodiversity, visitor management and environmental education have recently improved. These themes previously relied on short-term and often discontinued support from donors providing initial investments and technical assistance, but no funding to sustain the operations in the long-run. With PONT's long-term co-financing the protected area managers are able to recruit and retain new staff and gradually retrain the existing ones to develop and implement the key programmes. Capacities to mobilize, manage and implement additional funding from external sources for non-recurrent activities that have a more flexible timeline of implementation has increased. Several rangers, biologists, communication and education experts have joined the protected area management staff in the Prespa-Ohrid Ecoregion over the past two years filling in long-vacant positions of critical importance for their basic operations.

One of the roles of PONT is to facilitate and support the establishment of partnerships between the protected area management authorities and the environmental actors working on issues where the protected areas have insufficient capacity such as habitat/wetland mapping, wetland restoration, biodiversity monitoring, environmental education, tourism development, etc. Due to the improved planning by protected area management authorities it is more clear for the managers what can be done by themselves and where there is need for resources from third parties such as NGOs, scientific institutions or local people. Several formal and informal arrangement have been established over time such as the employment of temporary local workers helping Prespa National Park in Albania with the maintenance of hiking trails, removal of alien species and fire management. Already four formal partnership agreements/MoUs have evolved over time between protected area management authorities and environmental actors. Often the partnerships started informally by working together and after a certain period of cooperation these partnerships were acknowledged through MoUs. For example the Public Institution Galicica National Park established partnerships for nature-based tourism with the local Alpine club PATAGONIA Ohrid and Association of Sports "Sport for all – All for sport". Resen Municipality established partnerships with the Public Scientific Institution Hydrobiological Institute Ohrid and the Macedonian Ecological Society. Three other MoUs are currently being considered based on the good experiences of cooperation i.e., one by the Public Institution Galicica National Park with the Macedonian Academy of Sciences and Arts (MASA) in North Macedonia; one by the Regional Administration of Protected Area Korçe in Albania with the NGO PPNEA, and one between PPNEA and the University of Korçe (signed on 22 May 2021).

Highlighted should also be the more complex cooperation on a transboundary level taking place in Prespa and recently formalized and implemented by the three governments (29–30 June 2021). The transboundary 'Prespa Park' was created in 2000 with a declaration by the Prime Ministers of Albania, Greece and North Macedonia stating the importance of the Prespa basin and recognizing the preliminary work done by environmental NGOs. To institutionalize the operations of the 'Prespa Park' an agreement was signed by the three Environmental Ministers and the EU in 2010. This agreement stipulates the need to make a management plan and the development of Integrated River Basin Management Plans in line with EU and international standards. The agreement was signed by all parties in 2010, followed by a ratification process which was only finalized in 2019. Recently, the Prespa Park Management Committee has been established with representatives of administrations, protected areas, NGOs, and local municipalities to coordinate the work on environmental protection and sustainable development of Prespa. The process of coming from this 'de jure' transboundary cooperation on paper to a 'de facto' implementation has evolved over a period of 21 years.

In 2018 PONT won the first Pathfinder Award [33, 34]. Encouraged and supported by PONT, several of the stakeholders involved in these processes are currently developing their first PANORAMA solutions to identify the challenges and benefits of their successful management strategies, with a focus on building and maintaining partnerships among local protected area stakeholders [35].

5. Concluding remarks

The two case studies illustrate different but converging paths in the evolvement of local partnerships aiming at more effective protected area management. The core to the success is to build sufficient capacity within the protected area management authorities for them to understand the priorities and the resources needed to fund, manage and implement these priorities. Specialized skills and capacities in most countries in several subjects important for effective protected area management are limited and it would be impossible and too expensive to try to build this capacity in-house. Having a clear vision on what needs to be done and building a strong cooperation between partners through effective communication is the key to success to come to more effective protected area management (either on a national, regional or transboundary level).

The Protected Area Management Effectiveness (PAME) framework, developed by the IUCN World Commission for Protected Areas [36] provides a means to assess contributions of the solutioning approach for addressing challenges in protected area management [21]. The METT, which is built around the PAME framework, was applied in a participatory manner in both case studies, opening ways for building partnerships among major protected area stakeholders. A closer look at the METT scores for protected areas in the Prespa-Ohrid Ecoregion, reveals that the most significant progress since 2018 was made with respect to 'Inputs' and 'Processes'. The latter was mostly related to improved implementation of management-oriented surveys and research, as well as advancement of environmental education. The partnership agreements between protected area authorities and locally present NGOs, underpinned by the long-term PONT co-financing, directly contributed to these advancements. Similarly, the 2016 METT assessment for Hin Nam No showed that the management effectiveness score had increased by 13 per cent since 2014, accompanied by a 15 per cent increase in good governance score, as measured by the IUCN Indicators for Governance Quality [2].

The involvement of stakeholders in the METT assessments was instrumental to improving both management and governance aspects of conserving biodiversity in protected areas in both case studies. Many of the issues and challenges discussed and agreed during the METT assessments have both management and governance aspects and the solutions and approaches agreed upon are subsequently integrated into the strategic and operational planning. In both case studies a range of institutional mechanisms and processes (e.g. Management Boards, advisory councils, Strategic Environmental Assessment (SEA) procedures, MoUs, METT assessments, participatory monitoring and law enforcement, etc.) provide a diverse and complementary ways of sharing authority and responsibility among protected area stakeholders.

The case studies demonstrate that METT can be useful in evaluating the success of adapting and uptake of the building blocks of PANORAMA solutions to protected areas in different contexts and geographies. On the other hand, by offering a systematic and comprehensive approach to developing and sharing lessons learned regarding the challenges and successes in protected area management, the PANORAMA methodology encourages learning and experimentation among protected area stakeholders.

Acknowledgements

The authors would like to thank all relevant stakeholders involved in the work around Hin Nam No in Laos and the Prespa-Ohrid Ecoregion in the Western Balkans. We also would like to thank Marie Fischborn of the IUCN for her comments and edits. Without these inputs this publication would not have been possible.

Author details

Mirjam de Koning[1*] and Oliver Avramoski[2]

1 Prespa Ohrid Nature Trust, Frankfurt, Germany and Tirana, Albania

2 Prespa Ohrid Nature Trust, Tirana, Albania

*Address all correspondence to: mirjamdekoning@yahoo.com

References

[1] Borrini-Feyerabend G, Hill R. Governance for the conservation of nature. In: Worboys GL, Lockwood M, Kothari A, Feary S, Pulsford I, editors. Protected Area Governance and Management. Canberra: ANU Press; 2015. p. 169-206.

[2] Borrini-Feyerabend G, Dudley N, Jaeger T, Lassen B, Broome NP, Phillips A, Sandwith T. Governance of protected areas: from understanding to action. Gland: IUCN; 2013. 124 p.

[3] Ghimire KB, Pimbert MP. Social change and conservation: environmental politics and impacts of national parks and protected areas. London: Earthscan; 1997. 342 p.

[4] Stevens S. Conservation through cultural survival: indigenous peoples and protected areas. Washington: Island Press; 1997. 361 p.

[5] Anderson TL, James A. The politics and economics of park management. Lanham: Rowman and Littlefield; 2001. 224 p.

[6] Brechin SR, Wilshusen PR, Fortwangler CL, West PC. Beyond the square wheel: toward a more comprehensive understanding of biodiversity conservation as a social and political process. Society and Natural Resources. 2002;15:41-64.

[7] Phillips A. Turning ideas on their head: the new paradigm for protected areas. In: Jaireth H, Smyth D, editors. Innovative governance: indigenous peoples, local communities and protected areas. New Delhi: Ane Books; 2003. p. 1-27.

[8] Cumming D. Performance of parks in a century of change. In: Child B, editor. Parks in transition: biodiversity, rural development and the bottom line. London: Earthscan; 2004. p. 105-124.

[9] McCarthy J. States of nature: theorizing the state in environmental governance. Review of International Political Economy. 2007;14:176-194.

[10] Lockwood M. Good governance for terrestrial protected areas: a framework, principles and performance outcomes. Journal of Environmental Management. 2010;91(3):754-766.

[11] Munthali M. Trans-frontier conservation areas: integrating biodiversity and poverty alleviation in Southern Africa. Natural Resources Forum. 2006;31:51-60.

[12] Folke C, Hahn T, Olsson P, Nordberg J. Adaptive governance of social-ecological systems. Annual Review of Environment and Resources. 2005;30(1): 441-473.

[13] Armitage DR, Plummer R, Berkes F, Arthur RI, Charles AT, Davidson-Hunt IJ, Diduck AP, Doubleday NC, Johnson DS, Marschke M, McConney P, Pinkerton EW, Wollenberg EK. Adaptive co-management for social-ecological complexity. Frontiers in Ecology and the Environment. 2009;7(2):95-102.

[14] de Koning M. Additional local manpower improves protected area management effectiveness. PANORAMA: Inspiring Protected Area Solutions [Internet]. 2015. Available from: https://panorama.solutions/en/solution/additional-local-manpower-improves-protected-area-management-effectiveness [Accessed: 2021-05-28].

[15] de Koning M, Dobbelsteijn R. 2015. Participatory zonation, management and monitoring of Hin Nam No National Protected Area in Laos. In: XIV World Forestry Congress; 7-11 September 2015, South Africa. Durban: FAO.

[16] de Koning M, Parr J, Sengchanthavong S, Phommasane S. Collaborative governance improves management effectiveness of Hin Nam No National Protected Area in Central Lao PDR. Parks. 2016;22(2):27-40.

[17] de Koning M, Nguyen T, Lockwood M, Sengchanthavong S, Phommasane S. Collaborative governance of protected areas: success factors and prospects for Hin Nam No National Protected Area, Central Laos. Conservation and Society. 2017;15(1):87-99.

[18] de Koning M. Prespa Ohrid Nature Trust (PONT) – an innovative partnership enhancing conservation and cooperation. PANORAMA: Inspiring Protected Area Solutions [Internet]. 2018. Available from: https://panorama.solutions/en/solution/prespa-ohrid-nature-trust-pont-innovative-partnership-enhancing-conservation-and [Accessed: 2021-05-28].

[19] Avramoski O. Closing the gap between strategic and operational planning for protected areas. PANORAMA: Inspiring Protected Area Solutions [Internet]. 2020. Available from: https://panorama.solutions/en/solution/closing-gap-between-strategic-and-operational-planning-protected-areas [Accessed: 2021-05-28].

[20] PANORAMA. Solutions for a healthy Planet. 2018. ISSN 2511-7475. Available from: https://panorama.solutions/en/. [Accessed: 2021-05-28].

[21] Mattsson BJ, Fischborn M, Brunson M, Vacik H. Introducing and evaluating a knowledge transfer approach to support problem solving in and around protected areas. AMBIO. 2019;48(1):13-24. DOI: 10.1007/s13280-018-1048-5.

[22] Baird I. Integrating community-based fisheries co-management and protected areas management in Lao PDR: opportunities for advancement and obstacles to implementation. Evaluating Eden Series, IIED Discussion Paper No. 14. London: IIED; 2000. 19 p.

[23] Southammakoth S, Craig I. Participatory conservation co-management: a component description. Division of Forest Resources Conservation, Department of Forestry. Lao-Swedish Forestry Programme/Department of Forestry, Ministry of Agriculture and Forestry. Vientiane: Lao PDR. 2000. 9 p.

[24] Corbett J. Paper parks and paper partnerships: lessons for protected areas and biodiversity corridors in the Greater Mekong Sub-region. Paper synthesising lessons learnt about livelihoods, biodiversity, collaborative management initiatives and governance through the GMS/ BCI. Safeguarding Biodiversity for Poverty Reduction Project. Bangkok: IUCN. 2008. 19 p.

[25] Department of Forest Resource Management (DFRM), Ministry of Natural Resources and Environment (MoNRE). Hin Nam No National Protected Area Co-Management Plan 2016-2020. Thakhek: DFRM and MoNRE. 2015. 115 p.

[26] Mardiastuti A, Simorangkir AR, Kusrini MD, Buchori D, Suryadi I. Management effectiveness of ASEAN heritage parks: a study report. Los Baños, Philippines: GIZ and ASEAN Centre for Biodiversity. 2013. viii+94 p.

[27] Phommasane P. Scaling up Co-management from Hin Nam No to Village Forest Areas in Lao PDR. PANORAMA: Inspiring Protected Area Solutions [Internet]. 2017. Available from: https://panorama.solutions/en/solution/scaling-co-management-hin-nam-no-village-forest-areas-lao-pdr [Accessed: 2021-05-28].

[28] Dudley N, editor. Guidelines for applying protected area management

categories. Gland, Switzerland: IUCN. 2008. x+86 p. WITH Stolton S, Shadie P, Dudley N. IUCN WCPA Best Practice Guidance on recognising protected areas and assigning management categories and governance types, Best Practice Protected Area Guidelines Series No. 21. Gland, Switzerland: IUCN. 2013. xx p.

[29] Society for the Protection of Prespa (SPP), WWF-Greece, Protection and Preservation of Natural Environment in Albania (PPNEA), Macedonian Alliance for Prespa (MAP). Strategic action plan for the sustainable development of the Prespa Park: Executive summary. Agios Germanos: SPP. 2005. 75 p.

[30] United Nations Educational, Scientific and Cultural Organization (UNESCO). Natural and Cultural Heritage of the Ohrid Region (Albania, North Macedonia). Nomination Text [Internet]. 2018. Available from: https://whc.unesco.org/document/166324 [Accessed: 2021-05-28].

[31] Ministry of Culture (MoC), Ministry of Environment and Physical Planning (MoEPP). Management Plan for World Natural and Cultural Heritage of the Ohrid region (2020 - 2029) with an Action Plan. Skopje: MoC and MoEPP. 2019. 145 p. Available from: https://ohrid.gov.mk/wp-content/uploads/2017/08/Management-Plan-Ohrid-Region.pdf [Accessed: 2021-05-28].

[32] World Heritage Centre (WHC), International Council on Monuments and Sites (ICOMOS), International Union for Conservation of Nature and Natural Resources (IUCN). Joint WHC/ICOMOS/IUCN Reactive Monitoring mission to the World Heritage property Natural and Cultural Heritage of the Ohrid region (North Macedonia / Albania), 27 - 31 January 2020: Mission Report. Paris: WHC. 2020. 125 p. Available from: http://whc.unesco.org/document/187543

[33] Wildark. PONT Wins Pathfinder Award 2018 - Innovation & excellence in protected and conserved area financing and resourcing. Available from: https://wildark.org/journals/pont-wins-wildark-pathfinder-award/ [Accessed: 2021-06-23].

[34] Conservation Finance Alliance. Winner of 2018 Pathfinder Awards for Innovative Solutions in Protected Area Finance, 27 February 2019. Available from: https://www.conservationfinancealliance.org/news/2019/2/27/winner-of-2018-pathfinder-awards-for-innovative-solutions-in-protected-area-finance [Accessed: 2021-06-23].

[35] Buxhaku M. Promoting sustainable tourism through increasing municipal and entrepreneurship capacities in Wider Prespa Area. PANORAMA: Inspiring Protected Area Solutions [Internet]. 2021. Available from: https://https://panorama.solutions/en/solution/promoting-sustainable-tourism-through-increasing-municipal-and-entrepreneurship-capacities?fbclid=IwAR32lxXTG2ZmCqWVyMLdzMdqymAh9miRQXLB_VbpObx5X3bG4-kszym-w8w [Accessed: 2021-06-11].

[36] Hockings M, Stolton S, Leverington F, Dudley N, Corrau J. Evaluating effectiveness: a framework for assessing management effectiveness of protected areas, 2nd ed. Gland: IUCN; 2006. 105 p.

Section 5

Protected Area Conservation and Monitoring

Chapter 14

Novel Technologies and Their Application for Protected Area Management: A Supporting Approach in Biodiversity Monitoring

Daniel T. Dalton, Kathrin Pascher, Vanessa Berger, Klaus Steinbauer and Michael Jungmeier

Abstract

State-of-the-art tools are revolutionizing protected area (PA) manager approaches to biodiversity monitoring. Effective strategies are available for test site establishment, data collection, archiving, analysis, and presentation. In PAs, use of new technologies will support a shift from primarily expert-based to automated monitoring procedures, allowing increasingly efficient data collection and facilitating adherence to conservation requirements. Selection and application of appropriate tools increasingly improve options for adaptive management. In this chapter, modern biodiversity monitoring techniques are introduced and discussed in relation to previous standard approaches for their applicability in diverse habitats and for different groups of organisms. A review of some of today's most exciting technologies is presented, including environmental DNA analysis for species identification; automated optical, olfactory, and auditory devices; remote sensing applications relaying site conditions in real-time; and uses of unmanned aerial systems technology for observation and mapping. An overview is given in the context of applicability of monitoring tools in different ecosystems, providing a theoretical basis from conceptualization to implementation of novel tools in a monitoring program. Practical examples from real-world PAs are provided.

Keywords: protected area management, biodiversity monitoring system, environmental DNA, camera trapping, electronic nose, passive acoustic monitoring, remote sensing

1. Introduction

1.1 Recent history of biodiversity loss

Biodiversity is declining globally at an unprecedented rate, a trend that has proceeded unabated since the early 20th century [1–3]. Recognition of the importance and conservation needs of global biodiversity resulted in the proposal of the Convention on Biological Diversity (CBD) in Rio de Janeiro in 1992 [4]. More than 190 nations have since ratified the treaty. At the turn of the millennium, several international initiatives were started with the aim to change the trajectory of biodiversity conservation. Through the United Nations (UN) Millennium Ecosystem Assessment initiative (2001–2005), research was conducted with the goal to identify conservation priorities and set benchmarks for future actions [5]. At the time, the initiative provided a comprehensive summary of ecosystem changes and their effects on human well-being and linked to economic activities. The UN Millennium Development Goals (2000–2015) aimed to mitigate the extent of biodiversity loss. These goals are now addressed by the UN Sustainable Development Goals (SDGs) containing benchmarks for marine and terrestrial biodiversity [6]. In 2012, at the Tenth Meeting of the Conference of the Parties to the Convention on Biological Diversity, a strategic plan for the protection of biodiversity was formulated. The plan included 20 so-called Aichi targets to be addressed during the period 2011–2020. Ultimately, none of the Aichi targets were met on time (**Figure 1**) [7].

Looking forward to 2030, the SDGs provide a global framework toward sustainable development on economic, social, and environmental levels [8]. SDGs 14 and 15 are particularly relevant for biodiversity conservation. Goal 14 aims to protect life below water with a focus on marine pollution, protection, and restoration of ecosystems, reduction of ocean acidification, and sustainable fishing. Goal 15 targets terrestrial biodiversity, with a focus on protection, restoration, and promotion of sustainable forest management while reversing land degradation. To track evidence-based achievement of SDGs, far-reaching state-of-the-art monitoring capacities must be advanced.

1.2 Drivers of biodiversity loss

Despite the formation of the CBD, biodiversity has continued on a downward trajectory for vertebrate and insect species, while trends for many other taxa are unquantified [9, 10]. At least 900 species have gone extinct since 1500, and to date 1,145 species are listed as critically endangered or possibly extinct [11]. Given the considerable knowledge gap, these numbers are likely higher. The Living Planet Report noted a global decline in vertebrate abundance by 60% from the period 1970–2014 [12]. Main causes of biodiversity loss in the past century were associated with human population growth and economic development [13]. In its recent Global Assessment Report, the Intergovernmental Science-Policy Platform on

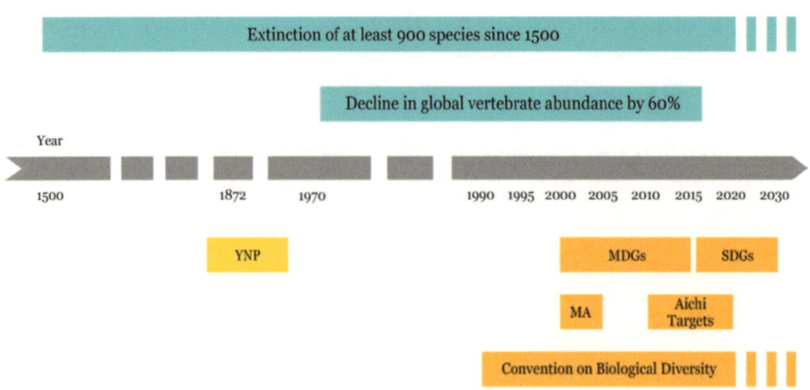

Figure 1.
Global conservation trends over the past 500 years (blue bars) and implementation of conservation treaties (orange bars). MA = millennium ecosystem assessment; MDGs = millennium development goals; SDGs = sustainable development goals; YNP = Yellowstone National Park established in 1872 (yellow bar). Timeline not drawn to scale.

Biodiversity and Ecosystem Services (IPBES) highlighted that terrestrial biodiversity losses were primarily linked to land-use changes caused by agricultural practices, whereas in maritime ecosystems overexploitation of fisheries caused major declines of biodiversity [14]. Other threats for biodiversity include climate change and proliferation of invasive alien species (IAS).

Biodiversity is under pressure due to human activities, and species extinctions will have severe negative feedbacks on human society in the future [15]. The impacts of biodiversity loss on global environmental change are comparable to climate change and need urgent attention. In its recent assessment, IPBES identified major drivers for current biodiversity losses: human-induced land-use changes, climate change, and IAS [16]. A separate study found that climate change, biodiversity loss and biogeochemical flows have already exceeded safe operating space [17]. Rising mean annual temperatures are linked to anthropogenic emissions of greenhouse gases. Temperatures have increased globally by about 0.2°C per decade since the 1970's [18], and climate change-driven impacts on biodiversity are documented across the globe [14]. Projections forecast further changes in the future [19–22].

1.3 Protected areas and biodiversity

The concept of protected areas (PA) may be as old as civilization itself [23]. Throughout the 20th century until today, the number of PAs has grown considerably to over 265,000 sites [24]. The CBD emphasized the importance of PAs for conservation of biodiversity and encouraged further PA establishment to mitigate ongoing biodiversity losses [4].

Some 76 years following the establishment of the world's first national park, Yellowstone, USA, the establishment of the International Union for Conservation of Nature (IUCN) occurred in 1948 and marked a landmark change in global biodiversity conservation [25]. Today, six commissions within the IUCN, including the World Commission on Protected Areas (WCPA) and Species Survival Commission (SSC), actively address environmental and socioeconomic issues related to conservation [23]. The importance of PAs is well-documented, but sufficient data on effectiveness of governance and management status for a majority of PAs are still lacking [26]. Recent studies additionally emphasize that biodiversity is on the decline in many PAs due to persistently high human pressures [27–29]. However, the advent of new technologies, with the possibility to provide fast and highly automated species identification and analysis across large spatial areas, points toward new perspectives in nature conservation [30].

True measurement of conservation outcomes requires effective and meaningful biodiversity monitoring systems (BMS). To foster best practice standards in governance and management of PAs, the WCPA released the Green List in 2016 [31]. In it, four components to evaluate the performance of PAs are described: good governance; sound design and planning; effective management; and successful conservation outcomes [32]. The SSC provides updated information on species and the status of ecosystem conservation in the IUCN Red List [11]. In 2009, the Joint Task Force on Biodiversity and Protected Areas was established by the WCPA and SSC. Their work focuses on two major objectives, determining best predictors of success for biodiversity conservation in PAs, and evaluating of key standards to identify sites that contribute significantly to biodiversity conservation.

1.4 Approaches to biodiversity monitoring

Monitoring of biodiversity is a challenge for many reasons, including deficits in the conception, methodologies, and technologies of BMS. Monitoring is expensive and demands significant human effort. Multiple species may require monitoring, but within the framework of data collection only a limited set of indicators can be selected. A sufficient number of specialists must be available to document taxa of expertise. Human resources can be limited by scheduling conflicts, poor weather, and inaccessible or hazardous field sites. BMS must additionally be reliable, reproducible, flexible, and comparable across sites, as well as applicable to different management questions. Perhaps most importantly, BMS should reflect the current state of the habitat or an organism group, providing key metrics to the manager in a timely and comprehensive manner. Solutions should take these limitations into consideration through application of effective technologies.

Novel approaches are now available to complement, or in some cases replace, classical monitoring methodologies. These exciting approaches are in different stages of maturity. In the following sections, we review digital monitoring techniques that are still under development or have become increasingly standardized in PA management in recent years.

Advances in computational technology over the past half century have revolutionized scientific capacity for monitoring of biodiversity. Digital methodologies that seemed unfathomable just a few years ago are now practical to enable rapid and automated collection of species data [33]. Primary among these state-of-the-art approaches are metagenomics through environmental DNA (eDNA) collection, camera trapping (CT) using digital trail cameras, environmental sampling of volatile organic compounds (VOCs) using digital sensors, passive acoustic monitoring (PAM), and earth-based remote sensing (RS) approaches [34]. In the field of biodiversity conservation, digital collection of big data is accomplished through use of data storage platforms such as GBIF; a lagging element is adequate analysis of these often-unstructured data [33, 35].

2. Advanced tools facilitating biodiversity monitoring

2.1 Applications of environmental DNA

Practical considerations constrain a BMS. One challenge is that due to time and cost considerations, often only limited selections of taxa can be monitored. To improve ecological assessments, metagenomics could be used to address sampling deficiencies. Molecular analysis could support a rapid survey of a wide range of taxa, quantify species richness, and measure diversity across different trophic levels of the ecosystem. Analysis of eDNA is increasingly becoming part of PA monitoring and management programs and can contribute to ensuring that conservation measures are implemented in a targeted manner.

Barcoding is a DNA-based taxonomic identification technique that allows a living organism to be identified on a genetic level through molecular analysis of skin, mucus, feces, or other biological samples [36]. Hair sample collection from the elusive European wild cat *Felis silvestris silvestris*, for example, can contribute to conservation activities by documenting species genetic composition across migration routes [37]. DNA metabarcoding combines barcoding and high-throughput DNA sequencing [38] and is applied for eDNA samples from diverse media such as soil, sediment, fresh water and seawater, and even air [39]. The sampling approach of eDNA collection is non-invasive, operator-independent, and flexible in its application for different taxonomic groups. Moreover, Herder and colleagues [40] highlight improved detection probability for rare and secretive species, including higher reliability of negative results, cost efficiency especially for species difficult to monitor with traditional methods, and species specificity without mismatch in identification. These features make metabarcoding attractive to fulfill PA monitoring goals [41]. Whereas morphological identification of immature aquatic insects is particularly challenging, eDNA analysis provides an objective way to differentiate species independent of life stage [42, 43].

Taxon-specific primers targeting highly conserved regions of the genome are used to amplify sample DNA in a thermocycler [44]. The sample is then sent to a Next Generation Sequencer. Species identification is based on output of nucleic acid sequences. Very short DNA primers, so-called mini-barcodes [45], allow amplification of degraded DNA, for example from soil samples [46].

DNA metabarcoding offers diverse applications to conservation, paleobiology, biomonitoring, and invasion biology. Metagenomics technologies under development could in the future provide more comprehensive biodiversity assessment in PAs using bulk samples from the environment. Moreover, interactions between taxonomic groups could be investigated, and detection of changes in these interactions could optimize adaptive management decisions [47]. For instance, aquatic eDNA sample collections are suited to detect pathogens in the environment including the fungus *Batrachochytrium dendrobatidis* in its host frog species [39, 40]. Discovery of incipient pathogens could help guide adaptive measures to limit spread of disease in the environment.

A coarse application of molecular diagnostics is the application of (molecular) operational taxonomic units, or (M)OTUs [48]. These are distinct clusters of reads whose nucleic acid sequences differ by less than a fixed threshold and can be applied as an initial survey of diversity. These OTUs are of particular value for soil biodiversity assessment in PAs, as no taxa of microorganisms need to be known to benchmark the diversity of different soil samples relative to one another.

Although DNA metabarcoding may have a highly supportive function in PA management, several challenges remain [40]. Reproducibility of results is a primary issue. For example, species composition of replicate samples taken from a freshwater stream may provide conflicting results. DNA detection in fresh water may be possible at a distance of 9 to 12 km away from the genetic source [49]. Species determination is influenced by the primers used and is highly dependent on the quality of available reference databases. Additionally, most designs are customized for the particular research question because there is no uniform approach for all applications. Another disadvantage includes limitations on accurate species density estimates. Furthermore, no information can be provided on the life stages or demographic structures of identified organisms, as eDNA analysis typically generates presence/absence data. Concerns exist that rare and endangered species could be reduced to numbers on a species list. But for their respect and protection, they would need support from society.

However, successful applications of eDNA analysis promote further usage of this novel approach in PAs. Much expectation is placed on future application of metabar-coding in a BMS. Favorable comparability of DNA-based and classical approaches has been demonstrated in the context of the European Union Water Framework Directive [50]. For the PA manager, several prerequisites for the workflow must be assessed. When using eDNA, the analytical procedure, which in most cases is carried out in an external laboratory, is not as important as the evaluation of conservation questions of interest. For this purpose, the manager must be familiar with the range of conclusions that could result from metagenomic analyses. Consideration must be given to whether eDNA collection would be the appropriate technique to answer the monitoring question. The next critical step for the manager is to acquire expert interpretation of the data. Yet, with appropriate research questions, analytical approaches using eDNA sampling have great potential to detect target species and contribute valuable insights to a BMS.

2.2 Camera trapping

Nature photography provides an archivable, permanent record on the in-situ occurrence of plants and animals. As a biodiversity research technique, photography dates back to the late 19th century [51]. In the early period of CT development, photographic approaches utilized cumbersome hardware and explosive compounds to create a flash [52–54]. Technological developments including remote triggering of the shutter, improved flash mechanisms, improvements to battery life, and digitization of images have enhanced cameras since the mid-20th century [51, 55]. With trail cameras, social media platforms, and dozens of smartphone apps, scientists and enthusiasts can now contribute to real-time photo documentation of species (**Figure 2**) [33, 56]. As a biodiversity research tool, CT compares favorably to many previously standard methodologies [57].

Formal CT studies for biodiversity monitoring came into existence a century ago [58]. Approaches have since undergone a dramatic evolution, with a wide selection of wildlife cameras now commercially available [55]. Use of remote photography has become standard for documenting species distributions over broad spatio-temporal scales [59]. Photographic approaches are suitable for examination of species occupancy or abundance in aquatic and terrestrial biomes [34] and are suitable for targeting a range of animal species [60–65]. Robust statistical methodologies are available for data analysis, including spatially explicit capture-recapture techniques (SECR), multi-layered robust principal component analysis, occupancy modeling, and predator–prey co-occurrence analysis [66–69]. Photographic and video processing programs are undergoing continual refinement, providing an ever-improving framework for data analysis and allowing inferences into animal behaviors and spatial distribution [70].

The field of big data analytics is advancing rapidly, utilizing machine learning (ML) algorithms to provide automated analysis of digital imagery [35]. Applications include identification of animals in pictures and systematic behavioral descriptions [71]. Today, deep convolutional neural networks (CNN) are applied to

Figure 2.
Trail cameras are widely available, allowing citizen scientists to capture the movement of animals, such as this family of American black bears (Ursus americanus) in Colorado, USA. Photo courtesy of K. Dalton.

image libraries, allowing rapid processing of large datasets using standard computer operating systems and open-source software [70]. Yet, ML works only if the computer is trained using accurately tagged photographs, which demands significant human effort. CNN in the context of CT research can be applied to identify any properly annotated object, from animals in PAs to agricultural pest insects [72–74]. Interconnectivity of hardware with cloud-based software is poised to empower real-time remote data collection in agriculture [75]. A parallel approach could be applied to state-of-the-art CT systems in PAs to provide real-time monitoring of animals or vegetation [76].

Passive infrared sensors (PIR) are the dominant feature used to trigger the camera shutter, while time-lapse (TL) approaches and PIR + TL in combination are also utilized [77]. Sensitivity of PIR is modulated by the camera field of vision and speed of the passing animal. A major shortcoming to PIR-activated cameras is that they often fail to trigger upon encounter by insects or small animals. Modifications of PIR sensor sensitivity or camera focal point distance can be made to improve detection of small-bodied animals [55, 77]. One advancement to PIR sensors, the so-called HALT trigger, utilizes a near-infrared beam to increase camera trapping performance on arthropods and small vertebrates [63]. As an alternative to sensor-based CT activation, automated TL photography has application to document arthropods, squamates, and avian roosting sites [62, 65, 77–79]. In addition to PIR and HALT, infrared technology has been used to create a less invasive flash mechanism for night photography compared to use of xenon or LED flash [55].

The advantages of remote CT are myriad. Today's automated approaches largely eliminate the requirement of human presence at a study site, restricting visitation to plot establishment and removal, and thereby reducing activities that could bias animal behavior. Furthermore, cameras can be deployed in locations that are difficult to access [79, 80]. Traps can be programmed to function at optimal times to detect target species behaviors. Exclusion of empty pictures or videos is enabled through automated image pre-processing [81, 82]. While studies generally focus on one or a small set of animals, the bycatch of unanalyzed photographs additionally serves as a rich source for wider ML training applications or retrospective occupancy analyses [83].

Despite the advancements of CT methodologies, critical logistical challenges remain. Animals may be able to detect CT through sight or sound, even in the absence of field workers [84]. A network of CT, deployed for weeks at a time, is necessary to acquire a robust dataset. The cumulative sampling effort of all cameras in an array, termed CT days, needs to be approximately determined prior to deployment [55]. Data analysis is an obstacle to understanding the value of CT schematics [59]. Another critical hindrance is the lack of standardization of CT technologies due to the wide selection of cameras on the market today [55, 80], although open standards to promote uniform collection of CT images have been proposed [85]. Up-front material costs of CT surveys can be high but are attenuated the longer the camera traps are in place [57]. The photo archive of a single project typically numbers in the thousands of images but requires a rapid turn-around time to inform management decisions. This problem is addressed through ML, but photographs must first be annotated, requiring months or years of technical effort depending on size of the photographic archive [71]. While automated identification of common species is reliable, identification of rare or undescribed species is challenging because photographic archives may not contain enough pictures to effectively train the computer [58].

With the use of appropriate digital camera sampling methodologies, the researcher no longer needs to interact directly with animals in order to gain insights into their behaviors or population structures. Images are either analyzed manually, or with a computer through ML approaches. Large networks of cameras may capture a representative number of individuals or species, allowing scientific inferences. In general, deployment periods need not exceed more than a few weeks to result in acceptable data. Foresight should be made when investigating particular behavioral attributes such as migration phenology or hibernation, because seasonality can affect captures of certain animals.

2.3 Electronic noses

Automated sensing of airborne chemicals is an emerging area of environmental diagnostics with high potential transferability to PA management. The use of electronic noses, or e-noses, is an established technique with diverse industrial and agricultural functions, including determination of the presence of VOCs, volatile inorganic compounds, and heavy metal pollutants in the environment [86]. Applications of e-nose technologies in conservation include monitoring of IAS and pathogenic infection of plants and animals [86–89]. E-nose devices are even capable of identifying species-level differences in plants based on their VOC emission profiles [90]. As such, e-noses are intelligent instruments that have great potential toward plant health monitoring [91], including in PAs.

Communication in mammals is moderated through sensory modalities, including scent. VOC emissions can be acquired from body surfaces, glands, or breath of animals [89, 92]. Insect communication is impacted by antennal detection of semiochemical VOCs [93]. In integrated pest management, this serves as the basis of mating disruption [94]. E-noses are designed to mimic mammalian or insect olfactory systems [86, 93]. First developed in the 1980's [95], e-noses can be equipped today with a variety of sensors. Among the most common sensor types are conductive polymer biosensors [86]. Environmental analysis using these sensors is an established method for ecological, forestry, and taxonomic research [90]. E-noses can be paired with fluorescence technologies and ML algorithms to allow reliable identification or diagnosis of VOC profiles [96]. Miniaturization of next-generation e-nose devices will allow greater utility in the field [86, 97].

Plants and animals emit altered suites of VOCs under biotic or abiotic stresses [86, 89, 97]. Comparison of VOC emissions can be made between field-grown plants and reference electronic aroma signature patterns to determine plant infection or infestation status [90]. In a study of North American ash trees, healthy trees had higher diversity of VOCs compared to trees infested with emerald ash borer *Agrilus planipennis*, a devastating IAS. Analysis of VOC patterns could help managers identify infested trees more rapidly than by using baits or traps for confirmation of infestation [88]. In the case of IAS introductions, such knowledge could advance containment measures and guide further surveillance actions [87]. Early detection of IAS or pathogenic infections of keystone species in PAs could similarly help managers determine adaptive management interventions.

Utilization of e-nose devices suffers from considerable practical limitations. Their bulky size and high price, coupled with difficulties of aroma profile detection, limit their application in the field [97]. E-noses only display raw response unless they are paired with computer-based training datasets [91]. When working with previously uncharacterized species, new computer algorithms and VOC reference libraries must be generated [86]. Moreover, due to geographic variability of abiotic factors, source materials for reference libraries should come from the sampled region [90]. Periodic calibration of e-nose monitors is necessary to maintain accuracy [86]. Sensors must be replaced periodically due to degradation over time [87]. Yet, the objective identification of VOC profiles in the environment represents a clear opportunity for management of plant health in PAs.

2.4 Passive acoustic monitoring

Animals communicate with one another for a number of biologically important reasons including defense, mating, group interactions, and orientation [98, 99]. Sound is recognized as a common means of communication in insects, fish, birds, squamates, and mammals [98, 100]. Call count censusing has long been a standard practice to identify community assemblages [101, 102]. Initially conducted with expensive, cumbersome equipment, census techniques using recorders now allow ecologists to document a wide diversity of species at a far lower cost than continual deployment of field crews [98]. Today, PAM uses autonomous recording units (ARUs), representing a non-invasive means to collect species-level occupancy data, thereby minimizing behavioral impacts or animal stress [103, 104].

Modern ARUs have many advantages over previously standard field techniques, enabling research crews to conduct more site surveys with fewer site visits and allowing improved biodiversity estimation in remote areas [105, 106]. Digital recordings further serve as permanent data records that can be played back for verification of species identity [101, 107–109]. Rapid acoustic surveys using microphone arrays have application in conservation, identifying changes in community species assemblages or migration patterns, phenology, communication, or even presence of IAS [105, 110, 111]. This approach may help to identify environmental impacts of anthropogenic disturbance, for example the impacts of artisanal mining on the local avian community [112].

Methodologies for detection of vocal species are well established, including classic field approaches of physical trapping, playback of audio recordings, point counts, and timed area searches [105, 108, 113, 114]. Bats and birds have been recorded in proximity to wind turbines using radar tracking, infrared imagery, and radio telemetry, [61, 115]. First formalized nearly 20 years ago, SECR techniques provide the statistical framework to document species density across microphone arrays [69, 103, 114, 116]. For some taxa, effectiveness of manual calling surveys has been directly compared to results from ARU methodologies, with both methods providing synergetic benefits to a monitoring program [101]. Manual calling surveys and ARU approaches can support similar conclusions; however, ARUs may provide biodiversity data with dramatically reduced human effort [117]. Similar to CT studies, well-established statistical techniques are available for studies using PAM to provide estimates on animal abundance, density, and occupancy [105, 113, 118, 119].

Species-specific auditory signals can be identified by experienced personnel, or automatically using ML algorithms. Several automated ML techniques are described [99, 100, 107, 120]. Two crucial components of automated bioacoustics analysis are recognized. First, auditory signals are characterized visually through spectrograms; subsequently, signals are extracted from continuous recordings through pairing with a "recognizer" template segment [105]. Spectrograms assist in species identification [106, 115]. Automation coupled with cloud-based technologies now enable remote real-time identification, potentially providing up-to-the-minute conservation information to a PA manager [107, 121].

Expert-based field identification may compare favorably to findings generated from remote microphone arrays linked to species recognition algorithms [108]. Yet, surveys relying on human skill for identification of species are prone to error due to imperfect species detection, confirmation bias, or listener fatigue [102, 103, 119, 122]. Lack of objective classification is especially challenging when a reviewer is charged with identifying rare or unknown species, with animals that are known to employ mimicry, or in complex soundscapes [104, 111]. Multiple factors influence the soundscape, including relative abundance of species, caller density, and community acoustic diversity [123]. Analysis of soundscape profiles can be facilitated through reduction of background noise [104, 109]. Incorporating species time of arrival or activity into a survey using fixed-point microphone arrays can be an approach to reduce bias [102, 114]. Through application of sound filters, automated programs can eliminate sections of uninformative data, facilitating verification of acoustic signals by a reviewer [117].

Important limitations persist for auditory species identification. Use of automated computer recognition of animal calls is currently underutilized [102]. For effective ML, hundreds of labeled sound records are required [115, 120]. Recordings may miss very faint or distant calls and allow overrepresentation of calls by noisy species [115, 117, 122]. Depending on equipment, costs can be high for acquisition and maintenance of a microphone array [105]. Furthermore, effective sampling area is often imprecisely known due to landscape features, thus limiting inference on species occupancy [103]. An effective study design can help alleviate some limitations, for example through strategic placement of microphone arrays providing overlap within species habitat. Certain types of hardware are becoming less expensive, while many software programs and call libraries are deposited in open-source libraries [99].

The generation of large amounts of data is a common feature to many PAM programs [117]. While automated identification of acoustic calls is possible for certain species or analytical processes, big data processing challenges remain [35, 99, 121]. Solutions to data management should be transferrable to personnel of all skill levels, and in a way that acoustic data can be statistically compared across sites [117]. Nonetheless, the recent advancements of automated PAM hold great promise for the future of PA management.

2.5 Applications of remote sensing in protected area monitoring

Management of PAs can be supported by RS applications. A range of different datasets can be produced using RS, including information on climate, characteristics of vegetation, plant phenology, water budget, energy exchange, and terrain models [124]. In order to ensure efficient use of such data, a clear implementation strategy is essential. Analysis of satellite data is a cost-efficient extension to conventional in-situ monitoring in the field, particularly in remote and inaccessible areas. Moreover, analyses can be carried out retrospectively with historic satellite imagery [125]. To detect different ecosystems and habitats, structural and functional attributes can be determined based on various RS technologies [124]. For example, LiDAR- and radar-derived elevation models are often used for forest mapping to assess aboveground structure and biomass [126]. Some RS techniques also provide the possibility to compare different PAs worldwide based on the same dataset, enabling global estimates of habitat availability. Local and regional datasets are often more accurate than global datasets, in particular for the use of unmanned aerial systems, or drones [124]. Drones are flexible vehicles that can be equipped with imaging sensors including thermal vision cameras, visible red-green-blue, near infrared, multispectral, or hyperspectral sensors, as well as ranging sensors including laser scanners and synthetic aperture radars. Drones come in multi-rotor or fixed wing configurations and are used in many conservation-based fields: wildlife monitoring and management, ecosystem monitoring, law enforcement, ecotourism, environmental management, and disaster response [127].

To improve management and monitoring effectiveness in PAs, software programs like Spatial Monitoring and Reporting Tool (SMART) combine geographic information systems (GIS) with database tools and digital field assessment [128]. Through such tools, standardized results of conservation efforts or PA law enforcement activities can be generated in real-time. SMART output shows the spatial distribution of illegal activities while simultaneously tracking patrol efforts and providing a record of the violation [129].

The SMART approach streamlines the time required for quality assessment. A multilingual interface facilitates its implementation in PAs anywhere in the world. The use of pictograms can further simplify the generation of datasets. Preparation of data templates also provides an efficient way to produce standardized reports that can be expressed as a dashboard visualizing monitoring results with only a few clicks [130]. Using cloud-based technology, it is now possible to produce near real-time (NRT) alerts directly from the field [131]. This allows immediate action on incidents of conservation interest, thus improving management of the PA.

A study on the impact of NRT alert systems for conservation concluded that such systems are suitable for identifying fire impact and illegal forest activities [132]. The accuracy and availability of NRT alerts are affected through different factors including spatial resolution or time lag due to cloud cover. Despite these limitations, RS datasets provide an important indication of potential threats [133].

Diverse methodologies and thresholds are used to assess key variables in forest inventories, making data comparison a challenge [134]. In particular, use of

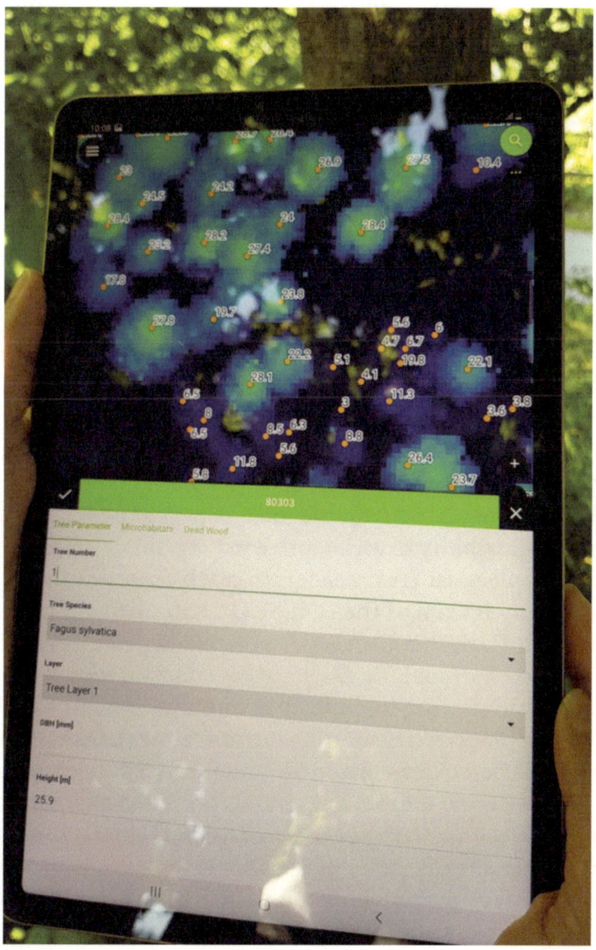

Figure 3.
In-situ single tree assessment with QField. The points represent single tree detections from a remote sensing approach and allow a linkage of tree parameters to the tree itself. Coloration indicates the tree height and crown structure.

subjective techniques can lead to faulty measurements. One solution is to compare parameters using RS such as above-ground woody biomass across national borders [135]. In this instance, generation of cross-comparable information could play an important role in understanding carbon sequestration dynamics of different forests [136]. By identifying, such datasets enable a comparison of individual tree characteristics at the landscape level [137]. The applicability of different methodologies and datasets for single tree detection has been studied for more than three decades [138] and is becoming more accurate. For laser scanner datasets, the point density to detect tree parameters can vary from 2 points m^{-2} up to more than 25,000 points m^{-2} [139, 140]. Furthermore, analysis of datasets with repeat survey dates allows detection of single missing trees. These so-called change detection approaches are already possible using consumer-level drones without post-processing effort, based on multi-temporal ultra-high-resolution orthomosaics (5 cm pixel resolution with a flight altitude of 100 m) and three-dimensional point clouds. The use of such technologies can thus increase the comparability and repeatability of monitoring datasets. With a combination of pre-processed single tree detection it is possible to ground-truth tree parameters or quantify microhabitats directly in the field based on the position of the trees [141].

Applications like QField further allow PA managers to establish digital assessments in the field based on GIS (**Figure 3**). Such applications promote effective workflows encompassing whole data assessment, data input, and digitization, thereby enabling data quality control. The availability of actual RS data in the field can further increase the quality of digitization [142].

3. Conclusion

In this chapter, a review of some of the most exciting technological advances to improve BMS is provided. To meet the urgent demands of international biodiversity conventions, state-of-the-art monitoring approaches must be quickly adopted on a broad scale. In some cases, completely new work flows will be required. Yet, in order to retain the value of historical data, utility of new technologies must be evaluated, compared with previously standard approaches, and visualized for interpretation. In other words, while application of individual novel technologies may be beneficial, no method alone provides a singular solution to improve conservation metrics. Instead, PA managers must select suitable tools as part of a toolkit to allow large-scale assessment and flexibility in an adaptive management program. Using such an integrated approach will assist PA managers to reach conservation goals. Currently, the BioMONITec research team of the UNESCO Chair on Sustainable Management of Conservation Areas Carinthia University of Applied Sciences, Austria, is constructing an online decision-making assistant, or configurator, to guide development of site-specific monitoring toolkits. In coordination with the IUCN WCPA, a comprehensive global biodiversity monitoring guideline that shall be applicable in PAs across the world is being developed (M. Jungmeier, *pers. comm.*).

Implementation of digital monitoring tools is poised to augment biodiversity monitoring programs, economizing both human capital and natural resources. Where monitoring data already exist, usage of new tools must allow valid comparison of data to permit identification of trends. High-throughput DNA metabarcoding techniques using eDNA sampling have proven to be invaluable for rapid and comprehensive biodiversity assessments in PAs. Advances in cloud-based computer frameworks and ML will allow sensor-based technologies to convey data in real-time to a manager. Drones and satellites can already provide NRT data from above the earth's surface, and these capabilities are continually improving. In this context,

PA managers of the future should not only be competently qualified scientists, excellent communicators and mediators, but must also be up-to-date technology enthusiasts.

Acknowledgements

The authors would like to thank Isabella Clemens, Laura Schnabl, and Hannes Egger for providing critical feedback on figures. Funding was provided through FFG COIN project number 884138, "Biodiversity Monitoring Technologies – test, development and transfer of disruptive engineering technologies into conservation practice" and the Austrian Federal Ministry for Digital and Economic Affairs (BMDW)." Mention of the BMDW is required by our funding sources. All opinions, findings, conclusions, and recommendations expressed in this publication are those of the authors and do not necessarily reflect the view of funding agencies.

Conflict of interest

The authors declare no conflict of interest.

Abbreviations

ARU	autonomous recording unit
BMS	biodiversity monitoring system
CBD	Convention on Biological Diversity
CNN	convolutional neural network
CT	camera trapping
eDNA	environmental DNA
GIS	geographical information system
IAS	invasive alien species
IPBES	Intergovernmental Science-Policy Platform on Biodiversity and Ecosystem Services
IUCN	International Union for the Conservation of Nature
ML	machine learning
(M)OTU	(molecular) operational taxonomic unit
NRT	near real-time
PA	protected area
PAM	passive acoustic monitoring
PIR	passive infrared
RS	remote sensing
SDGs	Sustainable Development Goals
SECR	spatially explicit capture-recapture
SMART	Spatial Monitoring and Reporting Tool
SSC	Species Survival Commission
TL	time-lapse
UN	United Nations
VOC	volatile organic carbon
WCPA	World Commission on Protected Areas

Author details

Daniel T. Dalton[1*], Kathrin Pascher[1], Vanessa Berger[1,2], Klaus Steinbauer[1] and Michael Jungmeier[1]

1 UNESCO Chair on Sustainable Management of Conservation Areas, Carinthia University of Applied Sciences, Villach, Austria

2 E.C.O. Institute of Ecology, Klagenfurt, Austria

*Address all correspondence to: d.dalton@fh-kaernten.at

References

[1] Pimm SL, Alibhai S, Bergl R, Dehgan A, Giri C, Jewell Z, et al. Emerging technologies to conserve biodiversity. Trends in ecology & evolution. 2015;30(11):685-696. doi:10.1016/j.tree.2015.08.008.

[2] Ceballos G, Ehrlich PR, Raven PH. Vertebrates on the brink as indicators of biological annihilation and the sixth mass extinction. Proceedings of the National Academy of Sciences of the United States of America. 2020;117(24): 13596-13602. doi: 10.1073/pnas. 1922686117.

[3] Barnosky AD, Matzke N, Tomiya S, Wogan GOU, Swartz B, Quental TB, et al. Has the Earth's sixth mass extinction already arrived? Nature. 2011;471(7336):51-57. doi: 10.1038/ nature09678.

[4] United Nations. The Convention on Biological Diversity. June 5, 1992. Rio de Janeiro, 30 p.

[5] Corvalán C, Hales S, McMichael AJ, Millennium Ecosystem Assessment. World Health Organization, editors. Ecosystems and human well-being: health synthesis. Geneva, Switzerland: World Health Organization; 2005. 53 p.

[6] Obrecht A, Pham-Truffert M, Spehn E, Payne D, Altermatt F, Fischer M, et al. Achieving the SDGs with biodiversity. Swiss Academies Factsheet 16(1) 2021. doi:10.5281/ zenodo.4457298.

[7] Xu H, Cao Y, Yu D, Cao M, He Y, Gill M, et al. Ensuring effective implementation of the post-2020 global biodiversity targets. Nat Ecol Evol. 2021;5(4):411-418. doi: 10.1038/ s41559-020-01375-y.

[8] United Nations. The future we want. June 20-22, 2012. Rio de Janeiro, 72 p.

[9] Pereira HM, Navarro LM, Martins IS. Global biodiversity change: the bad, the good, and the unknown. Annu Rev Environ Resour. 2012;37(1):25-50. doi:10.1146/annurev-environ-042911-093511.

[10] Hallmann CA, Sorg M, Jongejans E, Siepel H, Hofland N, Schwan H, et al. More than 75 percent decline over 27 years in total flying insect biomass in protected areas. PLoS ONE. 2017;12(10):e0185809. doi:10.1371/ journal.pone.0185809

[11] IUCN. The IUCN Red List of Threatened Species. Version 2021-1. 2021. Available from: https://www. iucnredlist.org [Accessed 2021 Jul 21].

[12] Grooten M, Almond REA, editors. Living planet report - 2018: aiming higher. Gland Switzerland: WWF; 2018.

[13] Pereira HM, Rosa IMD, Martins IS, Kim H, Leadley P, Popp A, et al. Global trends in biodiversity and ecosystem services from 1900 to 2050. doi:10.1101/2020.04.14.031716.

[14] IPBES. Global assessment report on biodiversity and ecosystem services of the Intergovernmental Science-Policy Platform on Biodiversity and Ecosystem Services. Diaz S, Settele J, Brondízio ES, Ngo HT, Guèze M, Agard J, et al., editors. 2019; doi:10.5281/zenodo. 3831674.

[15] Arneth A, Shin Y-J, Leadley P, Rondinini C, Bukvareva E, Kolb M, et al. Post-2020 biodiversity targets need to embrace climate change. Proc Natl Acad Sci USA. 2020 Dec 8;117(49):30882-30891. doi:10.1073/ pnas.2009584117.

[16] Balvanera P, Pfaff A, Viña A, Garcia Frapolli E, Hussain SA, Merino L, et al. Chapter 2.1 Status and trends –drivers of change. Lambin E, Jayalaxshni M,

editors. Zenodo; 2019. doi:10.5281/zenodo.3831881.

[17] Rockström J, Steffen W, Noone K, Persson A, Chapin FS, Lambin EF, et al. A safe operating space for humanity. Nature. 2009;461(7263):472-475. doi:10.1038/461472a

[18] IPCC. Climate change 2014: synthesis report. contribution of working groups i, ii and iii to the fifth assessment report of the intergovernmental panel on climate change [Core Writing Team, R.K. Pachauri and L.A. Meyer (eds.)]. Geneva, Switzerland; 2014 [cited 2021 Jan 15] 151 p. Available from: https://www.ipcc.ch/report/ar5/syr/ [Accessed 2021 Jul 28].

[19] Sweatman H, Delean S, Syms C. Assessing loss of coral cover on Australia's Great Barrier Reef over two decades, with implications for longer-term trends. Coral Reefs. 2011;30(2):521-531. doi:10.1007/s00338-010-0715-1.

[20] Bellard C, Bertelsmeier C, Leadley P, Thuiller W, Courchamp F. Impacts of climate change on the future of biodiversity. Ecology Letters. 2012;15(4):365-377. doi:10.1111/j.1461-0248.2011.01736.x.

[21] Anadón JD, Sala OE, Maestre FT. Climate change will increase savannas at the expense of forests and treeless vegetation in tropical and subtropical Americas. J Ecol. 2014;102(6):1363-1373. doi:10.1111/1365-2745.12325.

[22] Newbold T. Future effects of climate and land-use change on terrestrial vertebrate community diversity under different scenarios. Proceedings Biological Sciences. 2018;285(1881). doi:10.1098/rspb.2018.0792.

[23] IUCN. 50 years of working for protected areas: a brief history of IUCN World Commission on Protected Areas. 2010; 28 p.

[24] UNEP-WCMC and IUCN. Protected Planet Report 2020. UNEP-WCMC and IUCN: Cambridge UK; Gland, Switzerland; 2021.

[25] Holdgate M. The green web: a union for world conservation. London: Earthscan; 1999. 308 p.

[26] Secretariat of the Convention on Biological Diversity. Protected areas in today's world. Janishevski L, Noonan-Mooney K, Gidda SB, Mulongoy KJ, editors. Montreal, Quebec; Beaconsfield, Quebec. 2014; 96 p. (CBD Technical Series No. 36).

[27] Watson JEM, Dudley N, Segan DB, Hockings M. The performance and potential of protected areas. Nature. 2014;515(7525):67-73. doi:10.1038/nature13947.

[28] Geldmann J, Manica A, Burgess ND, Coad L, Balmford A. A global-level assessment of the effectiveness of protected areas at resisting anthropogenic pressures. Proceedings of the National Academy of Sciences of the United States of America. 2019;116(46):23209-23215. doi:10.1073/pnas.1908221116.

[29] Johnson CN, Balmford A, Brook BW, Buettel JC, Galetti M, Guangchun L, et al. Biodiversity losses and conservation responses in the Anthropocene. Science. 2017;356(6335):270-275. doi:10.1126/science.aam9317.

[30] Joppa LN. Technology for nature conservation: An industry perspective. Ambio. 2015;44(S4):522-526. doi:10.1007/s13280-015-0702-4.

[31] IUCN, World Commission on Protected Areas. IUCN Green List of Protected and Conserved Areas: Standard, Version 1.1. 2017. 43 p.

[32] IUCN, World Commission on Protected Areas, Assurance Services

International. IUCN Green List of Protected and Conserved Areas: User Manual, Version 1.2. Gland, Switzerland: IUCN; 2019. 110 p.

[33] Arts K, van der Wal R, Adams WM. Digital technology and the conservation of nature. Ambio. 2015;44(S4):661-673. doi:10.1007/s13280-015-0705-1.

[34] Stephenson P. Technological advances in biodiversity monitoring: applicability, opportunities and challenges. Current Opinion in Environmental Sustainability. 2020;45:36-41. doi:10.1016/j.cosust.2020.08.005.

[35] Gandomi A, Haider M. Beyond the hype: big data concepts, methods, and analytics. International Journal of Information Management. 2015;35(2):137-144. doi:10.1016/j.ijinfomgt.2014.10.007.

[36] Hebert PDN, Cywinska A, Ball SL, deWaard JR. Biological identifications through DNA barcodes. Proc R Soc Lond B. 2003;270(1512):313-321. doi:10.1098/rspb.2002.2218.

[37] Steyer K, Kraus RHS, Mölich T, Anders O, Cocchiararo B, Frosch C, et al. Large-scale genetic census of an elusive carnivore, the European wildcat (*Felis s. silvestris*). Conserv Genet. 2016;17(5):1183-1199. doi:10.1007/s10592-016-0853-2.

[38] Taberlet P, Bonin A, Zinger L, Coissac E. Environmental DNA – for biodiversity research and monitoring. Vol. 1. Oxford University Press; 2018. doi:10.1093/oso/9780198767220.001.0001.

[39] Ficetola GF, Miaud C, Pompanon F, Taberlet P. Species detection using environmental DNA from water samples. Biol Lett. 2008;4(4):423-425. doi:10.1098/rsbl.2008.0118.

[40] Herder J, Valentini A, Bellemain E, Dejean T, Van Delft JJCW, Thomsen PF, et al. Environmental DNA - a review of the possible applications for the detection of (invasive) species. 2014 doi:10.13140/RG.2.1.4002.1208.

[41] Ji Y, Ashton L, Pedley SM, Edwards DP, Tang Y, Nakamura A, et al. Reliable, verifiable and efficient monitoring of biodiversity via metabarcoding. Holyoak M, editor. Ecol Lett. 2013;16(10):1245-1257. doi:10.1111/ele.12162.

[42] Uchida N, Kubota K, Aita S, Kazama S. Aquatic insect community structure revealed by eDNA metabarcoding derives indices for environmental assessment. PeerJ. 2020;8:e9176. doi:10.7717/peerj.9176.

[43] Odero J, Gomes B, Fillinger U, Weetman D. Detection and quantification of *Anopheles gambiae* sensu lato mosquito larvae in experimental aquatic habitats using environmental DNA (eDNA). Wellcome Open Res. 2018;3:26. doi:10.12688/wellcomeopenres.14193.1.

[44] Kress WJ, Erickson DL. DNA barcodes: methods and protocols. Methods Mol Biol. 2012;858:3-8. doi:10.1007/978-1-61779-591-6_1.

[45] Hajibabaei M, Smith MA, Janzen DH, Rodriguez JJ, Whitfield JB, Hebert PDN. A minimalist barcode can identify a specimen whose DNA is degraded. Mol Ecol Notes. 2006;6(4):959-964. doi:10.1111/J.1471-8286.2006.01470.X.

[46] Bienert F, De Danieli S, Miquel C, Coissac E, Poillot C, Brun J-J, et al. Tracking earthworm communities from soil DNA. Mol Ecol. 2012;21(8):2017-2030. doi:10.1111/j.1365-294X.2011.05407.x.

[47] Seymour M, Edwards FK, Cosby BJ, Kelly MG, de Bruyn M, Carvalho GR, et al. Executing multi-taxa eDNA ecological assessment via traditional

metrics and interactive networks. Science of The Total Environment. 2020;729:138801. doi:10.1016/j.scitotenv.2020.138801.

[48] Blaxter M, Mann J, Chapman T, Thomas F, Whitton C, Floyd R, et al. Defining operational taxonomic units using DNA barcode data. Phil Trans R Soc B. 2005;360(1462):1935-1943. doi:10.1098/rstb.2005.1725.

[49] Deiner K, Altermatt F. Transport distance of invertebrate environmental DNA in a natural river. PLoS ONE. 2014;9(2):e88786. doi:10.1371/journal.pone.0088786.

[50] Zizka VMA, Leese F, Peinert B, Geiger MF. DNA metabarcoding from sample fixative as a quick and voucher-preserving biodiversity assessment method. Genome. 2019;62(3):122-136. doi:10.1139/gen-2018-0048.

[51] Kucera TE, Barrett RH. A history of camera trapping. In: O'Connell AF, Nichols JD, Karanth KU, editors. Camera traps in animal ecology: methods and analyses. Tokyo: Springer Japan; 2011;p. 9-26. doi:10.1007/978-4-431-99495-4_2.

[52] Woodford L. The earliest known camera trapping in Australia: a record from Victoria. Victorian Naturalist. 2015;132(6):171-176.

[53] Sanderson JG, Trolle M. Monitoring Elusive Mammals. American Scientist. 2005;93(2). doi:10.1511/2005.2.148.

[54] Wood D. The importance of artificial light in the development of night photography. Carbondale: Southern Illinois University; 1975. Available from: https://files.eric.ed.gov/fulltext/ED130267.pdf [Accessed: 2021-07-01].

[55] Rovero F, Zimmermann F, Bersi D, Meek P. "Which camera trap type and how many do I need?" a review of camera features and study designs for a range of wildlife research applications. Hystrix, the Italian Journal of Mammalogy. 2013;24(2):148-156. doi:10.4404/hystrix-24.2-8789.

[56] Luna S, Gold M, Albert A, Ceccaroni L, Claramunt B, Danylo O, et al. Developing mobile applications for environmental and biodiversity citizen science: considerations and recommendations. In: Joly A, Vrochidis S, Karatzas K, Karppinen A, Bonnet P, editors. Multimedia Tools and Applications for Environmental & Biodiversity Informatics. Cham: Springer International Publishing; 2018;p. 9-30. doi:10.1007/978-3-319-76445-0_2.

[57] Wearn OR, Glover-Kapfer P. Snap happy: camera traps are an effective sampling tool when compared with alternative methods. R Soc open sci. 2019;6(3):181748. doi:10.1098/rsos.181748.

[58] Delisle ZJ, Flaherty EA, Nobbe MR, Wzientek CM, Swihart RK. Next-generation camera trapping: systematic review of historic trends suggests keys to expanded research applications in ecology and conservation. Front Ecol Evol. 2021;9:617996. doi:10.3389/fevo.2021.617996.

[59] Ahumada JA, Fegraus E, Birch T, Flores N, Kays R, O'Brien TG, et al. Wildlife Insights: a platform to maximize the potential of camera trap and other passive sensor wildlife data for the planet. Envir Conserv. 2020;47(1):1-6. doi:10.1017/S0376892919000298.

[60] Tobler MW, Carrillo-Percastegui SE, Leite Pitman R, Mares R, Powell G. An evaluation of camera traps for inventorying large- and medium-sized terrestrial rainforest mammals. Animal Conservation. 2008;11(3):169-178. doi:10.1111/j.1469-1795.2008.00169.x.

[61] Kunz TH, Arnett EB, Cooper BM, Erickson WP, Larkin RP, Mabee T, et al. Assessing Impacts of wind-energy development on nocturnally active birds and bats: a guidance document. J Wildl Manag. 2007;71(8):2449-2486. doi:10.2193/2007-270.

[62] Neuharth DB, Ryberg WA, Adams CS, Hibbitts TJ, Walkup DK, Frizzell SL, et al. Searching for rare and secretive snakes: are camera-trap and box-trap methods interchangeable? Wildl Res. 2020;47(6):476. doi:10.1071/WR19230.

[63] Hobbs MT, Brehme CS. An improved camera trap for amphibians, reptiles, small mammals, and large invertebrates. PLoS ONE. 2017 Oct 5;12(10):e0185026. doi:10.1371/journal.pone.0185026.

[64] Marini S, Fanelli E, Sbragaglia V, Azzurro E, Del Rio Fernandez J, Aguzzi J. Tracking fish abundance by underwater image recognition. Sci Rep. 2018;8(1):13748. doi:10.1038/s41598-018-32089-8.

[65] Yoshioka A, Shimizu A, Oguma H, Kumada N, Fukasawa K, Jingu S, et al. Development of a camera trap for perching dragonflies: a new tool for freshwater environmental assessment. PeerJ. 2020;8:e9681. doi:10.7717/peerj.9681.

[66] Giraldo-Zuluaga J-H, Gomez A, Salazar A, Diaz-Pulido A. Camera-trap images segmentation using multi-layer robust principal component analysis. Vis Comput. 2019;35(3):335-347. doi:10.1007/s00371-017-1463-9.

[67] Steenweg R, Whittington J, Hebblewhite M, Forshner A, Johnston B, Petersen D, et al. Camera-based occupancy monitoring at large scales: power to detect trends in grizzly bears across the Canadian Rockies. Biological Conservation. 2016;201:192-200. doi:10.1016/j.biocon.2016.06.020.

[68] Sweitzer RA, Furnas BJ. Data from camera surveys identifying co-occurrence and occupancy linkages between fishers (*Pekania pennanti*), rodent prey, mesocarnivores, and larger predators in mixed-conifer forests. Data in Brief. 2016;6:783 792. doi:10.1016/j.dib.2016.01.032.

[69] Borchers D. A non-technical overview of spatially explicit capture–recapture models. J Ornithol. 2012;152(S2):435-444. doi:10.1007/s10336-010-0583-z.

[70] Young S, Rode-Margono J, Amin R. Software to facilitate and streamline camera trap data management: a review. Ecol Evol. 2018;8(19):9947-9957. doi:10.1002/ece3.4464.

[71] Norouzzadeh MS, Nguyen A, Kosmala M, Swanson A, Palmer MS, Packer C, et al. Automatically identifying, counting, and describing wild animals in camera-trap images with deep learning. Proc Natl Acad Sci USA. 2018;115(25):E5716-25. doi:10.1073/pnas.1719367115.

[72] Suckling DM, Stanbury ME, Lennon O, Colhoun KM, Chinellato F, El-Sayed AM. Kairomone and camera trapping New Zealand flower thrips, *Thrips obscuratus*. Insects. 2020;11(9):622. doi:10.3390/insects11090622.

[73] Böckmann E, Pfaff A, Schirrmann M, Pflanz M. Rapid and low-cost insect detection for analysing species trapped on yellow sticky traps. Sci Rep. 2021;11(1):10419. doi:10.1038/s41598-021-89930-w.

[74] Preti M, Verheggen F, Angeli S. Insect pest monitoring with camera-equipped traps: strengths and limitations. J Pest Sci. 2021;94(2):203-217. doi:10.1007/s10340-020-01309-4.

[75] Chiwamba SH, Phiri J, Nkunika POY, Sikasote C,

Kabemba MM, Moonga MN. Automated fall armyworm (*Spodoptera frugiperda*, J.E. Smith) pheromone trap based on machine learning. J Comp Sci. 2019;15(12):1759-1779. doi:10.3844/jcssp.2019.1759.1779.

[76] Marvin DC, Koh LP, Lynam AJ, Wich S, Davies AB, Krishnamurthy R, et al. Integrating technologies for scalable ecology and conservation. Global Ecology and Conservation. 2016;7:262-275. doi:10.1016/j.gecco.2016.07.002.

[77] Welbourne DJ, Claridge AW, Paull DJ, Ford F. Improving terrestrial squamate surveys with camera-trap programming and hardware modifications. Animals. 2019 Jun 25;9(6):388. doi:10.3390/ani9060388.

[78] Collett RA, Fisher DO. Time-lapse camera trapping as an alternative to pitfall trapping for estimating activity of leaf litter arthropods. Ecol Evol. 2017;7(18):7527-7533. doi:10.1002/ece3.3275.

[79] Jones FM, Allen C, Arteta C, Arthur J, Black C, Emmerson LM, et al. Time-lapse imagery and volunteer classifications from the Zooniverse Penguin Watch project. Sci Data. 2018;5(1):180124. doi:10.1038/sdata.2018.124.

[80] Beaudrot L, Ahumada JA, O'Brien T, Alvarez-Loayza P, Boekee K, Campos-Arceiz A, et al. Standardized assessment of biodiversity trends in tropical forest protected areas: the end is not in sight. PLoS Biol. 2016;14(1):e1002357. doi:10.1371/journal.pbio.1002357.

[81] Swinnen KRR, Reijniers J, Breno M, Leirs H. A novel method to reduce time investment when processing videos from camera trap studies. PLoS ONE. 2014;9(6):e98881. doi:10.1371/journal.pone.0098881.

[82] Tekeli U, Baştanlar Y. Elimination of useless images from raw camera-trap data. Turk J Elec Eng & Comp Sci. 2019;27:2395-2411. doi:10.3906/elk-1808-130.

[83] Green SE, Rees JP, Stephens PA, Hill RA, Giordano AJ. Innovations in camera trapping technology and approaches: the integration of citizen science and artificial intelligence. Animals. 2020;10(1):132. doi:10.3390/ani10010132.

[84] Meek PD, Ballard G-A, Fleming PJS, Schaefer M, Williams W, Falzon G. Camera traps can be heard and seen by animals. PLoS ONE. 2014;9(10):e110832. doi:10.1371/journal.pone.0110832.

[85] Forrester T, O'Brien T, Fegraus E, Jansen P, Palmer J, Kays R, et al. An open standard for camera trap data. Biodiversity Data Journal. 2016;4:e10197. doi:10.3897/BDJ.4.e10197.

[86] Wilson A. Diverse applications of electronic-nose technologies in agriculture and forestry. Sensors. 2013;13(2):2295-2348. doi:10.3390/s130202295.

[87] Poland TM, Rassati D. Improved biosecurity surveillance of non-native forest insects: a review of current methods. J Pest Sci. 2019;92(1):37-49. doi:10.1007/s10340-018-1004-y.

[88] Wilson AD, Forse LB, Babst BA, Bataineh MM. Detection of emerald ash borer infestations in living green ash by noninvasive electronic-nose analysis of wood volatiles. Biosensors. 2019;9(4):123. doi:10.3390/bios9040123.

[89] Doty AC, Wilson AD, Forse LB, Risch TS. Assessment of the portable C-320 electronic nose for discrimination of nine insectivorous bat species: implications for monitoring white-nose syndrome. Biosensors. 2020;10(2):12. doi:10.3390/bios10020012.

[90] Wilson AD, Lester DG, Oberle CS. Application of conductive polymer

analysis for wood and woody plant identifications. Forest Ecol Manag. 2005;209(3):207-224. doi:10.1016/j.foreco.2005.01.030.

[91] Ghaffari R, Laothawornkitkul J, Iliescu D, Hines E, Leeson M, Napier R, et al. Plant pest and disease diagnosis using electronic nose and support vector machine approach. J Plant Dis Prot. 2012;119(5-6):200-207. doi:10.1007/BF03356442.

[92] Wood WF, Weldon PJ. The scent of the reticulated giraffe (*Giraffa camelopardalis reticulata*). Biochem Syst Ecol. 2002;30(10):913-917. doi:10.1016/S0305-1978(02)00037-6.

[93] Brezolin AN, Martinazzo J, Muenchen DK, de Cezaro AM, Rigo AA, Steffens C, et al. Tools for detecting insect semiochemicals: a review. Anal Bioanal Chem. 2018;410(17):4091-4108. doi:10.1007/s00216-018-1118-3.

[94] Witzgall P, Kirsch P, Cork A. Sex pheromones and their impact on pest management. J Chem Ecol. 2010;36(1):80-100. doi:10.1007/s10886-009-9737-y.

[95] Persaud K, Dodd G. Analysis of discrimination mechanisms in the mammalian olfactory system using a model nose. Nature. 1982;299(5881):352-355. doi:10.1038/299352a0.

[96] Zhao P, Wu Y, Feng C, Wang L, Ding Y, Hu A. Conjugated polymer nanoparticles based fluorescent electronic nose for the identification of volatile compounds. Anal Chem. 2018;90(7):4815-4822. doi:10.1021/acs.analchem.8b00273.

[97] Cui S, Ling P, Zhu H, Keener H. Plant pest detection using an artificial nose system: a review. Sensors. 2018;18(2):378. doi:10.3390/s18020378.

[98] Obrist MK, Pavan G, Sueur J, Riede K, Llusia D, Márquez R. Bioacoustics approaches in biodiversity inventories. In: Eymann J, Degreef J, Häuser C, Monje JC, Samyn Y, VandenSpiegel D, editors. ABC Taxa. Manual on field recording techniques and protocols for all taxa biodiversity inventories. 2010. p. 68-99.

[99] Gibb R, Browning E, Glover-Kapfer P, Jones KE. Emerging opportunities and challenges for passive acoustics in ecological assessment and monitoring. Methods Ecol Evol. 2019;10(2):169-185. doi:10.1111/2041-210X.13101.

[100] Shiu Y, Palmer KJ, Roch MA, Fleishman E, Liu X, Nosal E-M, et al. Deep neural networks for automated detection of marine mammal species. Sci Rep. 2020;10(1):607. doi:10.1038/s41598-020-57549-y.

[101] Dorcas ME, Price SJ, Walls SC, Barichivich WJ. Auditory monitoring of anuran populations. In: Dodd K, editor. Conservation and ecology of amphibians. Oxford: Oxford University Press. 2009. p. 281-298.

[102] Dufourq E, Durbach I, Hansford JP, Hoepfner A, Ma H, Bryant JV, et al. Automated detection of Hainan gibbon calls for passive acoustic monitoring. Remote Sens Ecol Conserv. 2021; rse2.201. doi:10.1002/rse2.201.

[103] Measey GJ, Stevenson BC, Scott T, Altwegg R, Borchers DL. Counting chirps: acoustic monitoring of cryptic frogs. J Appl Ecol. 2017;54(3):894-902. doi:10.1111/1365-2664.12810.

[104] Priyadarshani N, Marsland S, Castro I. Automated birdsong recognition in complex acoustic environments: a review. J Avian Biol. 2018;49(5):jav-01447. doi:10.1111/jav.01447.

[105] Shonfield J, Bayne EM. Autonomous recording units in avian ecological research: current use and

future applications. ACE. 2017;12(1):art14. doi:10.5751/ACE-00974-120114.

[106] Sugai LSM, Silva TSF, Ribeiro JW, Llusia D. Terrestrial passive acoustic monitoring: review and perspectives. BioScience. 2019;69(1):15-25. doi:10.1093/biosci/biy147.

[107] Aide TM, Corrada-Bravo C, Campos-Cerqueira M, Milan C, Vega G, Alvarez R. Real-time bioacoustics monitoring and automated species identification. PeerJ. 2013;1:e103. doi:10.7717/peerj.103.

[108] Goyette JL, Howe RW, Wolf AT, Robinson WD. Detecting tropical nocturnal birds using automated audio recordings. J Field Ornithol. 2011;82(3):279-287. doi:10.1111/j.1557-9263.2011.00331.x.

[109] Pieretti N, Farina A, Morri D. A new methodology to infer the singing activity of an avian community: the acoustic complexity index (ACI). Ecological Indicators. 2011;11(3):868-873. doi:10.1016/j.ecolind.2010.11.005.

[110] Sueur J, Pavoine S, Hamerlynck O, Duvail S. Rapid acoustic survey for biodiversity appraisal. PLoS ONE. 2008;3(12):e4065. doi:10.1371/journal.pone.0004065.

[111] Gasc A, Pavoine S, Lellouch L, Grandcolas P, Sueur J. Acoustic indices for biodiversity assessments: analyses of bias based on simulated bird assemblages and recommendations for field surveys. Biol Cons. 2015;191:306-312. doi:10.1016/j.biocon.2015.06.018.

[112] Alvarez-Berríos N, Campos-Cerqueira M, Hernández-Serna A, Amanda Delgado CJ, Román-Dañobeytia F, Aide TM. Impacts of small-scale gold mining on birds and anurans near the Tambopata Natural Reserve, Peru, assessed using passive acoustic monitoring. Trop Cons Sci. 2016;9(2):832-851. doi:10.1177/194008291600900216.

[113] Rosenstock SS, Anderson DR, Giesen KM, Leukering T, Carter MF. Landbird counting techniques: current practices and an alternative. The Auk. 2002;119(1):46-53. doi:10.1093/auk/119.1.46.

[114] Stevenson BC, Borchers DL, Altwegg R, Swift RJ, Gillespie DM, Measey GJ. A general framework for animal density estimation from acoustic detections across a fixed microphone array. Methods Ecol Evol. 2015;6(1):38-48. doi:10.1111/2041-210X.12291.

[115] Andreassen T, Surlykke A, Hallam J. Semi-automatic long-term acoustic surveying: A case study with bats. Ecological Informatics. 2014;21:13-24. doi:10.1016/j.ecoinf.2013.12.010.

[116] Efford MG, Dawson DK, Borchers DL. Population density estimated from locations of individuals on a passive detector array. Ecology. 2009;90(10):2676-2682. doi:10.1890/08-1735.1.

[117] Marsland S, Priyadarshani N, Juodakis J, Castro I. AviaNZ: A future-proofed program for annotation and recognition of animal sounds in long-time field recordings. Methods Ecol Evol. 2019;10(8):1189-1195. doi:10.1111/2041-210X.13213.

[118] Dawson DK, Efford MG. Bird population density estimated from acoustic signals: population density from acoustic signals. J Appl Ecol. 2009;46(6):1201-1209. doi:10.1111/j.1365-2664.2009.01731.x.

[119] MacKenzie DI, Bailey LL, Nichols JD. Investigating species co-occurrence patterns when species are detected imperfectly. J Anim Ecology. 2004;73(3):546-555. doi:10.1111/j.0021-8790.2004.00828.x.

[120] Acevedo MA, Corrada-Bravo CJ, Corrada-Bravo H, Villanueva-Rivera LJ, Aide TM. Automated classification of bird and amphibian calls using machine learning: a comparison of methods. Ecological Informatics. 2009;4(4):206-214. doi:10.1016/j.ecoinf.2009.06.005.

[121] Brown A, Garg S, Montgomery J. AcoustiCloud: a cloud-based system for managing large-scale bioacoustics processing. Environmental Modelling & Software. 2020;131:104778. doi:10.1016/j.envsoft.2020.104778.

[122] Mortimer J, Greene T. Investigating bird call identification uncertainty using data from processed audio recordings. NZ J Ecol. 2017;41(1):126-33. doi:10.20417/nzjecol.41.10.

[123] Gasc A, Sueur J, Jiguet F, Devictor V, Grandcolas P, Burrow C, et al. Assessing biodiversity with sound: do acoustic diversity indices reflect phylogenetic and functional diversities of bird communities? Ecological Indicators. 2013;25:279-287. doi:10.1016/j.ecolind.2012.10.009.

[124] Martínez-López J, Bertzky B, Willcock S, Robuchon M, Almagro M, Delli G, et al. Remote sensing methods for the biophysical characterization of protected areas globally: challenges and opportunities. Int J Geo-Inf. 2021;10(6):384. doi:10.3390/ijgi10060384.

[125] Szantoi Z, Brink A, Buchanan G, Bastin L, Lupi A, Simonetti D, et al. A simple remote sensing based information system for monitoring sites of conservation importance. Remote Sens Ecol Conserv. 2016;2(1):16-24. doi:10.1002/rse2.14.

[126] Zolkos SG, Goetz SJ, Dubayah R. A meta-analysis of terrestrial aboveground biomass estimation using lidar remote sensing. Remote Sensing of Environment. 2013;128:289-298. doi:10.1016/j.rse.2012.10.017.

[127] Jiménez López J, Mulero-Pázmány M. Drones for conservation in protected areas: present and future. Drones. 2019;3(1):10. doi:10.3390/drones3010010.

[128] Hötte MHH, Kolodin IA, Bereznuk SL, Slaght JC, Kerley LL, Soutyrina SV, et al. Indicators of success for smart law enforcement in protected areas: A case study for Russian Amur tiger (*Panthera tigris altaica*) reserves. Integrative Zoology. 2016;11(1):2-15. doi:10.1111/1749-4877.12168.

[129] IUCN. Regional action plan for the conservation of western lowland gorillas and central chimpanzees 2015-2025. Maisels F, Williamson L, Strindberg S, Pokempne A, Greer D, Stokes E, et al., editors. International Union for Conservation of Nature; 2015. 54 p. doi:10.2305/IUCN.CH.2005.SSC-RAP.1.en.

[130] SMART. Spatial Monitoring and Reporting Tool [Internet]. 2021. Available from: https://smartconservationtools.org/ [Accessed: 2021-07-30]

[131] Anbaroğlu B, Coşkun İB, Brovelli MA, Obukhov T, Coetzee S. Educational material development on mobile spatial data collection using open source geospatial technologies. Int Arch Photogramm Remote Sens Spatial Inf Sci. 2020;XLIII-B4-2020:221-7. doi:10.5194/isprs-archives-XLIII-B4-2020-221-2020.

[132] Musinsky J, Tabor K, Cano CA, Ledezma JC, Mendoza E, Rasolohery A, et al. Conservation impacts of a near real-time forest monitoring and alert system for the tropics. Dubois G, Klerk H, editors. Remote Sens Ecol Conserv. 2018;4(3):189-196. doi:10.1002/rse2.78.

[133] Weisse M, Nogueron R, Vicencio REV, Soto DAC. Use of near-real-time deforestation alerts: a case

study from Peru. World Resources Institute. 2019. 36 p. Available from: https://www.wri.org/publication/use-near-real-time-deforestation-alerts [Accessed 2021 Jul 28].

[134] Winter S, Chirici G, McRoberts RE, Hauk E, Tomppo E. Possibilities for harmonizing national forest inventory data for use in forest biodiversity assessments. Forestry. 2008;81(1):33-44. doi:10.1093/forestry/cpm042.

[135] Gallaun H, Zanchi G, Nabuurs G-J, Hengeveld G, Schardt M, Verkerk PJ. EU-wide maps of growing stock and above-ground biomass in forests based on remote sensing and field measurements. Forest Ecology and Management. 2010;260(3):252-261. doi:10.1016/j.foreco.2009.10.011.

[136] Eggleston HS, Intergovernmental Panel on Climate Change, National Greenhouse Gas Inventories Programme, Chikyū Kankyō Senryaku Kenkyū Kikan. 2006 IPCC guidelines for national greenhouse gas inventories. 2006. Available from: https://www.worldcat.org/title/2006-ipcc-guidelines-for-national-greenhouse-gas-inventories/oclc/192005769 [Accessed 2021 Jul 9].

[137] Weinstein BG, Marconi S, Bohlman SA, Zare A, Singh A, Graves SJ, et al. A remote sensing derived data set of 100 million individual tree crowns for the National Ecological Observatory Network. eLife. 2021;10:e62922. doi:10.7554/eLife.62922.

[138] Pinz A. Final results of the vision expert system VES: Finding trees in aerial photographs. Wissensbasierte Mustererkennung (Knowledge-based Pattern Recognition). 1989;49:90-111.

[139] Kaartinen H, Hyyppä J, Yu X, Vastaranta M, Hyyppä H, Kukko A, et al. An international comparison of individual tree detection and extraction using airborne laser scanning. Remote Sensing. 2012;4(4):950-974. doi:10.3390/rs4040950.

[140] Saarinen N, Kankare V, Vastaranta M, Luoma V, Pyörälä J, Tanhuanpää T, et al. Feasibility of terrestrial laser scanning for collecting stem volume information from single trees. ISPRS Journal of Photogrammetry and Remote Sensing. 2017;123:140-158. doi:10.1016/j.isprsjprs.2016.11.012.

[141] Thiel C, Müller MM, Berger C, Cremer F, Dubois C, Hese S, et al. Monitoring selective logging in a pine-dominated forest in central Germany with repeated drone flights utilizing a low cost RTK quadcopter. Drones. 2020;4(2):11. doi:10.3390/drones4020011.

[142] Nowak MM, Dzióbk K, Ludwisiak Ł, Chmiel J. Mobile GIS applications for environmental field surveys: a state of the art. Global Ecology and Conservation. 2020;23:e01089. doi:10.1016/j.gecco.2020.e01089.

Chapter 15

Development of Eco-Tourism Vis-à-Vis Conservation Measures for Protecting the Endangered Species of Flora and Fauna: A Study on Bhitarkanika National Park and Wildlife Sanctuary, Odisha, India

Dillip Kumar Das

Abstract

Bhitarkanika is the second largest mangrove ecosystem in India after the Sundarbans National Park and is situated in the state of Odisha. It is the home of diverse flora and fauna that are intricately linked with each other. Bhitarkanika comprises lots of endangered species of flora and fauna that needs immediate attention by different stakeholders especially the government and NGOs for conservation and preservation of these diverse resources. The present study is focused on identifying various eco-tourism resources available at this destination and also analyse various protection measures undertaken to conserve and preserve the flora and fauna inside the sanctuary. This study also examines the role of government for conservation and preservation of eco-tourism resources inside Bhitarkanika. The current study reveals that Bhitarkanika Wildlife Sanctuary have immense tourism potentials to become a perfect eco-tourism destination. However, in spite of that, this destination has not been extensively promoted and marketed as a perfect eco-tourism destination in this region. Since this area provides a huge employment opportunity for the local community, so there is an urgent need to conserve and preserve this sanctuary for a better eco-tourism destination in the years to come.

Keywords: ecosystem, flora, fauna, Ramsar sites, endangered species, mangrove Forest

1. Introduction

Bhitarkanika is one of the finest wildlife sanctuaries in Asia, the second-largest mangrove ecosystem in India after the Sundarbans National Park and is situated in the Kendrapada district of Odisha. It is also the home of diverse flora and fauna, which are intricately linked with each other and are the integral components of eco-tourism destination. The conservation and preservation of flora and fauna in the study area show an improvement in the vegetation and increase the livelihood of the local people resulting in an increase in the tourist flow to this destination. Bhitarkanika, as an eco-tourism destination, comprises lots of endangered species of flora and fauna that needs immediate attention by different stakeholders especially the government and other stakeholders to conserve and protect. Bhitarkanika is situated in-between the river Brahmani and Baitarani and forms the deltaic region in between these two rivers. This area was originally belonging to the wrest while ruler of Kanika and a large portion of this forest land of this delta was encroached in the later phase due to expansion of agriculture in this region by the local people [1]. From tourism perspective, Bhitarkanika consists of areas including Dangmal, Bagagahan, Gahiramatha, Ekakula, Havalikhati, and mangrove forests, rivers, cricks, mudflats, mammals, reptiles, crocodiles, snakes, birds, etc. It has become a global tourist attraction due to the pride possession of white crocodile [2].

2. About Bhitarkanika National Park and Bhitarkanika wildlife sanctuary

This National Park has been declared by the Department of Forest and Environment; Government of Odisha vide Notification No.19686/F & E dated 16.9.1998 with an area Covering 145 sq. km area that is notified as Bhitarkanika National Park. The national park is popular because of its ecological significance coupled with a biological background. It mainly comprises estuarine crocodile, mangrove forest, backwaters, river creeks etc. Bhitarkanika Wildlife Sanctuary was declared vide notification No.6958/FF AH dated. 22.04.1975 with an area of 672 sq. km. of mangrove forest & wetland that provides a home to well over 215 species of birds including winter migrants from central-Asia and Europe. Giant saltwater crocodiles and a variety of other wildlife inhabitants in this ecosystem make this place unique as Asia's one of the most spectacular wildlife areas [3].

The major resources of Bhitarkanika Wildlife sanctuary are the popular mangrove forest and the endangered saltwater crocodile. Apart from the above, the mangrove forest is also a good habitat for the king cobra, Indian python and water monitored lizard. In the year 2002, the Bhitarkanika mangroves with an area of 2672 sq. km. has been declared as a *Ramsar site* being a place of having wetland international importance by Wetland International, South Asia [4].

3. The major objectives of the study

3.1 The major objectives of the studies are as follows

The first objective is to study the tourism potentials of Bhitarkanika. The second objective is to identify different flora and fauna of this place with special emphasis on conserving endangered species. The next objective is to examine the role of government in the conservation and preservation of eco-tourism resources. The last but not the least objective is to analyse how to promote and market Bhitarkanika as a perfect eco-tourism destination.

4. Scope of the study

The current study on Bhitarkanika Wildlife sanctuary and National Park does not only provide you with the scope for getting knowledge on different flora and fauna available inside the sanctuary but also study the scope of promoting Bhitarkanika as an important eco-tourism destination in entire eastern India with special emphasis to the state of Odisha. Here in this study, the author has considered Estuarine crocodile (Salt Water Crocodile) and Olive Reedley Turtle as major endangered species inside the sanctuary in terms of conservations of fauna, which needs to be conserved and protected. However, in terms of flora, the author considers the mangrove trees as an endangered species in Bhitarkanika, which needs to be protected in order to save the sanctuary and market this destination as a perfect eco-tourism place for the state of Odisha.

5. Methodology used

For preparing this article, the author adopted a checklist and participant observation method. Sample size was restricted to 50. A total of 50 checklists were filled in by the students on their field study visit to Bhitarkanika in the year 2020. Apart from those, interactions were also made with the staff of Bhitarkanika Wildlife Sanctuary, staff of the accommodation units inside the wildlife sanctuary and the guides and boatmen including the District Forest Officer Bhitarkanika through an in-depth interview. The author also used the observation method to discuss and analyse the data. A major limitation was that the dwellers in the Sanctuary could not be contacted to elicit their perceptions/concerns. Since the District Forest Officer himself along with other government officials were present with the team inside the Sanctuary, the author could able to authenticate the information in order to arrive at specific suggestions and conclusion given in this study.

6. Tourism potentials of Bhitarkanika

Being the only state in India, the state of Odisha, one can find all the three varieties of crocodile species mainly Gharial, Mugger and the saltwater crocodile. The first crocodile project was launched in the state of Odisha.

The term Bhitarkanika is formed of two words, "Bhitar" and "Kanika". In Odia language, Bhitar means interior and Kanika means extraordinarily beautiful. So Bhitarkanika is popularly known as the place having eternal beauty. The Odisha Tourism is developing Bhitarkanika as a destination mainly for ecotourism purposes [5]. Bhitarkanika is a hot spot of bio-diversity and home to giant saltwater crocodiles. One more place inside Bhitarkanika is the Gahiramatha, which is a very important nestling place of Olive ridley sea turtles making Bhitarkanika famous in the entire world.

Inside Bhitarkanika forest block, Bagagahan is the place where mostly as well as different migratory birds used to come and used the mangrove forest for their nestling place. Tourists can see these birds with the help of a watchtower, which can be accessed by foot getting down from the boat to the watchtower [6].

6.1 DANGMAL: saltwater crocodile project at Dangmal

The quietness and scenic atmosphere of Dangmal make it one of the most popular places for tourists where the salt water crocodile project is made. Other attractions at Dangmal include Captive breeding & Research Centre for estuarine crocodile. Here one can see a huge female white crocodile named Gauri. Tourists can see here Python and King Cobra staying together without harming each other in an enclosure. One interpretation hall is constructed in Dangmal, where film show on

Bhitarkanika is made available on demand. This beautiful place can be reached by road from Rajnagar via Khola Check gate.

The boat ride from Khola to Dangmal and return is one of the major activities for the tourists. Khola is one of the gateways into the park. This is along an artificial creek and it passes through dense mangrove forest providing a glimpse into the estuarine ecosystem and its wealth of fauna. Sightseeing facilities are provided with the help of boats that are having a valid licence from the forest department.

7. Saltwater crocodile

It is regarded as the Earth's largest living crocodile species. This variety of crocodilesis basically known as man-eater. In India apart from Bhitarkanika, one can find this variety in The Sundarbans (West Bengal) and Andaman and Nicobar Island. This variety is considered as a threatened species as per IUCN list and it is the IUCN List of Threatened Species that is included in schedule 1 of the Wildlife Protection Act of India 1972.

Saltwater crocodiles are also called Indian muggers. The saltwater crocodile, as evident by its name, can tolerate saline environments very well, that is why it is typically found in brackish water around coastal areas and rivers. The saltwater crocodile (*Crocodylus porosus*) is the largest of all crocodilians and the largest reptile in the world. Bhitarkanika is a very good place to sight the giant Saltwater Crocodile, some of them are growing to 23 feet in length. They are also known as man-eater [7].

One can find the saltwater of estuarine crocodile in the Brahmani and Baitarani River delta region popularly known as Bhitarkanika national park of Odisha. Being an endangered species, the estuarine crocodile population was sharply declined due to extreme poaching, hunting, as well as due to exploitation. So, keeping in mind the above, a conscious effort was first initiated by the forest department of Odisha in form of launching a conservation project popularly known as project Baula.

This project was mainly funded by the FAO of UNDP initially [8]. As a result of this project, several crocodiles were released and also some rare varieties of crocodiles were supplied to other projects launched in different states of India. As a result of this project, illegal trapping and killing of crocodiles were stopped.

Efforts have been made every year to count the crocodile population inside Bhitarkanika. It was also observed that night-time is considered to be the best time in comparison to the daytime for counting crocodile numbers because of hatching and yearling. The study was mainly conducted to make sure that all classes of crocodiles should be present in different places showing a better sign of variable population resulting in a positive sign of showing an increase in crocodile population.

7.1 Major objectives of crocodile conservation projects in Odisha/Bhitarkanika

To protect the remaining population of crocodilians in their natural habitat by creating sanctuaries, to rebuild natural population quickly through 'grow and release' or 'rear and release' technique that involves the following phases of operation, to promote captive breeding to increase crocodile population, to take-up research to improve management as apart of scientific study on crocodile population and their behaviour, to build up the skills of the personnel for better continuity of the project through trainings imparted at the project-sites and the Central Crocodile Breeding and Management Training Institute, in different states of India including Odisha and to involve the local people intimately to maintain ecological balance and improve economic conditions [9].

The Project aimed at maintaining the ecological disbalance caused by deaths of crocodiles in the river basins. To make this happen, three research units on crocodile conservation were established at Dangmal, Tikarpara and Nandankanan Biological parks in Odisha.

7.2 Baula project at Dangmal

In Oriya literature, Estuarine or saltwater crocodiles are popularly known as 'Baula'. Accordingly, a Baula Crocodile Project has been introduced in Dangmal inside Bhitarkanika Wildlife sanctuary. Under this project, several tests have been conducted successfully in a phase wise manner. In this process, eggs of saltwater crocodiles were collected and kept under supervision for breeding to increase the crocodile population inside Bhitarkanika. This project was found to be successful in increasing the population but also balancing the ecological set-up inside the national park.

The Crocodile Conservation Project was launched in 1975 in different States. As a result of the programme, the estimated number of saltwater crocodiles increased from 96 in 1976 to 1640 in 2012 in India (**Table 1**) [10].

Year	No of saltwater Crocodile Population In Bhitarkanika
1975	Program introduced
1976	96
2004	1308
2012	1640
2019	1742
2020	1757
2021 Jan	1768

Source: Compiled from forest department Statistics.

Table 1.
Saltwater crocodile population in Bhitarkanika.

7.3 Gahirmatha sanctuary

Gahiratha is a popular marine sanctuary situated in the state of Odisha famous for being the world's largest mating and nestling place for the most endangered Olive ridley turtles. Established in 1997, it covers an area of 1435 sq. km. Extending from Dhamara River from the north to Brahmani River form the south of Kendrapada districts of Odisha. Apart from Gahirmatha, one can find mass nesting and mating centres for Olive ridley turtles in Rushikulya and Devi River [11, 12].

The Project aimed at maintaining the ecological balance caused by deaths of crocodiles in the river basins. To make this happen, three research units on crocodile conservation were established at Dangmal, Tikarpara and Nandankanan Biological parks in Odisha.

8. Olive Ridley turtles

This is the smallest and most abundant of all available sea turtles found in the entire world. This species is listed as one of the most vulnerable species in the IUCN Red List, Appendix 1 in CITES, and listed in schedule 1 in Wildlife Protection Act, 1972. This variety of turtle is characterised by mass nesting where thousands of

Year	No of turtle arrivals to Bhitarkanika/ Gahiramatha Marine Sanctuary (In Lakh)
2020	407,000
2019	450,000
2018	470,000
2014	400,000

Source: Compiled from Forest department statistics.

Table 2.
Arrival details of olive Radley turtles to Bhitarkanika/Gahiramatha sanctuary.

female turtles move together to specific places on a regular basis again and again to lay eggs. Normally, a female Olive Ridley Turtle lays eggs up to 100–150 eggs each generally during the night in the small pits they dig, leaving the beach after covering the holes with sand. The hatchling process starts after 45–60 days and then these small turtles crawl into the sea in the absence of their parents [13].

8.1 Major threats to turtles in Bhitarkanika are

Unfriendly turtle fishing practice by the fishermen, development of tourist activities at the nesting places especially Dangmal and Ekakula, more fishing activities in the specified route in which turtles are generally coming to Gahiramatha for laying eggs, excessive use of speed boats and trollers, development of new ports alongside the nesting ground and presence of wild animals in the nesting areas.

Role of government, especially the forest department is taking concrete measures to ensure a smooth mating period followed by laying of eggs in the Gahiramatha area. The major activities in this regard include:

The Forest Department's intensive patrolling, including at night, had helped decrease turtle mortality especially during the mating season, banning of use of fishing and use of fishing net in the turtle route in which they are coming to lay eggs, banning the use of trawlers in the area by earmarking the route, The Forest Department also keeps a watch on fishing trawlers venturing into the area up to 10 km into the sea from different entry point to Bhitarkanika. Especially from Balasore from north side to Andhra Pradesh from the south side, The forest Department had also started fencing the beach between Gokharakuda and Bateswar, to pave the way free from predators and facilitate smooth mass nesting by the turtles. Last but not the least, the local community-based social organisations were also helping in cleaning activities by sensitising the locals as well as the tourists (**Table 2**).

9. Endangered flora of Bhitarkanika: mangrove ecosystem

No doubt, the Saltwater Crocodile Conservation Programme has ensured the survival of crocodiles in the Bhitarkanika Sanctuary that has been in operation for more than four decades but the major issue to give a new lease of life, the possibilities lie in managing the area as a Biosphere Reserve containing the mangrove ecosystem.

Recent studies show that the mangrove forest is diminishing due to lack of freshwater supply in the mangrove areas and anthropogenic pressures from the surrounding areas including illegal encroachment for shrimp culture, agricultural activity etc. Another important issuethatis responsible for reducing the mangrove forest is due to regular occurrence of cyclones in the state of Odisha. Because of its adjacent proximity to the sea, most of the times the cyclones are hitting this area especially causing devastating the mangrove forest ecosystem and as a result, it is also observed that the migratory bird populations are also reducing slowly to these areas because of loss of mangrove trees inside the biosphere reserve.

Bhitarkanika wildlife sanctuary has become a floristic composition, nature of distribution, sociability, rarity present in this region for better tourism. Natural area tourism has a major effect on the economy of the country since it provides direct and indirect employment opportunities to raise the standard of living of the host population. In view of this fact, it is felt that there is an urgent need to conserve this sanctuary for a better tourism point of view [14].

10. Conservation measures by government

Mangrove plantation and Nalia grass (*Myriostachya wigstiana*) plantation has been taken up under this programme on riverside of 5 km. as a pilot programme in Bhitarkanika Ramsar site to prevent soil erosion and also natural conservation with livelihood enhancement.

The women are the most vulnerable during disasters for collecting drinking water and sanitation. So, the platforms of 150 nos. Existing tube wells have been raised to the high flood level to avail the drinking water facility during flood. For better sanitation management during the time of flood, 850 nos. of high raised plinth toilets have been installed in coastal flood-affected villages. To manage the ecosystem in water logged areas, proper drainage facility has been created. The creeks of 8 km have been renovated as a pilot programme to release the flood water fast from the project villages.

The government is now encouraging participation in the implementation of Govt. owned programmes, the community has been capacitated by providing training as well as facilitating the action in the field. The Govt. programmes like sanitation, insurance, MGNREGS, horticulture, agriculture, fishery, forest, child welfare etc. has been included in the project villages through convergence.

A massive awareness program has been undertaken by the district administration supplemented with the formation of an anti-poaching camp at different strategic points to stop the poachers away from poaching. To encourage eco-tourism, training camps for eco-tourism guides and boat-man associations are being organised in a regular time interval.

Ministry of Environment and Forest, Government of India provides funds related to digging, plantation programs, including renovation of creeks and ponds to improve habitat inside the sanctuary.

11. Management strategy for conservation program to protect the biodiversity

Several measures have been taken care of by the state government as well as the Ministry of Environment and Forest, Government of India for conservation and preservation of unique bio-diversity of Bhitarkanika. The most important measures to create database regarding the destination, implementation conservation program for saltwater crocodiles, control to protect the water bodies from weeds, control poaching of migratory birds and other animals, sensitising the local community regarding the importance of wetlands and mangrove forest, awareness about the community participation and the need of capacity building program for promotion and development of eco-tourism in the study area i.e., Bhitarkanika.

12. Major findings at different tourist destinations inside Bhitarkanika

a. The interpretation centre is found to be very small with limited facilities.

b. Basic amenities like adequate, safe and pure drinking water is not available inside.

c. No provision of refreshments and food facilities either at the entrance or inside the sanctuary, causing highly inconvenience to the visitors/tourists.

d. The waste bins/cans are not installed at the entrance and inside the park premises.

e. Accommodation units both outside and inside the park are found to be inadequate.

f. The watchtowers are not in sufficient number and are also not sufficient in height.

g. Many trails were noticed on both sides of the itinerary inside the park.

h. The guides are mainly locals and less trained but found to be inadequately conversant with languages other than Oriya and Bengali.

Some general findings are as follows:

- Production of a large number of captive stocks without adequate suitable places for release causes ecological imbalance inside the sanctuary.
- Due to an increase in population in and around the sanctuary, resulting in reduction in undisturbed habitats for released crocodilians back in the wild.
- The Financial assistance received from both the state, central and other international organisations is found to be insufficient and irregular causing hindrance in the conservation process.
- During the study, it was also revealed that international organisations like FAO/UNDP that were earlier associated with this project have stopped giving funds to this project.
- Increased and unplanned encroachment by the fishermen to establish Gheris for prawn culture is found to be one of the major threats to the destination.
- It was worthy to mention that the number of nesting grounds for crocodiles is declining every year due to floods and cyclones, which are occurring almost every year.
- Transportation cost for the tourist from the nearest railway station to the entry point of Bhitarkanika is found to be substantially high.

13. Suggestions and recommendations

The suggestions proposed in this study are purely based on observations inside park during the field visit and taking into the theoretical considerations. However, these suggestions are quite relevant and can consider as important values for sustained growth of ecotourism in Bhitarkanika.

The tourism potentials, as well as tourism resources of Bhitarkanika, are beyond doubt and they can attract mainly the eco-tourists from different parts of the country as well as from the world. The strengths of the park are relatively easy accessibility and abundance of natural resources including its beauty and tranquillity and natural set-up.

Considering the diverse nature of tourism resources of Bhitarkanika, the authorities should think of introducing nature-based tourism activities like bird-watching, crocodile breeding, health tourism, camping and trekking etc., which can be planned in a sustainable manner to provide a variety of tourism resources to the tourist.

A major strength of Bhitarkanika is the abundance of medicinal plant varieties. This, combined with the rich ayurvedic tradition of Odisha, the author suggests that health tourism should be considered as a priority segment for future development. The State of Odisha can certainly be benefited in future if planned in a proper manner.

The interpretation centre, which is constructed inside the sanctuary, is ill-equipped and is found to be inadequate for sensitising the tourist about the park and the significance of nature tourism to their life. There should be provisions for audio-visual systems so that short films and documentary films can be shown to the tourist for sensitization and better education related to their behaviour and movement inside the sanctuary.

Provision of dustbins and bio-degradable carry bags inside the sanctuary especially different entry points at the destinations must be introduced to avoid an unhealthy and uncleaned environment.

Being an eco-tourist destination, special experienced tourist guides, particularly from the local trained youth should be introduced inside the destination for a better visitor management system inside the destination.

Keeping in pace with the tourist traffic, few restaurants including a few refreshment centres should be opened inside the parks mainly at the entrance and terminal point of important tourist points.

Provision of watchtowers inside the park are found to be inadequate in number and are not above the tree line. So, a greater number of watchtowers with bigger heights should be constructed to see the wild animals, especially at night.

The tourist points should be properly planned and built in a synchronised way so that the tourist can enjoy the destination based on a specific time frame with a proper visitor management facility.

Provision for fresh and pure drinking water must be provided for the tourists in specific destinations inside the sanctuary. This will help in discouraging the tourists from carrying bottled water, a potential source of plastic pollution.

The vehicle movements inside the park should be restricted and monitored. Caution needs to be taken about the colour and speed of the vehicles moving inside the park including the sound pollution made by the vehicles for avoiding disturbance to animals.

To optimise the benefits of eco-tourism as well conservation of the sanctuary the people living in both the core and buffer areas may be sensitised properly. The local government shall come up with a suitable agenda for providing better employment opportunities and also create entrepreneurial skills. Different schemes associated with the development of eco-tourism projects like start-up grants, poverty alleviation programs, rural employment opportunities guarantee schemes etc. should be implemented properly to maximise the benefits to the local community.

A proper mechanism for calculating the number of tourist arrivals inside the destination including their demographic profiles and purpose of visit should be made. Occasional surveys must be carried out to find out the activities undertaken inside the park, their expectations and levels of satisfaction. This will help the facility development/improvement, regulation and monitoring activities inside the park.

Modern boating facilities with proper safety management facilities must be ensured while providing boating service to the tourists. Strict law enforcement is necessary to prevent unauthorised entry into Bhitarkanika and handle armed criminals, especially for hunting. The exact and scientific reasons for declining trends of nestling ground for crocodiles must be examined and thoroughly studied to know the accurate fact.

14. Conclusion

As the name suggests the destination Bhitarkanika is definitely having eternal beauty for which a large number of tourists are now visiting this destination every year. No doubt the Bhitarkanika Wildlife Sanctuary and National Park is a paradise for all the wildlife lovers with its rich bio-diversity, but due to the lack of certain facilities and amenities, the flow of tourists to this spot is not so encouraging in comparison to its tourism resources. If the following points to be taken care of seriously like: facilitating the development of conservation measures both for flora and fauna inside the sanctuary, creation of more employment opportunities through tourism development inside the study area, use of local agricultural and household products including marketing of local handicrafts and souvenir to the tourists who are coming to visit the destination and encouragement of use local facilities like catering, transport, guides, fishing etc.

Then, the place can be very easily marketed as a perfect eco-tourism place of the state of Odisha. If the sanctuary is developed according to the biological and physical tolerance level by maintaining the carrying capacity, then a day will come when this sanctuary will become the main attraction for the tourists of National and International level of the global tourists' market. It is also worthy here to mention that in the year 2001, Bhitarkanika has been identified as the most unexplored eco-tourism destination of India. The scope for marketing Bhitarkanika as an eco-tourism destination is very high. What the destination presently needs is proper marketing, development of infrastructure, especially setting up of accommodation units inside the sanctuary, enhancing the conservation projects for better survival of endangered species, better promotion, maintaining the carrying capacity of the destination and above all planning for a better visitor management system with highest monitoring facilities will definitely make Bhitarkanika a world-class eco-tourism destination in the state of Odisha.

DOI: http://dx.doi.org/10.5772/intechopen.102502

Author details

Dillip Kumar Das
Department of Tourism Management, The University of Burdwan,
Burdwan, West Bengal, India

*Address all correspondence to: das.dillipkumar@gmail.com

References

[1] Behura BK. Bhitarkanika–A biological paradox. In: Behura BK, editor. Bhitarkanika–The Wonderland of Orissa. Bhubaneswar: Nature and Wildlife Conservation Society of Orissa; 1999. pp. 44-68

[2] Hossain F, Ibrahim Beg I. Economic impact of tourism—A case study of Bhitarkanika Wildlife Sanctuary. Jamshedpur Research Review. Year IV. Volume IV. Issue XVIII. ISSN: 2320-2750. JRR. December 2016–February 2017; Jamshedpur, Jharkhand, India

[3] Pathy SP. Interpreting eco-tourism: The scenario of Odisha. International Journal of Current Research in Life Sciences. 2018;7(2):874-876

[4] Banerjee LK. Vegetation of Bhitarkanika sanctuary in Cuttack District, Orissa, India. Journal of Economics Taxes Bot. 1984;**5**(5): 1065-1079

[5] Home (Public Relations) Department. Odisha, India: Government of Orissa; 1995. p. 14

[6] Pandav B. Birds of Bhitarkanika mangroves. Eastern India Forktail. 1996;**12**:9-20

[7] Bustard HR, Kar SK. Attacks on domestic livestock by the Saltwater Crocodile (Crocodylus porosus) in Orissa. India British Journal of Herpetology. 1981;**6**(4):137

[8] FAO (1975): India: Gharial and Crocodile conservation management in Orissa. Based on the work of H.R. Bustard. FAO, Rome. FAO, Rome. (FO:IND/71/033). pp. 15

[9] Singh DK, Panda GK. Bhitarkanika and its Environs – a Geographical Appraisal. In: Bhitarkanika: The Wonderland of Orissa. Bhubaneswar, India: Nature and Wildlife Conservation Society of Orissa; 1999. pp. 10-18 123

[10] Singh LAK, Kar SK. Status of the saltwater crocodile in Orissa: An overview. Journal of the Bombay Natural History Society. 2006; **103**(2-3):274-285

[11] Chadha S, Kar CS. Bhitarkanika: Myth and Reality. Dehradun: Natraj Publishers; 1999. pp. 388

[12] Chadha SK. The mangrove swamps of Devi River mouth, Puri District (Orissa). Bhubaneswar: Paper presented at the National Workshop on Mangrove Conservation and Restoration; 2001 October 30 - November 1, 2001. Mimeo. pp. 4

[13] Biswas S. A report on the olive ridley, Lepidocheysolivacia (Eschscholtz) (Testudines: Chelonidae) of Bay of Bengal. Records of the Zoological Survey of India. 1982;**79**:275-302

[14] Choudhury BP. The unique mangrove forest of Bhitarkanika. Orissa Review; 1990

Chapter 16

Use of Phytosociology and Remote Sensing to Classify and Map the Vegetation in Protected Areas, Botswana

Tsholofelo Lori

Abstract

In a natural environment, the vegetation is organized into different plant communities. The vegetation maps produced through phytosociological and remote sensing techniques can be used in the conservation, management, and monitoring of wildlife habitats in protected areas. A desk study was conducted to review studies conducted by various peer-reviewed researchers that used phytosociology and remote sensing methods to classify and map the vegetation in Botswana's protected areas from 2000 to 2021. Seven studies were carried out in the last two decades, and four out of these studies were conducted in Northern Botswana. Even though a variety of satellite imagery was used, Landsat was the most commonly used. Maximum-likelihood supervised classification and random forest were the most common classification methods used to classify and map the vegetation. Vegetation maps are crucial in knowing which plant species occur in which protected areas, and they are used to manage effectively the vegetation in protected areas. It is important to incorporate phytosociology and remote sensing technology with the management of protected areas to conserve effectively and monitor the vegetation in these areas.

Keywords: phytosociology, remote sensing, protected areas, plant communities, classification, vegetation map, conservation, Botswana

1. Introduction

Vegetation is organized into different plant communities in a natural environment. According to Brown et al. [1], "vegetation is a collective term for all the plant communities." Clements [2] describes a plant community as a discrete and natural organism, whereas Gleason [3] states that a plant community is a collection of individual plants. It is important to integrate phytosociology with remote sensing when mapping the vegetation in protected areas. Phytosociology is a subsection of vegetation science, that focuses on existing plant communities and emphasizes their classification [4]. It concentrates on classifying plant communities based on their species composition and how different plant species relate to each other [5]. During the era of climate change, phytosociological studies are more crucial and necessary in the conservation of plant communities as well as in understanding the past and future changes occurring to these plant communities since in most cases, only vegetation data are accessible for comparisons [1, 6]. Computer technology has allowed the improvement of new methods to semi-automatically classify big datasets of vegetation and this has removed vegetation classification from just assigning the vegetation types to more organized data analysis [7]. Plant ecologists had generally agreed that the vegetation consisted of natural plant communities, which can be recognized as distinct formations with real boundaries [8]. Modern remote sensing products are likely to offer much more thorough arrangements of plant diversity than maps drawn by experts that subjectively assigned vegetation types in the olden days [7].

The classification, description, and mapping of plant communities are the important initial steps in constructing a basis in understanding, protecting, conserving, and management of natural resources in protected areas [9]. The International Union for the Conservation of Nature (IUCN) defines a protected area as "a clearly defined geographical space, recognized, dedicated and managed, through legal or other effective means, to achieve the long-term conservation of nature with associated ecosystem services and cultural values" [10]. Even though most of the protected areas are located in very remote areas, it is very crucial to manage and monitor the vegetation in these areas. Field-based sampling using phytosociological methods for monitoring the vegetation in these remote areas is logistically challenging, costly, labor-intensive, and time consuming [11, 12]. In contrast, remote sensing monitoring is cheap, requires less labor, and is more objective than field-based methods, and it allows mapping of the vegetation in the remote areas to be efficient, effective, and economical [7, 11, 12]. Remote sensing in savanna landscapes is complicated because the landscapes are heterogeneous and there is a likelihood of spectral confusion between a shrub and a tree [13].

The vegetation maps produced through phytosociological and remote sensing techniques can be used in conservation and monitoring of wildlife habitats in protected areas. Vegetation maps are crucial in knowing which plant species occur in which protected areas, and they are used to effectively manage the vegetation in protected areas. Furthermore, vegetation maps are important in defining seasonal habitat use of collared wild animals, which cannot easily be tracked in huge wilderness areas with little road access more especially in Northern Botswana [14]. Mosugelo et al. [15] performed 36-year study on vegetation changes in Chobe National Park and they found that the reduction of woodland cover near Chobe river could be due to heavy browsing by elephants and impala in dry seasons. Still in Chobe National Park, Herrero et al. [13] found that increased elephant population has increased the amount of degradation in the riverfront area. The aim of this chapter is to review the phytosociological and remote sensing methods used by various peer-reviewed researchers to produce vegetation maps in Botswana's protected areas. The literature for these studies is from 2000 to 2020. It is important to conduct a local review because it can give details on the main concerns and monitoring methods of protected areas in different environments together with providing specific information on the management of each protected area [12]. The current review focuses on information concerning the location of the study area, the study aim, satellite imagery used, and the classification method used to map the vegetation in each protected area.

2. Protected areas in Botswana

Botswana is a landlocked country located in Southern Africa and shares borders with South Africa, Namibia, Zimbabwe, and Zambia. There are 22 protected areas in Botswana [16]. A total of 245, 244 km^2 of Botswana (over 37%) is committed to the conservation of wildlife, with >17% of the country being designated as protected national parks and game reserves, and 20% is utilized as wildlife management areas [17]. "Protecting such large areas of pristine wilderness across a wide variety of habitats has ensured that much of the biodiversity within Botswana is intact" [17]. Out of the 22 protected areas, there are 3 national parks, 1 transfrontier park, 7 game reserves, 6 forests reserves (located in Chobe District), and 4 sanctuaries in Botswana. **Table 1** gives the names of protected areas found in Botswana, their sizes, and the years in which they were declared protected areas. These areas comprise national parks, game reserves, forest reserves, and sanctuaries (**Figure 1**). The Botswana National Conservation Strategy was developed in 1990 because the Botswana government acknowledged the importance of its natural resources and the goal of the strategy is sustainable development and conservation of natural resources [17]. According to DWNP [22], there is a policy framework in place which guides the management of the national parks and game reserves and this is done through the Wildlife Conservation Policy of 1986, the Tourism Policy of 1990, and National Development Plan No. 9 of 2003, whereas the Wildlife Conservation and National Parks Act of 1992 and National Parks and Game Reserves Regulations of 2000 provide the legislation. The Ministry of Environment, Natural Resources and Tourism (MENT), through the Department of Wildlife and National Parks (DWNP), is responsible for the management of protected areas in Botswana.

Protected areas in Botswana		Size (km^2)	Declaration Year
National Parks	Chobe National Park	15,400	1960
	Kalahari Transfontier Park	35,551	2000
	Nxai Pan National Park	1500	1971
	Makgadikgadi Pans National Park	1500	1970
Game Reserves	Central Kalahari Game Reserve	52,800	1961
	Khutse Game Reserve	2600	1971
	Moremi Game Reserve	4871	1963
	Gaborone Game Reserve	3	1980
	Nnywane Dam Game Reserve	10	1969
	Mannyelanong Game Reserve	3	1985
	Northern Tuli Game Reserve	780	1964
Forest Reserves	Chobe Forest Reserve	1432	1976
	Maikaelelo Forest Reserve	543	1981
	Kasane Forest Reserve	149	1968
	Kasane Forest Extension	641	1981
	Kazuma Forest Reserve	195	1981
	Sibuyu Forest Reserve	1194	1981
Sanctuaries	Maun Game Sanctuary	85	1975
	Nata Bird Sanctuary	961	1993
	Mogobane Bird Sanctuary	9	1940
	Bathoen Dam Bird Sanctuary	5	1992
World Heritage and Ramsar Site	Okavango Delta System	55,374	1996

Table 1.
Protected areas in Botswana [18–20].

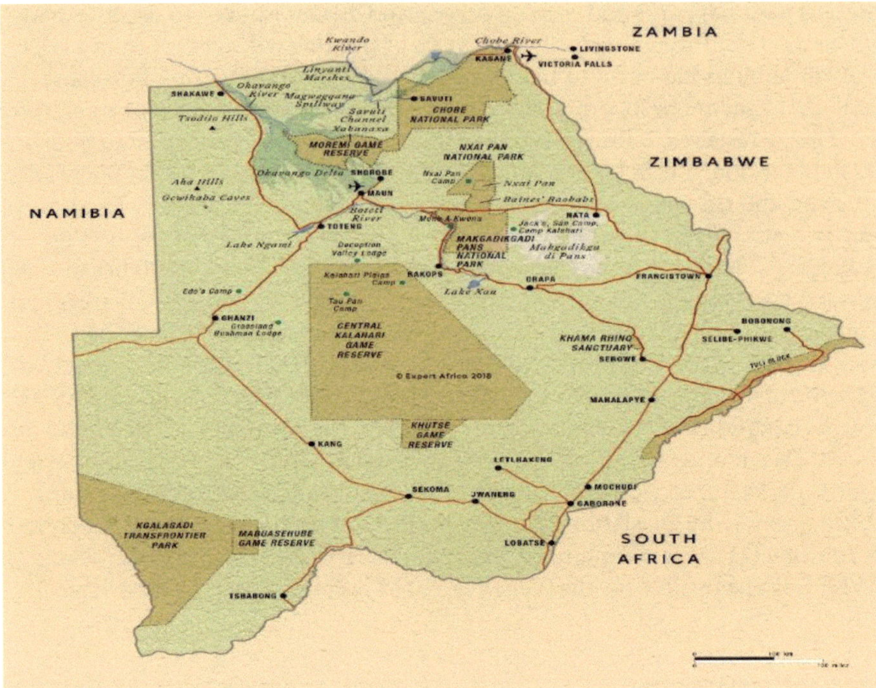

Figure 1.
Map of main national parks and game reserves in Botswana [21].

Its sister department, the Department of Forestry and Range Resources (DFRR) is responsible for the management and conservation of forests through Forest Act 1968, Forest Reserves and State Land, Herbage Preservation Act, and Forest Policy 2011 [18]. In addition to these acts and policies, there are management plans of the protected areas, which offer guidance in their management.

Chobe National Park is considered one of the most important national parks in Africa [23] and it hosts the largest elephant (*Loxodonta africana*) population in Africa. Makgadikgadi Pans National Park is located in northeastern Botswana and it contains pans that host one of the most important breeding sites for flamingos. Nxai Pan National is found on the northern side of Makgadikgadi Pans National Park. Central Kalahari Game Reserve (CKGR) is the largest game reserve in Botswana which is located in Ghanzi District and it shares the border with Khutse Game Reserve that is in Kweneng District. Moremi Game Reserve is the second largest game reserve and it is found in Ngamiland District. The Kgalagadi Transfrontier Park (KTP) is the first transboundary park in Africa and is located between Botswana and South Africa. It was formed by the amalgamation of the former Kalahari Gemsbok National Park (proclaimed in 1931) in South Africa and the Gemsbok National Park (proclaimed in 1971) in Botswana [24]. In addition to the protected areas, there are wildlife management areas surrounding the protected areas and private game reserves around the country. Non-consumptive utilization of wildlife is permitted in the protected areas, whereas both sustainable consumptive and non-consumptive utilization of wildlife are allowed in the wildlife management areas.

3. Study approach

This chapter presents a desk study that was conducted to review studies that used phytosociological and remote sensing methods to classify and map the vegetation in the protected areas in Botswana. Phytosociological methods include going to the field to study and collect vegetation data, whereas remote sensing methods involve using satellite imagery to study and map the vegetation. The current study used Google, ScienceDirect, and Web of Science to search for keywords such as phytosociology, remote sensing, national park, game reserve, plant community, classification, mapping, conservation, protected areas, Botswana. English literature published from 2000 to 2020 from peer-reviewed journal articles, books, edited book chapters, electronic academic thesis, and technical reports were selected for review. The full texts of the studies were downloaded, and the information on the study area, study objective, satellite imagery used, and classification type used to map the vegetation was extracted.

4. Vegetation description, classification, and mapping

According to this review, seven vegetation description, classification, and mapping studies have been conducted in Botswana's protected areas in the last two decades. Most of the studies were carried out in Northern Botswana. The results of the review on the phytosociological and remote sensing methods used by researchers to produce vegetation maps in Botswana's protected are summarized in **Table 2**. The table provides information on the study area, satellite imagery used classification method, and the reference of the researchers who conducted the studies. Van Rooyen [25] used Landsat ETM+ to classify and map the entire Kgalagadi Transfrontier Park (KTP). This produced a vegetation map consisting of 13 major plant communities that were found on Botswana side of the KTP (**Figure 2**). The study found that the vegetation varies from open to dense tree savanna.

In Chobe National Park, Herrero et al. [13] mapped vegetation changes in Chobe riverfront using Landsat TM and AVHRR. The study used random forest because it is a good classification method in spatially and temporally complex heterogeneous savanna landscapes [13]. The overall classification accuracy was 79.8% for 1989–1990 and 78.5% for 2008–2009 Fox et al. [26] used Landsat 5TM, 7ETM+, and 8OLI to study land cover change (LCC) in Northern Botswana which included Chobe National Park and the six forest reserves. The study found that LCC processes in semi-arid savannas in Southern Africa are influenced by environmental and anthropogenic factors. Interactive self-organizing (ISO) clustering was the classification

Study area	Satellite imagery	Classification method	Source
Kalahari Transfontier Park	Landsat ETM+		[25]
Chobe National Park	Landsat TM & AVHRR	Random Forest	[13]
Savuti-Mababe-Linyanti	RapidEye & Landsat	Maximum Likelihood	[14]
Northern Botswana	Landsat 5TM, 7ETM+, 8OLI	ISO Clustering	[26]
Kasane Forest Reserve	Landsat 5TM	Support Vector Machine	[27]
Central Kalahari Game Reserve	MODIS	Random Forest	[28]
Khutse Game Reserve	Sentinel-2A	Maximum Likelihood	[30]

Notes: AVHRR, Advanced Very High Resolution Radiometer; ETM, Enhanced Thematic Mapper; OLI, Operational Land Imager; ISO, Interactive Self-Organizing; MODIS, Moderate Resolution Imaging Spectroradiometer; TM, Thematic Mapper.

Table 2.
Satellite imagery and classification methods used to map the vegetation in different protected areas in Botswana.

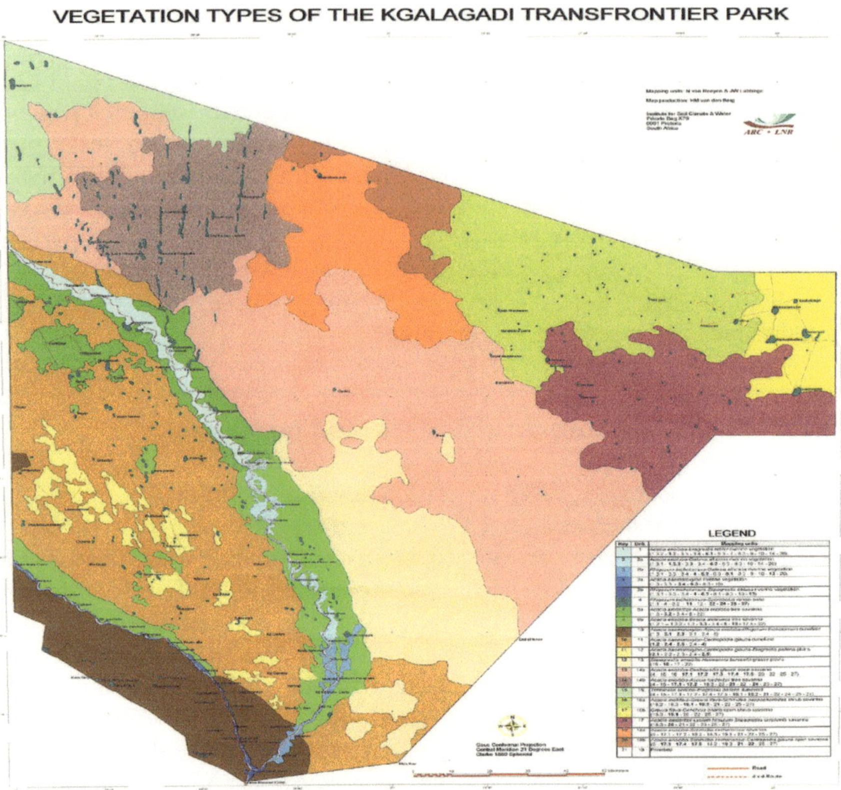

Figure 2.
Vegetation map of Kgalagadi Transfrontier Park [25].

method used resulting in 86.7% overall accuracy and a Kappa coefficient of 0.832, with the highest confusion coming from woodland and shrubland [26]. In Northern Botswana, Sianga and Fynn [14] conducted a study in Savuti-Mababe-Linyanti ecosystem, which also covers Chobe National Park and the forest reserves. The authors used RapidEye &and Landsat to classify and map 15 plant communities in this ecosystem. The study used maximum-likelihood supervised classification and concluded that vegetation map will provide an important database for research in wildlife habitat selection and monitoring of plant communities [14]. Basalumi et al. [27] classified four carbon classes with Landsat 5TM and produced above ground carbon stock map of Kasane Forest Reserve (**Figure 3**). The supervised classification method used was Support Vector Machine and it yielded 97.8% overall classification accuracy. The study suggested that in miombo woodlands, the use of Landsat was ideal for monitoring biomass and carbon stock [27].

Mishra et al. [28] used MODIS to broadly and physiognomically map six vegetation morphology types in Central Kalahari Game Reserve and Khutse Game Reserve. The random forest classification method was used for this study and overall accuracy was 91.9% and Kappa coefficient was 0.88. Lori et al. [29] classified and described nine plant communities in Khutse Game Reserve. Lori [30] has the details of this study which include the mapping of these plant communities using Sentinel-2A imagery (**Figures 4** and 5). **Figure 6** shows one of these nine plant communities, that is, *Heliotropium lineare-Enneapogon desvauxii* community. Maximum-likelihood supervised classification method resulted in overall classification accuracy of 61.67% and overall Kappa coefficient of 58.18%. The heterogeneous savanna vegetation in the study area might have contributed to the optimal

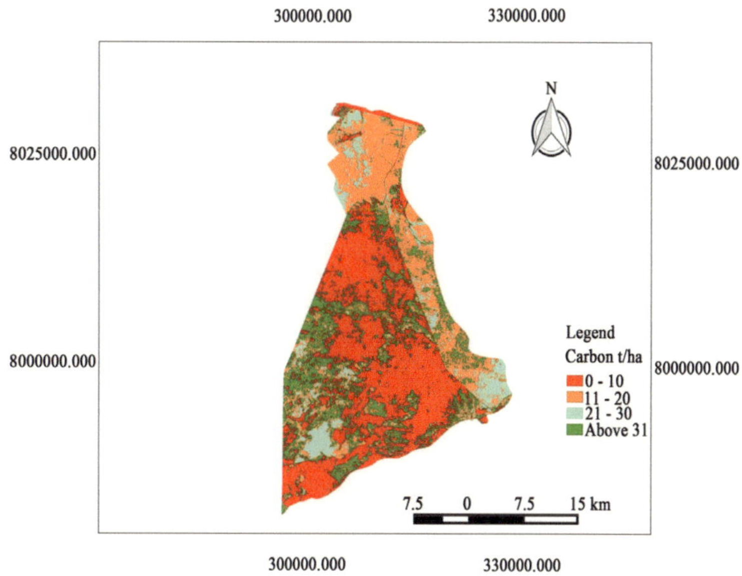

Figure 3.
Above ground carbon stock map of Kasane Forest reserve [27].

Figure 4.
Sentinel-2A natural color RGB (red, green, and blue) imagery with red squares representing the sampling plots in Khutse Game Reserve [30].

overall accuracy and medium Kappa value [30]. This study differs from the one by Mishra et al. [28] because it used Sentinel-2A imagery that has a high spatial scale (i.e., 10 m) to indicate fine-scale spatial heterogeneity of the area, as compared to MODIS with a low spatial resolution (i.e., 232 m) [28].

In this review, Landsat satellite imagery was the most commonly used. This might be due to the fact that Landsat is the most advanced, free, and easy to access online. The results indicate that maximum-likelihood supervised classification and random forest were the most common classification methods used to classify and map the vegetation and each of the seven studies used different satellite imagery. The results show that there is still a lot that needs to be done in terms of mapping

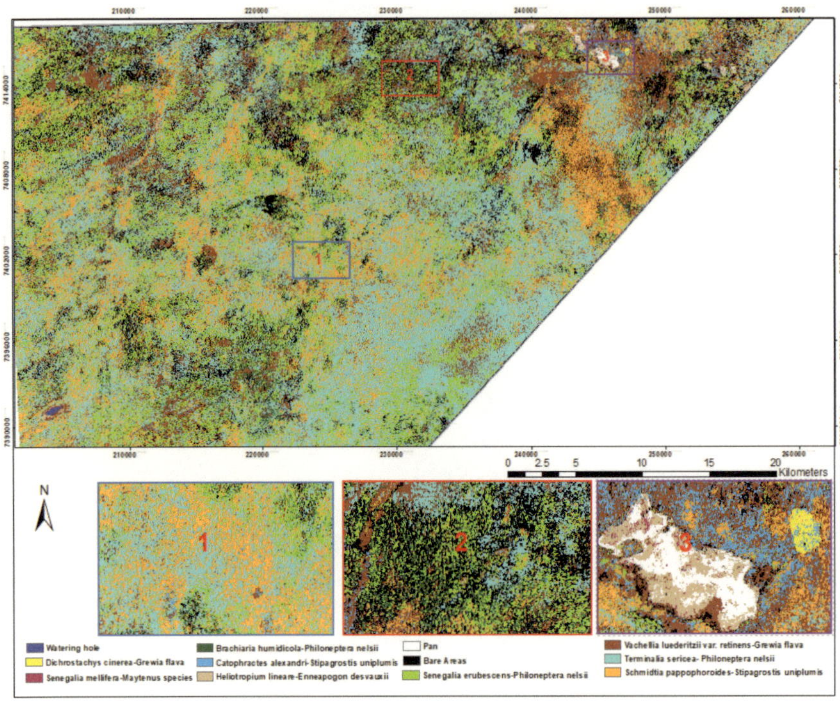

Figure 5.
Plant community map of Khutse game reserve [30].

Figure 6.
A pan habitat consisting of Heliotropium lineare-Enneapogon desvauxii plant community in Khutse Game Reserve. Photo credit: Tsholofelo Lori.

and monitoring vegetation in Botswana's protected areas using remote sensing. Even though different researchers use different satellites with different spatial resolutions, there is a general agreement in methods used between different studies in remote sensing of protected areas in Botswana.

5. Conclusion

A review of the literature on the phytosociological and remote sensing methods used by researchers to produce vegetation maps in Botswana's protected areas was performed and found that there is still a lot that needs to be done in terms of producing up-to-date vegetation maps for the protected areas in Botswana. There is currently a limited number of published works on the use of remote sensing data to map the vegetation in the protected areas. Due to their remoteness, some protected areas in the country are still understudied and there is a lack of in situ vegetation data for these areas. Vegetation classification and mapping are crucial because the vegetation maps can be used to detect vegetation change over time caused by climate change. Researchers used similar methods in remote sensing of the protected areas in Botswana. It is recommended that remote sensing technology should be incorporated with the management of protected areas to effectively conserve and monitor the vegetation in these areas. Research institutions with resources and capacity should work closely with the Ministry of Environment, Natural Resources and Tourism on remote sensing of vegetation in the protected areas.

Acknowledgements

The author appreciates and thanks an anonymous person for lending her an Internet modem which enabled her to write this book chapter. The author is also grateful to an anonymous reviewer for his valuable encouragement and comments.

Conflict of interest

The author declares no conflict of interest.

Author details

Tsholofelo Lori
Hope Nature Conservation Centre, Molepolole, Botswana

*Address all correspondence to: tshololori@gmail.com

References

[1] Brown LR, du Preez PJ, Bezuidenhout H, Bredenkamp GJ, THC M, Collins NC. Guidelines for phytosociological classifications and descriptions of vegetation in southern Africa. Koedoe. 2013;**55**(1) #1103;10 p. DOI: 10.4102/koedoe.v55i1.1103

[2] Clements FE. Plant Succession: An Analysis of the Development of Vegetation. Washington, DC, US: Carnegie Institute; 1916 Publication #242

[3] Gleason HA. The individualistic concept of plant association. Bulletin of the Torrey Botanical Club. 1926;**53**:7-26

[4] Dengler J, Chytrý M, Ewald J. Phytosociology. In: Jorgensen SE, Fath BD, editors. Encyclopedia of Ecology. 2nd ed. Oxford: Elsevier; 2008. p. 516-527. DOI: 10.1016/ B978-0-444- 63768-0.00533-3

[5] Ismail IM, Elawad A. Phytosociological analysis and species diversity of herbaceous layer in Rashad and Alabassia localities, South Kordofan State, Sudan. Jordan Journal of Biological Sciences. 2015;**8**(2):151-157

[6] Qose A, Proko A. Remote Sensing techniques and phytosociological methods on the inventory of medicinal plants (Case study on the Skrapari's Municipality). Albanian Journal of Agricultural Sciences. 2018:604-620

[7] Revermann R, Finckh M. Vegetation Survey, Classification and Mapping in Angola. In: Huntley BJ, Russo V, Lages F, Ferrand N, editors. Biodiversity of Angola Science & Conservation: A Modern Synthesis. Vol. 2019. Switzerland AG: Springer Nature; 2019. pp. 97-108. DOI: 10.1007/978-3-030-03083

[8] Nicolson M. Community concepts in plant ecology: from Humboldtian plant geography to the super organism and beyond. Web Ecology. 2013;**13**:95-102

[9] Van Rooyen MW, van Rooyen N, Jacobus du P, van den Berg HM. Landscapes in the Kalahari Gemsbok National Park, South Africa. Koedoe. 2008;50(1):99-112. DOI: 10.4102/ koedoe.v50i1.154

[10] Dudley N, Stolton S (editors). Defining protected areas: In: Proceedings of the International Conference in Almeria, Spain. Gland, Switzerland: IUCN; 2008. 220 p. ISBN: 978-2-8317-1132-4

[11] Oldeland J, Dorigo W, Lieckfeld L, Lucieer A, Jürgens N. Combining vegetation indices, constrained ordination and fuzzy classification for mapping semi-natural vegetation units from hyperspectral imagery. Remote Sensing of Environment. 2010;**114**:1155-1166

[12] Mao L, Li M, Shen W. Remote Sensing Applications for Monitoring Terrestrial Protected Areas: Progress in the Last Decade. Sustainability. 2020;**12**:5016. DOI: 10.3390/su12125016

[13] Herrero HV, Southworth J, Bunting E. Utilizing Multiple Lines of Evidence to Determine Landscape Degradation within Protected Area Landscapes: A Case Study of Chobe National Park, Botswana from 1982 to 2011. Remote Sensing. 2016;**8**(623):1-17. DOI: 10.3390/rs8080623

[14] Sianga K, Fynn R. The vegetation and wildlife habitats of the Savuti-Mababe-Linyanti ecosystem, northern Botswana. Koedoe. 2017;**59**(2):1-16. DOI: 10.4102/koedoe. v59i2.1406

[15] Mosugelo DK, Moe SR, Ringrose S, Nellemann C. Vegetation changes during a 36-year period in Northern Chobe National Park, Botswana. African Journal of Ecology. 2002;**40**:232-240

[16] IUCN ESARO. The state of protected and conserved areas in Eastern and Southern Africa. In: , State of Protected and Conserved Areas Report Series No. 1. Nairobi, Kenya: IUCN ESARO; 2020

[17] Maude G, Reading R. The role of ecotourism in biodiversity and grassland conservation in Botswana Great Plains Research: Paper 1077. A Journal of Natural and Social Sciences. 2010;**20**(1):109-119

[18] Centre for Applied Research. Forest management and use in Botswana: brief situation analysis and options for the Forest Conservation Strategy. Forest Conservation Botswana; Gaborone, Botswana; 2013. p. 36

[19] Campbell A. Establishment of Botswana's National Park and Game Reserve System. Botswana Notes and Records. 2004;**36**:55-66

[20] UNEP-WCMC Protected Area Profile for Botswana from the World Database of Protected Areas [Internet]. June 2021. Available from: www.protectedplanet.net [Accessed: 23-06-2021]

[21] Expert Africa [Internet]. 2018. Available from: https://www.expertafrica.com/botswana/reference-map [Accessed: 23-06-2021]

[22] DWNP. Central Kalahari Game Reserve and Khutse Game Reserve Management Plan. Gaborone, Botswana: Department of Wildlife and National Parks; 2003 132 p

[23] Nellis MD, Bussing CE. Spatial variation in elephant impact on the Zambezi Teak Forest in the Chobe National Park, Botswana. Geocarto International. 1990;**2**:55-57

[24] Thondhlana G, Shackleton S, Muchapondwa E. Kgalagadi Transfrontier Park and its land claimants: a pre- and post-land claim conservation and development history. Environmental Research Letters. 2011;**6**;024009.12p. DOI: 10.1088/1748-9326/6/2/024009

[25] Van Rooyen N. Vegetation survey of the Gemsbok National Park, Botswana and mapping of the Kgalagadi Transfrontier Park. Stellenbosch: Peace Parks Foundation; 2000 Final Project Report. Project number PPF/P/23

[26] Fox JT, Vandewalle ME, Alexander KA. Land Cover Change in Northern Botswana: The Influence of Climate, Fire, and Elephants on Semi-Arid Savanna Woodlands. Land. 2017;**6**(73):1-23. DOI: 10.3390/land6040073

[27] Basalumi L, Kilawe CJ, Mauya EW. Linking Ground Forest Inventory and NDVI in Mapping above Ground Carbon Stock in Kasane Forest Reserve, Botswana. Open Journal of Forestry. 2018;**8**:429-438. DOI: 10.4236/ojf.2018.83027

[28] Mishra NB, Crews K, Miller J, Meyer T. Mapping vegetation morphology types in southern Africa savanna using MODIS time-series metrics: a case study of central Kalahari, Botswana. Land. 2015;**4**(1):197-215

[29] Lori T, Ditlhogo MK, Setshogo MP, Koosaletse-Mswela P. Classification, description and mapping of the vegetation in Khutse Game Reserve, Botswana. Botswana Journal of Agriculture and Applied Sciences. 2019;**13**(2):8-23. DOI: 10.37106/bojaas.2019.45

[30] Lori T. Classification, description and mapping of the vegetation in Khutse Game Reserve, Botswana [PhD Thesis]. Gaborone: University of Botswana; 2019

Chapter 17

Human Settlement Encroachment in Kainji Lake National Park, Nigeria

Adelakun Kehinde Moruff, Shafiu Kilishi Halidu, Azeez Olalekan Ibrahim and Olorunfemi Boye Oyediran

Abstract

Land encroachment is severely degrading and destroying many of Nigerian protected area as a result of high population pressure caused by high population growth and immigrations trends. Therefore, this study aimed at assessing human settlement encroachment in Kainji Lake National Park in order to established the extent to which this threat have been upheld and therefore call for park management to seek for proper approaches to deal with it. Data for this study were collected through the administration of questionnaire to the villages adjacent to Borgu sector of Kainji Lake National Park. Seven (7) randomly selected villages, namely Luma, Kuble, Audu Fari, Kali, Malale, Leshegbe and Gada Oli were surveyed. The result indicates gender of respondents where 65.09% of the responses were received from males while 34.91% from females suggesting that responses are the views from both gender parties who are mostly (33.96%) within the age range of 31–40 years and are predominantly farmer (43.42%). 88% of the respondents' indicated that increased need of land for developments are the major causes of encroachment around the park while other human activities such as farming, deforestation and grazing of domesticated animals, by encroaching to protected areas have led into competition over natural resources. It is also established by 92% of the respondents agree that migration of people for livelihood support has led to increased competition between people and wildlife hence, this increase in population has consequently led to encroachment into the park as attested by a whopping 100% of the respondents. Kainji Lake National Park management need to do more in sensitizing the local communities on importance of wildlife conservation as most (54%) locals disagreed to awareness of conservation education and engages in intense vigilance against encroachment into the park land.

Keywords: encroachment, human activities, population pressure, protected area, resources

1. Introduction

Protected areas have long been recognized as the single most important method of conserving wildlife and preserving biological diversity [1]. They protect the fertility and stability of soils, play a key role in watershed management, and are the habitats of countless species of wildlife. Hence, it is important as a result of its significant contribution to economic and social status of their host country. Protected areas are popularly associated with large areas of 'undisturbed wilderness' [2]. Many of these protected areas including the parks and game reserves especially in the developing countries are affected by degradation of the ecosystem which involves hunting, logging, livestock keeping, cultivation, wildfire and this has led to establishment of conservation programmes for sustainable management of these protected areas [3]. Another view sees protected areas as social spaces; that is, they are socially conceived and preserved [2]. For instance National park concept involves the exclusion of people from wildlife areas apart from visitors and employees concerned with management [4], it also conserve many of the world's habitats and species. Despite the high productivity of National Parks, and provision of many benefits, it has been found that these protected areas natural features have been destroyed everyday as a result of encroachment [5, 6]. Human encroachment, especially in the tropics, is severely degrading and destroying many of these areas [7] as a result of high population pressure caused by high population growth and immigrations trends [3].

Encroachment on public property is defined as:" the existence of any structure or item of any kind under, upon, in, or over the project lands or waters and/or the destruction, injury, defacement, removal or any alteration of public property including natural formations, historical and archeological features, and vegetative growth [8]. It also "denotes an illegal activity as one where the person who encroaches is not deemed to have any legal right to do so" [9]. The above two definitions suggest that encroachment results when there is an unlawful activity/entry on forest (gradually and without permission).

Encroachment in the protected areas is one of the major causes of degradation of ecosystem in many parts of the world [10]. Human encroachment into wildlife areas, which has increased almost exponentially over the past few decades, has usually resulted in the elimination of the larger species, particularly the large mammals (e.g. [4]).

Destruction of wildlife habitats through human encroachment has remained the leading threat to biodiversity. This destruction, taking different forms, for example degradation, fragmentation or outright loss, is a function of the growing human activities prompted mainly by such factors as poverty, demographic factors, land tenure systems, inadequate conservation status, development policies and economic incentives [1].

The park could be subjected to encroachment through physical development which poses problems to sustainable resource utilization [11] and this may be connected to their importance to the livelihoods of local communities, especially indigenous people who live and/or depend on the resources available in the park for their survival [12]. However, the problem of encroachment caused by economic development and other human activities will exert pressure on biodiversity, resulting in the interference in the wildlife management approaches and make it difficult to protect Nigerian National Parks.

Therefore, this study assessed human settlement encroachment in Kainji Lake National Park in order to established the extent to which this threat have been upheld and therefore call for park management to seek for proper approaches to deal with it.

2. Materials and method

2.1 Study area

Kainji Lake National Park is located in the North central part of the country lies latitude 9'45 and 10'23 N and longitude 3'40 and 5'47E. It is made up of

Villages	Population size (households)	Sample size (10% of each households)
Luma	365	37
Kuble	50	5
Audu fari	66	7
Kali	71	7
Malale	308	31
Leshegbe	70	7
Gada oli	139	14
Total	1069	108

Source: Modified from [14] report.

Table 1.
Sampling population and size.

two sectors (Borgu and Zugurma) situated in Borgu and Kaima/Baruten Local Government Area of Niger and Kwara State respectively. It covers a total land area of 5,340.825q [13].

2.2 Method of data collection

2.2.1 Sampling techniques

The administration of questionnaire for this study was restricted to the villages adjacent to Borgu sector of Kainji Lake National Park. Seven (7) randomly selected villages, namely Luma, Kuble, Audu Fari, Kali, Malale, Leshegbe and Gada Oli were survey.

2.2.2 Household questionnaire and interview survey

To obtain information questionnaire will be prepare to correspond all the aspects of the study. Interviewing method will be used to collect information. Randomness will also be strictly ensured for better output (**Table 1**).

2.2.3 Data processing and analysis

Available data were processed, analyzed using Special Package for Social Science (SPSS 17) and interpret to find the result of the study. After completion of data collection the responses to the questions of livelihoods in the study area were transferred to a master sheet to facilitate tabulation. The analyzed data were then represented through tabular and graphical form.

3. Results

3.1 Demographic characteristic of the respondents

Table 2 indicates gender of respondents where 65.09% of the responses were received from males while 34.91% of the responses were from females suggesting that responses are the views from both gender parties. The age groups of respondents fall between $\leq$ 20 years with 3.77%, 21–30 years with 40.57%, 31–40 years with 33.96% and $\geq$ 40 are 21.70%.

Demographic	Categories	Frequency	Percentage (%)
Sex	Male	69	65.09
	Female	37	34.91
	Total	106	100
Age	≤ 20	4	3.77
	21–30	43	40.57
	31–40	36	33.96
	≥ 40	23	21.70
	Total	106	100
Year of residing in the area	1–10	56	52.83
	11–20	20	18.87
	21–30	13	12.26
	31–40	8	7.55
	40 Above	9	8.49
	Total	106	100
Major occupation	Civil servant	9	11.84
	Farming	33	43.42
	Trading	15	19.74
	Artisan	4	5.26
	Fishing	15	19.74
	Total	76	100

Table 2.
Demographic characteristic of the respondents.

Majority (52.83%) of these people have been residing in these areas between 1 to 10 years while small fractions of 7.55% of the respondents were occupant for 31–40 years. The bigger (43.42%) percentages of respondents was farmers, and therefore are likely to require land for settlement and agriculture. Other occupations prominent in the area are fishing (19.74%), trading (19.74%), civil service (11.84%) and artisan (5.26%).

3.2 Causes of human encroachment into wildlife corridors

From the **Table 3**, 88% of the respondents' indicated that increased need of land for developments are the major causes of encroachment around the park. It is also revealed that 74% of the respondents agreed that human activities such as farming, deforestation and grazing of domesticated animals, by encroaching to protected areas have led into competition over natural resources. The study also indicates that most respondents 83% do agree that natural factor like drought had led to encroachment in the study area. All these are indicators that there exist encroachment activities around Kainji Lake National Park as a result of competition for limited resources (**Table 3**).

3.3 Perceived indicators of human population increased and its impact on human-wildlife conflicts in wildlife corridors

Table 4 shows parameters that were used to evaluate indicators of human population increased and its impact on wildlife corridors. The result shows that

Variable	Strongly agreed		Agreed		Neutral		Disagreed		Strongly disagreed		Means	Std. Dev.
	f	%	F	%	F	%	F	%	f	%		
Need of land for human development	63	59	31	29	0	0	8	8	4	4	4.52	0.412
Impact of human activities such as farming, deforestation and grazing	48	45	31	29	19	18	0	0	8	8	4.54	0.408
Search of water for domestic purpose	27	26	38	36	27	26	9	9	5	5	4.05	0.090
Natural factor like drought that push human to wildlife corridor	4	4	36	34	13	12	8	9	41	42	2.50	1.416

Table 3.
Causes of human encroachment into wildlife corridors.

Variable	Strongly agreed		Agreed		Neutral		Disagreed		Strongly disagreed		Means	Std. Dev.
	F	%	F	%	F	%	F	%	F	%		
Migration of people for livelihood	67	63	31	29	0	0	8	8	0	0	4.45	0.454
Encroachment into protected area	106	100	0	0	0	0	0	0	0	0	5.00	0.000
Emergence of towns and trading centre next to park	98	92	8	8	0	0	0	0	0	0	4.93	0.090
Difficult to catch up with poachers whenever they strike due to high population	0	0	35	33	8	8	10	9	53	50	2.24	1.362

Table 4.
Perceived indicators of human population increased and its impact in Kainji Lake National Park.

human population contributes a lot in competition of resources between human beings and wildlife. 92% of the respondents agree that migration of people for livelihood support has led to increased competition between people and wildlife. The increase population has also led to encroachment into protected area is factor

Variable	Strongly agreed		Agreed		Neutral		Disagreed		Strongly disagreed		Means	Std. Dev.
	F	%	f	%	F	%	F	%	F	%		
Community awareness and education	12	12	16	15	20	19	5	5	52	49	2.37	1.51
K.L.N.P has developed voluntary relocation program for affected people	14	13	32	30	14	13	9	9	37	35	2.78	1.51
There is intense human vigilance by K.L.N.P ranger against attack by wild animals.	0	0	47	49	10	10	15	16	24	25	2.17	1.28
K.L.N.P has intensified its fencing to bar wild animals from freely moving to human habitat	0	0	10	9	24	23	5	5	67	63	1.78	1.10
K.L.N.P has corporate social responsibility/ community enterprise for the to prevent encroachment	71	100	0	0	0	0	0	0	0	0	5.00	0.000

Table 5.
Approaches in management of human encroachment in Kainji Lake National Park.

that a whopping 100% of the respondents agreed. Emergence of trading centres always attracts people close to it for essential services and as per the findings; most people (92%) agree that the growth of the centres next to Kainji Lake National Park has contributed to wildlife management challenges. Catching poachers in a crowded area has not pose a serious challenge in the study area as 50% and 9% of respondents strongly disagreed and disagreed respectively with any difficulty in this regard.

3.4 Approaches in management of human encroachment in Kainji Lake National Park

The results in **Table 5** show that KLNP need to do more in sensitizing the local communities on importance of wildlife conservation as most (54%) locals disagreed to awareness of conservation education.

Most respondents also disagreed and strongly disagreed (44%) to any voluntary relocation programmes for the affected people by the Park and 49% agreed to intense vigilance against encroachment into the park land though 100% respondents strongly agreed that KLNP has corporate responsibility to the affected communities.

4. Discussion

From the study, majority of settlers around Kainji Lake National Park are farmers, and therefore are likely to require land for settlement and agriculture may encroaches into protected area. Socio-economic factors has compel people to abuse the use of National Parks Adelakun et al. [15] and this may result to conflict because of the human overlap with wildlife requirements resulting in costs to both native residents and animals [16].

Increased need of land for developments are the major conflict agent between human and animal as well human activities such as farming, deforestation and grazing of domesticated animals, by encroaching to protected areas have led into competition over natural resources. This is consensus with Kate [17] who reported that human activities such as farming infrastructure development and tourism can radically alter wildlife habitat.

Increased human population on wildlife corridor and protected areas has influenced conflicts in the study area because conflict is most acute in areas in which a wide range of wildlife species co-exists with high density human populations [18].

Human population increased in wildlife corridors had earlier being envisaged [19]; that population increase may be witness as result of marital status in the study area and this will mount more pressure on the park resources. This study also reflected that people migration for security reasons as well as emergence of trading centres always attracts people close to it for essential services. This study further revealed that human settlement encroachment contributes a lot in competition of resources between human beings and wildlife hence leading to conservation challenges corroborates the Ijeomah and Akosim [20] who was of the assertion that there is a relationship between population growth and resource conservation.

Approaches in management of human settlement encroachment show that KLNP need to do more in sensitizing the local communities on importance of wildlife conservation as most locals disagreed to awareness of conservation education. This contradicts Akosim et al. [19] who reported that the park authority has expended a great deal of efforts in educating the local residents. Morrison et al. [21] pointed out that conservation strategies can be addressed using the proactive or reactive measures. Proactive measures are the same as preventive measures, these measures are crucial in wildlife conservation, reducing encroachment of park land and coming up with strategies to minimize these challenges. An example of a preventive measure is the education and awareness programs. These strategies increase the tolerance level towards wildlife, and can help improve the resource conservation. Studies have earlier shown that when fringe communities of protected areas are forced to absorb living with wildlife, local support for conservation may be seriously undermined [22]. In another report, Muller and Albers [23] confirms ecologically valued lands as economically valuable and so in the absence of development interventions that would provide the residents with alternative means of livelihood, illegal activities, which undermine wildlife conservation, would continue.

5. Conclusion

The research shows that Kainji Lake National Park has been encroached due to need for human settlement and farming which subsequently lead to wild animal raiding communities. Anthropogenic activities such farming, hunting, and fishing are the main activities responsible for the encroachment in the protected area. Human population contributes a lot in competition of resources between human beings and wildlife as people migrating towards protected area for livelihood support. These have led to increased competition between people and wildlife. Emergence of trading centres such as markets have also attracts people close to park for available and probably affordable essential services which consequently leading to encroachment and hence contributed to wildlife conservation challenges in the study area.

6. Recommendations

The following recommendations were made from the study:

With the manifestation of encroachment activities, Kainji Lake National Park needs to consider reviewing its policy for the minimization of human activities in and around the park; for instance enforcement of regulations and legislation on the safe distance on community settlement from the Park.

National Parks should re-strategies on the new ways of mitigating human encroachment and settlement through surveillance and monitoring of people from illegal entry to the park.

Farmers being the most affected in terms of farming activities in the park, the government should come up with an alternative way of livelihood that suits the farmers living around Kainji Lake National Park to ease competition over resources.

Author details

Adelakun Kehinde Moruff*, Shafiu Kilishi Halidu, Azeez Olalekan Ibrahim and Olorunfemi Boye Oyediran
Federal College of Wildlife Management of Forestry Research Institute of Nigeria, New Bussa, Nigeria

*Address all correspondence to: adelakunkehinde@gmail.com

References

[1] Kideghesho, J. R., Nyahongo, J. W., Hassan, S. N., Tarimo, T. C., and Mbije, N. E. (2006). Factors and ecological impacts of wildlife habitat destruction in the Serengeti ecosystem in northern Tanzania. African Journal of Environmental Assessment and Management, 11, 17-32.

[2] Twyman, C. (2001). Natural resource use and livelihoods in Botswana's wildlife management areas. Applied Geography, 21(1), 45-68.

[3] Olupot, W., and Chapman C. A. (2006). Human Encroachment and Vegation Change in Isolated Forest Reserves, the Case of Bwindi Impenetrable National Park in Uganda., http://chapmanresearch.mcgill.ca/Pdf/206_Olupot%20%20Chapman.pdf

[4] Eltringham, S. (1990). Wildlife carrying capacities in relation to human settlement. Koedoe, 33(2), 87-97.

[5] Binlinla, J. K., Voinov, A. and Oduro, W. (2014). Analysis of human activities in and around protected areas: case of kakum conservation area in Ghana, international journal of biodiversity and conservation Academic journals, vol. 6(7), pp. 541-554.

[6] Kpontsu E.A. (2011). Patterns of Woody Plant Species Richness, Diversity and Structure along a Disturbance Gradient in the Atiwa Range Forest Reserve, Eastern Region, Ghana: Thesis KNUST, Kumasi.

[7] Wells, M. and Brandon, K. (1992). People and Parks: Linking Protected Areas Management with Rural Communities. World Bank/World Wildlife Fund/US Agency for International Development, Washington DC.

[8] Agyen-Brefo, R. (2012). The Effects of Encroachment on Sustainable Public Land Management: A Case Study of the Owabi Catchment Area in Kumasi, KNUST, Ghana.

[9] Shitima, E.M. (2005). Forest Conservation and people's Livelihoods: Explaining Encroachment on Zambia's Protected Forest Landscapes-the Case of Mwekera National Forest, Kitwe, Copper Belt. Trondheim, Norway.

[10] Cavanagh C. J. (2012). Anthropocene Conservation: Governing Enviromental Change, Biodiversity and Local Resistence at Mt Elgon Uganda. Master Thesis. Noragric, UMB.

[11] Biojo Y. (2018). Factors Influencing Human-Wildlife Conflicts in Communities around Kainji Lake National Park, Unpublished HND project, Federal College of Wildlife Management, New Bussa, Nigeria. 68p.

[12] DeFries, R., Rudel, T., Uriarte, M., et al. (2010). Deforestation driven by urban population growth and agricultural trade in the twenty-first century. Nature Geosci 3: 178-18. DOI:10.1038/ngeo756

[13] Ayeni, J. S. O. (2007). Participatory Management Plan in Kainji Lake National Park. ENVIRON-CONSULT: Lagos. 156p.

[14] NPC (2015). Population and Housing Census 2015. Avalaible www.npc.org.ng/npcc_2015

[15] Adelakun K.M., Adeola A.J., Halidu S.K., Bakare K. and Omole E.B. (2015). Socio economic activities of rural community and their perception on conservation in Okomu National Park, Nigeria. Int. J. of Devt. Strat in Hum. Manage. And Soc. Sci. Vol. 5(1): 103-115.

[16] Distefano, E. (2004). Human-Wildlife Conflict Worldwide: Collection

of Case Studies, Analysis of Management Strategies and Good Practices. FAO.

[17] Kate K. (2012). Possible strategies/ practices in reducing wild animal (primate) crop raids in unprotected areas in Hoima District, Uganda. A Report to the PLCG-Uganda.

[18] Ogada, M. O., Woodroffe, R., Oguge, N. O., and Frank, L. G. (2003). Limiting depredation by African carnivores: The role of livestock husbandry. Conservation Biology, 17(6), 1521-1530.

[19] Akosim C., Bode A.S., Kwaga B.T. and Dishan E.E. (2010). Perceptions and involvement of neighbouring communities of Kainji Lake National Park towards the parks conservation programmes. Journal of Research Forestry, Wildlife and Environment. Vol. 2(1):44-59.

[20] Ijeomah, J.U. and Akosim, C. (2000). Elements of Biological Conservation 1 Ed. Trinity Graphics System Limited, Jos, Plateau State, Nigeria.

[21] Morrison, K., Victurine, R. and Mishra, C. (2009). Trans-Links Lessons Learned, Opportunities and Innovations in Human Wildlife Conflict Compensation and Insurance Schemes, Report prepared for WCS Trans Links Program.

[22] Ogra, M. and Badola, R. (2008). Compensating human–wildlife conflict in protected area communities: Ground-level perspectives from Uttarakhand, India. Human Ecology, 36, 717-729.

[23] Muller J, and Albers, H. J. (2001). Enforcement, payments, and development projects near protected areas: how the market setting determines what works where. Resource and Energy Economics 26: 185-204

Chapter 18

Polytrichum formosum and *Vaccinium myrtillus* as Phytoindicators of Pollutants from Long-Range Emissions of Environmentally Important Protected Areas (The Tatra National Park, the Central Western Carpathians, Poland)

Joanna Korzeniowska

Abstract

The study determined the influence of altitude on the content of heavy metals in selected plant species of the Tatra National Park (TNP). The metals (Cd, Cr, Cu, Ni, Pb, and Zn) were identified in two species of plants, i.e., in the moss (*Polytrichum formosum* Hedw.) and in the blueberry leaves (*Vaccinium myrtillus* L.). Plant samples were collected in two test areas every 100 meters of altitude of the area, starting from 1,000 m above sea level in the Lake Morskie Oko test area and from 1,100 m above sea level in the Kasprowy Wierch test area, and ending at 1,400 m above sea level for Lake Morskie Oko and 1,550 m above sea level for Kasprowy Wierch. The two test areas are different from each other in terms of natural and physico-geographical conditions (geological structure, landform, climatic conditions, etc.). The conducted research shows that the content of heavy metals in the studied species of plants increases with the altitude above sea level. Both *P. formosum* Hedw and *V. myrtillus* L. can be good phytoindicators in mountainous areas. In the tested plant species, the contents of heavy metals were also found to be higher than the natural contents, which is most likely related to long-range emission. Long-distance transport of pollutants causes that important natural protected areas, such as the Tatra National Park, are exposed to excessive pollution, including the accumulation of heavy metals in plants.

Keywords: heavy metals, plants, mountains, *P. formosum* moss, *V. myrtillus* blueberry, the Tatras, protected areas

1. Introduction

There are numerous places of natural value in the world. They are often protected areas, such as national parks or nature reserves. However, it should be remembered that such areas are not free from the influence of human activities. In protected areas, humans do not have a direct negative impact on the natural environment, but industrial and automotive emissions reach these areas. These are the so-called long-range emissions, as a result of which fine dust particles are transported over distances of several hundred kilometers [1–3]. Dust particles contain heavy metals, the spreading of which over long distances is related to the long duration of dust pollution in the atmosphere. The length of time during which dust particles remain in the atmosphere depends on the particle size, terrain configuration, and meteorological conditions. Low pressure, strong wind, significant cloud cover, and high precipitation contribute to the spreading of pollutants over long distances [2]. The length of time during which heavy metals remain in the environment differs for individual metals. Lead and cadmium are metals that remain in the atmosphere for a long time and are characterized by the very small diameter of their particles. They are easily transported over long distances and, therefore, contaminate the environment on a global scale [2, 3].

As a result of the transport of pollutants over long distances, areas considered to be of natural value and protected may have a problem with an increased amount of metals in soil or vegetation. An example of such an area is the Tatra National Park, where we are dealing with long-range emissions. The Tatra National Park is one of 23 national parks in Poland. It has the highest regime of all forms of nature protection in Poland. Together with the Slovak part of the Tatra National Park (Tatranský národný park), it constitutes a UNESCO biosphere reserve. Its natural value is evidenced by the fact that for many centuries a large part of this area has not been directly changed by man, and this condition has continued to this day. The small area of the national park (211 km^2) contains a wealth of flora and fauna, often endemic and relict species, as well as a variety of landscapes. This is the result of, among other things, the specific geographic location of the Tatra Mountains in Europe, which is influenced by, among other factors, transitional climate and overlapping ranges of various flora and fauna. Although human pressure on the natural environment of this area concerns the in situ impacts related to tourism, the entire area of the national park is affected by external influences.

Pollutants from the Czech Republic, Slovakia, and the Silesia region are transported to the TNP area, where they fall and cause increased metal content in soils and plants. In Slovakia and the Czech Republic, the industrial sector is dominated by metallurgy, and chemical, defense, electrical, and electronic industries, as well as by the production of aluminum, nickel, and copper. These industries are a source of heavy metals in the natural environment.

The metals particularly dangerous to living organisms include cadmium, chromium, copper, nickel, lead, and zinc. Currently, the content of these metals in soils and plants is higher than their natural content [4, 5]. The excessive amount of the mentioned heavy metals in the natural environment results from industrial activity and motorisation, and leads to irreversible changes in ecosystems [6, 7].

The phytoindication method is commonly used in environmental monitoring. Technical monitoring is reduced to the direct measurement of pollutants, while biomonitoring supplements technical monitoring and can be carried out within any number of stations. It provides direct information on the level of soil and plant contamination. In this study, the monitoring of contamination with metals was based on two plant species commonly found in the TNP (the European blueberry *V. myrtillus* L. and the moss *P. formosum* Hedw.).

Species recommended as bioindicators (phytoindicators) should meet, inter alia, the following criteria:

- show strong or selective accumulative properties,
- occur in large populations in various habitats of the natural environment [4].

The selection of indicator plants was based primarily on the prevalence of given species in the Tatra National Park, followed by the criterion of their ability to accumulate heavy metals. Literature data [8–11] as well as our own observations [12, 13] have provided some indication of the heavy metal accumulation capacity of selected plant species. The moss species *Pleurozium schreberi*, recommended in the European monitoring programme, has been abandoned owing to its wide geographical range. Despite the fact that it is a very good indicator of environmental pollution,, it nevertheless occurred within a too small number of sites in the studied mountainous area. In particular, it rarely appeared within dense spruce stands. On the other hand, samples of the moss *P. formosum* Hedw. as one of the most widespread species in the forest areas of the Tatra National Park were collected. The samples of the plant material were collected at the end of September, that is at the end of the growing season in the mountains. Markert and Weckert [14] provide some clue as to the date of moss sampling. They state that the heavy metal content in the *P. formosum* Hedw. is subject to significant seasonal fluctuations, and this variation is often more important than the variation between sites. Therefore, in order to obtain comparable results, it is recommended that moss is sampled for biomonitoring purposes in the same period of time, preferably at the end of the plant growing season, i.e. at the turn of September and October.

The main objective of the study is to determine the content of heavy metals in two plant species depending on the absolute altitude in protected mountain areas.

2. Materials and methods

2.1 Study area

The research area is located in the Tatra National Park (**Figure 1**). The park covers an area of 211.64 km^2, of which 149.84 km^2 is under strict protection, 34.69 km^2 is under active protection, and 27.14 km^2 under landscape protection. In 2019, the TNP was visited by 3.9 million people. The Tatras are the only alpine mountains in Poland, where mainly alpine relief is protected, as well as valuable species of plants and animals (including endemics and relics). The research area is located in the Polish part of the Central Western Carpathians, in the northern part of the Tatra Range macroregion [15] and it is the highest part of the entire Carpathians. The specificity of this area is the complex geological structure [16–18], land relief heterogeneity (fluvial-denudation, karst, and glacial) [19–21], climatic conditions changing with the increase in altitude above sea level (air temperature, total precipitation, etc.). The specificity of the climate of the Tatra Range is determined by the incidence of different air masses. Arctic maritime air masses (PPm) have the largest share in the formation of weather, i.e., 65% of days a year, while continental polar air masses (PPk) approximately 20% of days a year [22, 23]. The above elements determine the specificity of water circulation (spatially diversified possibility of water retention, the volume of runoff, water chemistry, etc.). The soil cover of the Tatra Mountains is strongly related to, among other features, their geological substrate, morphogenetic processes, and climatic conditions, and its

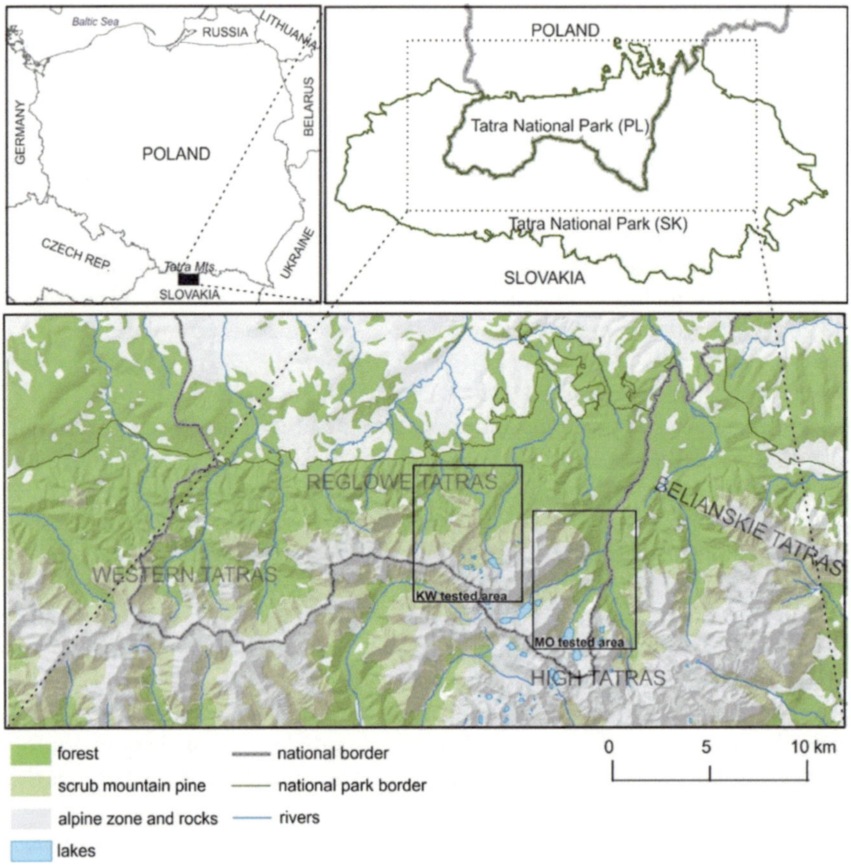

Figure 1.
Location of the study area on the background of the map of Poland and Tatra National Parks.

characteristic feature is openwork, as well as poorly developed soils (i.e., initial soils) [24]. All the physico-geographical zones, characteristic of high mountain areas, have developed in the Tatra Mountains [25]. Two test areas in the Tatra National Park in Poland, on the northern slope of the Tatra Mountains, were selected for the study. These areas were selected owing to the diversity of the natural environment, including the physico-geographical location, landscape zone, and geological structure. The test areas were given working names—Kasprowy Wierch (KW) and Morskie Oko (MO).

2.1.1 Kasprowy Wierch

The test area covers two physico-geographical mesoregions, i.e., the Reglowe Tatras (sampling points 1–4) and the Western Tatras (sampling point 5) [26], and ranges from the forest level to the alpine level (**Table 1**). The geological structure is strongly diversified in terms of lithology and tectonics. This affects, among other things, the incompatibility of the topographic watershed with the underground watershed. The area belongs to the Bystra catchment (with the sub-catchment of the Potok Jaworzynka) and the Sucha Woda Gąsienicowa catchment, which is part of the Dunajec basin. Depending on the altitude, the mean annual air temperature ranges from 0–6°C [27], the annual total of precipitation ranges from 800 mm to 1,800 mm, and the length of the snow cover deposition ranges from 100 to 200 days a year [28]. The soil cover is varied and dominated by the following soils:

DOI: http://dx.doi.org/10.5772/intechopen.99962

Sample no.	Altitude [m asl]	Geographical coordinates	Dominant area exposure	Terrain slope grade	Land cover features	Geological structure [17, 18]	Physico-geographical mesoregion [15]
1	1100	N49°15.572′ E19°59.322’	NE	20°-30°	Coniferous forest, spruce forest	Boulders, gravel, sand, and silts of stones and river terraces 0.5–3.0 m high, e.g., rivers (Holocene)	The Reglowe Tatras
2	1200	N49°15.424′ E19°59.645’	N	30°-40°	Coniferous forest, spruce forest	Dolomites, limestones, siltstones, and breccia (Lower Triassic)	The Reglowe Tatras
3	1300	N49°15.254′ E19°59.681’	W	20°-30°	Glade (area covered with grasses, sedges, herbaceous plants)	Dolomites, limestones, siltstones, and breccia (Lower Triassic)	The Reglowe Tatras
4	1400	N49°15.252′ E19°59.908’	NW	20°-30°	Rows and groups of the Norway spruce or the Swiss pine in the mountain pine, dense clumps of Norway spruce in the mountain pine	Dolomites and limestones, undivided (Middle Triassic)	The Reglowe Tatras
5	1550	N49°14.497′ E20°00.097’	N	0°-10°	Glade (area covered with grasses, sedges, herbaceous plants)	Boulders, moraine rock debris, clayey (Pleistocene)	Western Tatras
6	1000	N49°15.065′ E20°05.898’	SE	0°-10°	Coniferous forest, spruce forest	Boulders, gravel, sand, clayey sands and silts of cones, of fluvioglacial levels and terraces 12.0–15.0 m high, e.g., rivers (Pleistocene)	High Tatras
7	1100	N49°13.984′ E20°05.524’	NE	20°-30°	Coniferous forest, spruce forest	Granodiorites and tonalities, equal grained, gray (Carbon)	High Tatras
8	1200	N49°13.270′ E20°05.647’	NE	0°-10°	Young Norway spruce stand	Boulders, moraine rock debris, clayey (Pleistocene)	High Tatras
9	1300	N49°12.893′ E20°04.867’	NE	10°-20°	Coniferous forest, spruce forest	Boulders, rock debris and silts of dump and alluvial cones (Pleistocene–Holocene)	High Tatras
10	1400	N49°12.021′ E20°04.115’	E	10°-20°	Coniferous forest, spruce forest	Boulders, rock debris and silts of dump and alluvial cones (Pleistocene–Holocene)	High Tatras

Table 1.
Characteristics of sampling points in the Kasprowy Wierch (KW) test area - samples No. 1–5, and the Lake Morskie Oko (MO) test area - samples No. 6–10.

Fluvisols, Rendzic Leptosols, Folic Rendzic Leptosols, Cambic Rendzic Leptosols, Haplic Cambisols (Eutric), Haplic Podzols (Skeletic), Entic Podzols, Leptic Podzols, and Folic Leptosols [29].

2.1.2 Morskie Oko

The area is located within the High Tatras, in the Białka catchment (the Dunajec river basin) drained by the Rybi Potok, the Roztoka, and the Białka (**Table 1**). With regard to the zonation of the environment, it is entirely located within the forest level. It is part of one of the largest post-glacial grooves in the Tatras (a U-shaped valley). Depending on the altitude, the mean annual air temperature ranges from 2–4°C [27], the annual total of precipitation ranges from 1,000 mm to 1,400 mm, and the length of snow cover deposition ranges from 120 to 160 days a year [28]. The dominant soils in this part are, among others: Haplic Podzols (Skeletic), Haplic Cambisols (Dystric, Skeletic), Lithic Leptosols, and Regosols (Hyperskeletic) [29].

2.2 Sampling and analysis

2.2.1 Sampling

Plant samples (two species: the moss *Polytrichum formosum* Hedw. (green parts) and the European blueberry *Vaccinium myrtillus* L. (leaves) were sampled in the area of the Tatra National Park, from the Kasprowy Wierch (KW) test area and the Lake Morskie Oko (MO) test area. The samples were taken every 100 meters of altitude, starting from an altitude of: 1,100 m above sea level for KW and from 1,000 m above sea level for MO. Owing to the limited range of occurrence at higher altitudes, the plants were sampled up to 1,550 m above sea level for KW. The geographical coordinates of the sampling sites and the designations adopted are presented in **Table 1**.

2.2.2 Chemical analysis

According to the suggestions of the following authors: Maňkovska et al. [30] and Sawidis et al. [31] regarding the sample preparation procedure, the plant material was left unwashed. The samples were dried in an electric drier at a temperature of 400°C for 72 h. Needles were separated from branches. Equal amounts of biomass from primary samples from the same plot were combined. Dry and homogenized samples were pulverized in an electric grinder. Portions of 1 g dry weight material were placed in Teflon vessels. 5 cm3 of 65% HNO3 and 3 cm3 of 36% H2O2 were added to each vessel. The mixture was mineralized in a Berghof Speed Wave microwave at a temperature of 200°C and at a pressure of 4 MPa. After processing, the samples were diluted with deionized water to a total volume of 50 cm3 and filtered through a hard paper filter. The final solutions were analyzed for heavy metals (Cd, Cr, Cu, Ni, Pb and Zn) using the inductively coupled plasma mass spectrometry (ICP-MS) method in the Bureau Veritas laboratory. Such standards and reference materials (for plants) were used. The detection limits (μg/g dm) were as follows: for Cd: 0.01, Cr: 0.1, Cu: 0.01, Ni: 0.1, Pb: 0.01 and Zn: 0.1.

2.2.3 Statistical study

Statistical analyses were performed using the IBM SPSS program. Owing to the different conditions and differences in the altitude of sampling, the analysis was carried out in two groups depending on the location of the tests (MO Lake Morskie

Oko and KW Kasprowy Wierch). In order to assess the compliance of the distributions with the normal distribution, the Kolmogorov-Smirnow tests were performed. As the distributions differed from the normal distribution, non-parametric methods were used for further analyses. Spearman's coefficients were used to assess the relationship between the variables. In order to assess the significance of differences between the two groups, descriptive statistics were calculated and Mann–Whitney tests were performed. P = 0.05 was assumed as the limit of statistical significance below which the results were considered significant.

3. Results

The mean content of heavy metals in plants is presented in **Table 2**.

On the basis of the mean metal content in plants, calculated from all the collected samples, the following series of heavy metal concentrations were obtained:

Zn > Pb > Cr > Cu > Ni > Cd for the moss *P. formosum* Hedw.

Zn > Cr > Pb > Cu > Ni > Cd for the European blueberry *V. myrtillus* L.

The metal concentration series look similar for both plant species with the difference in lead and chromium content, where the moss *P. formosum* Hedw. accumulated more lead than chromium, while in the case of the European blueberry

Species	Study area	Altitude [m asl]	Heavy metals [μg/g d.m.]					
			Cd	Cr	Cu	Ni	Pb	Zn
Moss *P. formosum* Hedw.	MO	1000	2.1	24.9	18.7	16.7	30.2	70.8
		1100	2.1	24.8	18.8	17.0	30.8	70.6
		1200	2.3	25.1	19.5	17.7	32.3	71.9
		1300	2.4	26.6	20.3	18.1	33.7	74.2
		1400	2.5	27.9	20.9	18.8	35.5	76.9
	KW	1100	1.6	15.9	12.6	10.2	17.6	45.6
		1200	1.8	16.4	12.7	10.3	18.0	46.4
		1300	1.8	17.4	13.1	10.8	18.6	47.7
		1400	2.0	17.6	13.6	11.2	19.3	49.4
		1550	2.2	18.3	14.2	11.7	19.9	50.7
European blueberry *V. myrtillus* L.	MO	1000	1.5	17.0	11.8	12.2	15.9	42.4
		1100	1.6	17.3	12.2	12.3	15.9	43.8
		1200	1.6	17.7	12.5	12.7	16.6	44.5
		1300	1.7	18.7	13.2	13.2	17.3	45.6
		1400	1.8	19.7	14.1	14.3	18.2	48.1
	KW	1100	1.1	11.1	9.3	8.7	10.0	26.1
		1200	1.2	11.6	9.8	9.3	10.6	26.2
		1300	1.3	12.2	10.2	9.5	11.0	27.6
		1400	1.4	12.2	10.9	10.1	11.5	28.0
		1550	1.5	13.2	11.4	11.0	12.1	28.6

Table 2.
Mean concentrations of heavy metals in the moss species Polytrichum formosum *Hedw. And in the European blueberry* Vaccinium myrtillus *L.*

V. myrtillus L. the opposite was true (the European blueberry accumulated more chromium than lead). For both plant species, the last three metals presented in the series were similarly accumulated (higher amounts of copper compared to nickel and cadmium, and nickel compared to cadmium).

Comparing the mean metal contents in both plant species, it can be seen that the moss *P. formosum* Hedw. was characterized by a greater accumulation of metals (2.1 μg Cd/g dm, 21.5 μg Cr/g dm, 16.4 μg Cu/g dm, 14.2 μg Ni/g dm, 25.6 μg Pb/g dm, 60.4 μg Zn/g dm) compared to the European blueberry *V. myrtillus* L. (1.5 μg Cd/g dm, 15.1 μg Cr/g dm, 11.5 μg Cu/g dm, 11.3 μg Ni/g dm, 13.9 μg Pb/g dm, 36.1 μg Zn/g dm). Both the moss *P. formosum* Hedw. and the European blueberry *V. myrtillus* L. accumulated zinc in the highest amounts (60.4 and 36.1 μg Zn/g dm for the moss and the European blueberry, respectively), and cadmium in the smallest amounts (2.1 and 1.5 μg Cd/g dm).

Analyzing the data in **Table 2**, it can also be concluded that both plant species accumulated greater amounts of metals in the Morskie Oko test area than in the Kasprowy Wierch test area. For the same absolute altitudes, the content of heavy metals in plants, in particular of lead (1.8), nickel (1.7) and zinc (1.6) for the moss and of zinc (1.7) and lead (1.6) for the European blueberry, were almost twice as high in the Morskie Oko test area (**Table 3**). Smaller differences in the metal content in plants between the test areas were observed for cadmium, copper, and chromium.

The absolute altitude coefficient was calculated as the quotient of the heavy metal content in plants in the test areas for a given altitude (e.g. for Cd 1,100 m asl, the coefficient is the quotient of the Cd content in the MO test area to the Cd content in the KW test area for an altitude of 1,100 m asl).

The values of the heavy metals plant enrichment factor metals depending on the absolute height were determined as the second factor. This coefficient was calculated as the quotient of the heavy metal content for the lowest altitude to the highest metal content in the plants for a given test area, e.g. Cd 1,400 m asl to Cd 1,000 m asl for MO). The values of the enrichment coefficient are presented in **Table 4**.

The enrichment factors, calculated for all absolute heights, reach a value greater than or equal to 1.0. This indicates the presence of heavy metals accumulation in plants that is lower or similar to the highest altitude tested for a given test area (1,400 m asl for MO and 1,550 m asl for KW). The increase in the accumulation of metals in plants along with altitude was observed for the two plant species and for

Species	Study area	Altitude [m a.s.l.]	Absolute altitude coefficient					
			Cd	Cr	Cu	Ni	Pb	Zn
Moss *Polytrichum formosum* Hedw.	MO/KW	1100	1.3	1.6	1.5	1.7	1.8	1.5
		1200	1.3	1.5	1.5	1.7	1.8	1.6
		1300	1.3	1.5	1.5	1.7	1.8	1.6
		1400	1.3	1.6	1.5	1.7	1.8	1.6
European blueberry *Vaccinium myrtillus L.*	MO / KW	1100	1.4	1.6	1.3	1.4	1.6	1.7
		1200	1.4	1.5	1.3	1.4	1.6	1.7
		1300	1.3	1.5	1.3	1.4	1.6	1.7
		1400	1.3	1.6	1.3	1.4	1.6	1.7

Table 3.
Absolute altitude coefficients.

Species	Study area	Altitude [m a.s.l.]	Enrichment factor					
			Cd	Cr	Cu	Ni	Pb	Zn
Moss *Polytrichum formosum* Hedw.	MO	1400/1000	1.2	1.1	1.1	1.1	1.2	1.1
		1400/1100	1.2	1.1	1.1	1.1	1.2	1.1
		1400/1200	1.1	1.1	1.1	1.1	1.1	1.1
		1400/1300	1.0	1.0	1.0	1.0	1.1	1.0
	KW	1550/1100	1.3	1.2	1.1	1.1	1.1	1.1
		1550/1200	1.2	1.1	1.1	1.1	1.1	1.1
		1550/1300	1.2	1.0	1.1	1.1	1.1	1.1
		1550/1400	1.1	1.0	1.0	1.0	1.0	1.0
European blueberry *Vaccinium myrtillus* L.	MO	1400/1000	1.2	1.2	1.2	1.2	1.1	1.1
		1400/1100	1.1	1.1	1.2	1.2	1.1	1.1
		1400/1200	1.1	1.1	1.1	1.1	1.1	1.1
		1400/1300	1.1	1.1	1.1	1.1	1.1	1.1
	KW	1550/1100	1.3	1.2	1.2	1.3	1.2	1.1
		1550/1200	1.3	1.1	1.2	1.2	1.1	1.1
		1550/1300	1.2	1.1	1.1	1.2	1.1	1.0
		1550/1400	1.1	1.1	1.0	1.1	1.1	1.0

Table 4.
Heavy metals plant enrichment factors.

Species	Study area	Altitude [m asl]	Plant accumulation coefficient					
			Cd	Cr	Cu	Ni	Pb	Zn
Moss *Polytrichum formosum* Hedw./ European blueberry *Vaccinium myrtillus* L.	MO	1000	1.4	1.5	1.6	1.4	1.9	1.7
		1100	1.3	1.4	1.5	1.4	1.9	1.6
		1200	1.4	1.4	1.6	1.4	1.9	1.6
		1300	1.4	1.4	1.5	1.4	1.9	1.6
		1400	1.4	1.4	1.5	1.3	1.9	1.6
	KW	1100	1.5	1.4	1.4	1.2	1.8	1.7
		1200	1.5	1.4	1.3	1.1	1.7	1.8
		1300	1.4	1.4	1.3	1.1	1.7	1.7
		1400	1.4	1.4	1.3	1.1	1.7	1.8
		1550	1.5	1.4	1.2	1.1	1.6	1.8

Table 5.
Heavy metals plant accumulation coefficients.

each tested element. The increase in the content of all metals in plants occurs for two test areas.

Table 5 summarizes the plant accumulation coefficients calculated as the ratio of the content of a given metal in the moss to the content of the same metal in the European blueberry for the same absolute height and the same test area.

On the basis of the calculated accumulation coefficients it is clearly visible that the moss *P. formosum* Hedw. has a greater ability to absorb and accumulate heavy metals than the European blueberry *V. myrtillus* L. For each of the heavy metals determined, it was the moss that accumulated greater amounts. The largest differences in the accumulation of metals between the tested plants were observed for lead (a coefficient of 1.6–1.9) and zinc (a coefficient of 1.6–1.8), and the lowest for nickel (a coefficient of 1.1–1.4). The coefficients obtained for cadmium, chromium, and copper were similar and fell within the range of: 1.2–1.6.

Based on the Kolmogorov–Smirnov test (**Table 6**), distributions close to normal were recorded only for the content of Cd in the moss and Cu in the European blueberry. The other variables had distributions deviating from the normal distribution. Therefore, in order to determine the similarity, non-parametric tests were used in further analysis.

A statistically significant positive relationship was found between the Cd content and altitude in the moss (rho = 0.227; p = 0.023). There was no statistically significant relationship between the altitude and the content of other elements (**Table 7**).

Species		Altitude	Cd	Cr	Cu	Ni	Pb	Zn
Moss	N	100	100	100	100	100	100	100
	Test statistics	0.135	0.085	0.110	0.136	0.096	0.174	0.219
	p	0.000	0.070	0.005	0.000	0.025	0.000	0.000
European blueberry	N	100	100	100	100	100	100	100
	Test statistics	0.135	0.142	0.135	0.069	0.096	0.212	0.109
	p	0.000	0.000	0.000	0.200	0.024	0.000	0.005

Table 6.
Assessment of compliance with the normal distribution - Kolmogorov–Smirnov test for one sample.

Species	Spearman's rho	Cd	Cr	Cu	Ni	Pb	Zn
Moss	rho	0.227	−0.081	−0.064	−0.047	−0.069	−0.057
	p	0.023	0.422	0.525	0.642	0.493	0.573
European blueberry	rho	0.132	−0.002	0.085	0.088	0.065	−0.076
	p	0.190	0.988	0.403	0.386	0.521	0.455

Table 7.
Assessment of the relationship between the variables (metal-altitude) - Spearman's coefficients.

Species		Cd	Cr	Cu	Ni	Pb	Zn
Moss	Z	−4.798	−6.571	−5.726	−5.330	−6.132	−6.250
	p	0.000	0.000	0.000	0.000	0.000	0.000
European blueberry	Z	−4.962	−5.057	−2.479	−3.200	−2.775	−5.291
	p	0.000	0.000	0.013	0.001	0.006	0.000

Table 8.
Mann–Whitney test results for species.

There were statistically significant differences between the locations for both the moss and the European blueberry. Each element had a lower concentration in the measurements at Kasprowy Wierch than at Lake Morskie Oko (**Table 8** and **Figure 2**).

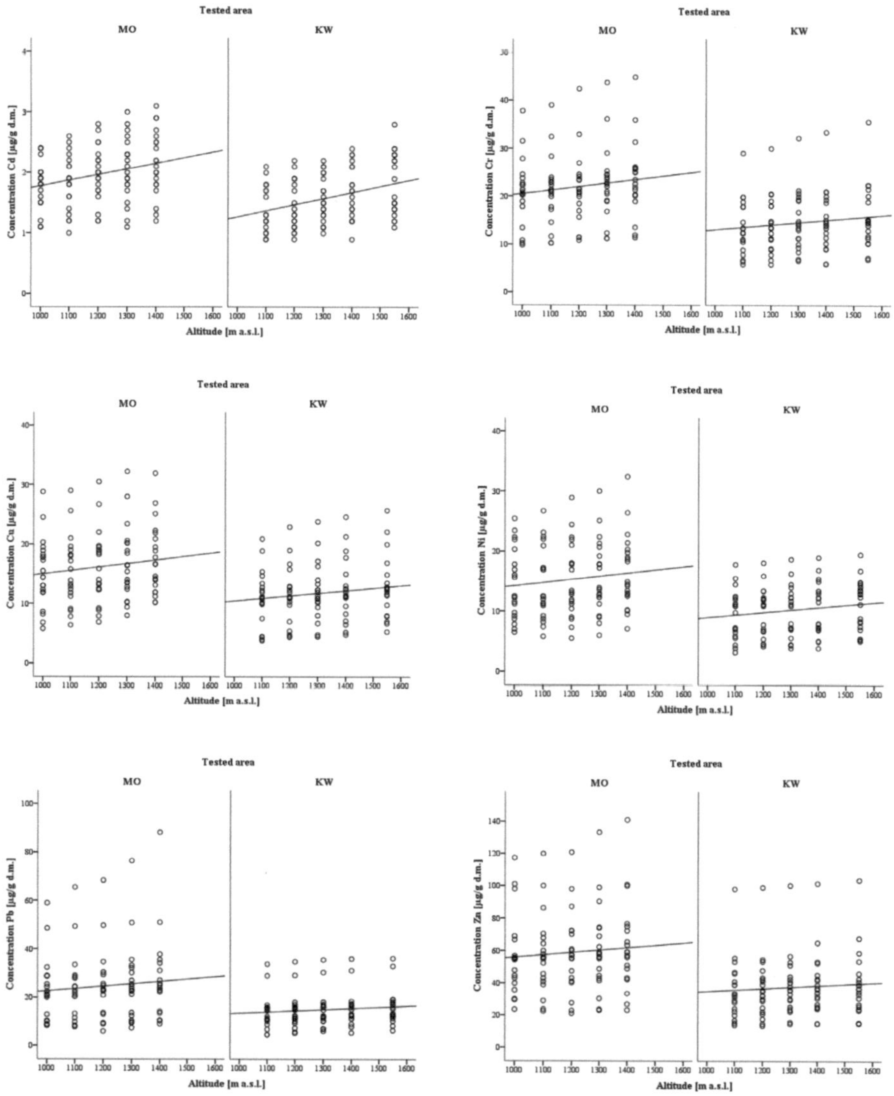

Figure 2.
Assessment of the relationships between variables (metal-altitude) - scatter dot plots for Cd, Cr, Cu, Ni, Pb and Zn.

Location		Cd	Cr	Cu	Ni	Pb	Zn
Morskie Oko	Z	−7.036	−5.106	−5.954	−3.709	−5.512	−5.653
	p	0.000	0.000	0.000	0.000	0.000	0.000
Kasprowy Wierch	Z	−6.390	−4.472	−2.096	−1.759	−5.937	−6.067
	p	0.000	0.000	0.036	0.079	0.000	0.000

Table 9.
Mann–Whitney test results for location.

Considering the differences in the accumulation of metals by the moss and the European blueberry, statistically significant differences were obtained for each location. Moss accumulated more of all the tested metals compared to the European blueberry. The only exception was the Ni content in the Kasprowy Wierch test area, for which no statistically significant differences were found in accumulation by the moss and the European blueberry (**Table 9**).

4. Discussion

The contents of heavy metals in the tested plants were compared to the natural and toxic contents of metals in plants provided by Kabata-Pendias and Pendias [4], which amount to, respectively: 0.05–0.2 and 5–30 μg Cd/g, 0.1–0.5 and 5–30 μg Cr/g, 5–30 and 20–100 μg Cu/g, 0.1–5.0 and 10–100 μg Ni/g, 5–10 and 30–300 μg Pb/g, 27–150 and 100–400 μg Zn/g dm. It was found that the contents of copper and zinc in plants are within the natural ranges. However, the contents of cadmium, chromium, nickel, and lead in both plant species for both test areas exceeded the natural values. They were exceeded by several, or several dozen times: more than 3 times for lead and nickel, several times for cadmium, and the highest - several dozen times for chromium. Such high exceedances in the content of heavy metals in the tested plant species in relation to the natural value of these metals in plants indicate the anthropogenic pollution of the selected research areas. Owing to their location (TNP) and great natural value, the research areas should be free from anthropogenic pollutants, however, long-range emission in this case has a large impact on the quality of the natural environment. It should be noted that in mountain areas, the content in plants of heavy metals is influenced by long-range emissions. Communication and industrial pollutants from areas with increased emissions are transported over long distances, even several dozen to several hundred kilometers. The transport of pollutants is related to the prevailing wind directions. For the TNP area, the dominant wind direction is to the south-west. Dusts containing heavy metal are transported from this direction, i.e., from industrial areas in Poland (Silesia) as well as from the Czech Republic and Slovakia. Long-range emissions result in such a high accumulation of metals in plants of the protected area [32–34]. Additionally, the high accumulation of metals in the higher parts of the mountains is also influenced by high wind velocity and a large amount of precipitation [34].

Determining the content of heavy metals in the plants in the tested test areas, an increase in metals was found with increasing altitude. The increase in the metal content concerned all of the tested metals and two plant species. However, this increase was slight and similar for the Kasprowy Wierch and Morskie Oko test areas. Similarly to the author of the present study, an increase in the content of metals in plants along with an increase in altitude was found in their research by Shetekauri et al. [35] in the western Caucasus Mountains for As, Cd, Ti, W in mosses, Sahin et al. [36] in the Kumalar Mountains for Cu, Zn in herbaceous plants, Zechmeister [32] in the Alps for As, Pb, Zn, and V in the *P. schreberi* and *Hylocomium splendens*, Šoltés [37] for the content of Pb in the *Sphagnum girgensohnii* in the Tatra Mountains in Slovakia, Samecka-Cymerman et al. [38] in the Tatra National Park for the content of Cd, Ni, and Zn in the *Athyrium distentifoliu*, Panek [39] in the Poland's Carpathian region for Pb in *P. formosum* and Kuklová et al. [40]. Kuklová et al. [40] found an increase in the content of Cu and Zn in three plant species (the *Dryopteris filix-mas*, *Rubus idaeus*, and *V. myrtillus*) with an increase in altitude. They observed an increase in the Cd content for two plant species: the *D. filix-mas* and the *V. myrtillus*. The re-search was carried out in the Slovak Paradise National Park (Slovenský Raj National Park), Eastern Slovakia, collecting samples of plants growing at an altitude of 750, 760, 950, 960, 1,000, and 1,110 m above sea level.

Comparing the accumulation of metals in the two tested plant species, it can be seen that the moss *P. formosum* Hedw. has higher amounts of metals compared to the European blueberry *V. myrtillus* L. The higher accumulation of metals in the moss results from its morphological structure and the ability to accumulate pollutants. In addition, the moss accumulated pollutants for longer than the European blueberry, because in the case of the moss, it was the green parts of the plant (stem and leaves), estimated to be about 3 years old, that were sampled for analysis, while in the case of the European blueberry, it was the leaves (about six months old) that were sampled for analysis. This gives several times longer accumulation time of pollutants.

5. Conclusions

In order to determine the content of heavy metals in the natural environment, indicator plants are used, the so-called phytoindicators. These include plant species that can absorb and accumulate toxic substances, such as heavy metals. The conducted research makes it possible to make the conclusion that both plant species showed a high ability to absorb air pollutants, which directly translated into high concentrations of heavy metals in plants. Both the moss *P. formosum* Hedw. as well as the European blueberry *V. myrtillus* L. can be good phytoindicators in mountainous areas. The conducted research showed that both plant species accumulated greater amounts of heavy metals in the Morskie Oko test area compared to the Kasprowy Wierch area.

Author details

Joanna Korzeniowska
Institute of Geography, Pedagogical University, Krakow, Poland

*Address all correspondence to: joanna.korzeniowska@up.krakow.pl

References

[1] Sörme L, Bergbäck B, Lohm U. Goods in the antroposphere as a metal emission source. Water Air Soil Pollut. 2001;1:213-227. DOI: 10.1023/A: 1017516523915

[2] Lehndorf E, Schwark L. Accumulation histories of major and trace elements on pine needles in the Cologne Conurbation as function of air Quality. Atmospheric Environment. 2008;42:833-845. DOI: 10.1016/j. atmosenv.2007.10.025

[3] Steinnes E, Friedland AJ. Metal contamination of natural surface soils from long-range atmospheric transport: existing and missing knowledge. Environmental Reviews. 2006;14: 169-186.

[4] Kabata – Pendias A, Pendias H. Trace elements in soils and plants. CRC Press, Boca Raton, Fla., London, UK; 2001. 204 p.

[5] Nguyen HTH, Nguyen BQ, Duong TT, Bui ATK, Nguyen HTA, Cao HT, Mai NT, Nguyen KM, Pham TT, Kim K-W. Pilot-Scale Removal of Arsenic and Heavy Metals from Mining Wastewater Using Adsorption Combined with Constructed Wetland. Minerals. 2019;9:379. DOI: 10.3390/min9060379

[6] García-Lorenzo ML, Crespo-Feo E, Esbrí JM, Higueras P, Grau P, Crespo I, Sánchez-Donoso R. Assessment of Potentially Toxic Elements in Technosols by Tailings Derived from Pb–Zn–Ag Mining Activities at San Quintín (Ciudad Real, Spain): Some Insights into the Importance of Integral Studies to Evaluate Metal Contamination Pollution Hazards. Minerals. 2019;9:346. DOI:10.3390/ min9060346

[7] Korzeniowska J, Krąż P. Heavy Metals Content in the Soils of the Tatra National Park Near Lake Morskie Oko and Kasprowy Wierch - A Case Study (Tatra Mts, Central Europe). Minerals. 2020;10:1120. https://doi.org/10.3390/ min10121120

[8] Szczepaniak K, Biziuk M. Aspects of the biomonitoring studies using mosses and lichens as indicators of metal pollution. Environmental Research. 2003;93:221-230. https://doi.org/ 10.1016/S0013-9351(03)00141-5

[9] Harmens H, Norris D, Sharps K, Mills G, Alber R, Aleksiayenak Y, Blum O, Cucu-Man S.-M, Dam M, De Temmerman L, et al. Heavy metal and nitrogen concentrations in mosses are declining across Europe whilst some “hotspots” remain in 2010. Environ Pollut. 2015;200:93-104. https://doi.org/ 10.1016/j.envpol.2015.01.036

[10] Suchara I, Sucharová J, Hola M, Reimann C, Boyd R, Filzmoser P, Englmaier P. The performance of moss, grass, and 1- and 2-year old spruce needles as bio indicators of contamination: A comparative study at the scale of the Czech Republic. Sci Total Environ. 2011;409:2281-97. DOI: 10.1016/j.scitotenv.2011.02.003

[11] Rühling Å, Tyler G. Changes in the atmospheric deposition of minor and rare elements between 1975 and 2000 in south Sweden, as measured by moss analysis. Environmental Pollution. 2004;131:417-423. https://doi.org/ 10.1016/j.envpol.2004.03.005

[12] Korzeniowska J, Panek E. Trace metal concentrations in *Pleurozium schreberi* and *Taraxacum officinale* along the road no. 7. Ecological Chemistry and Engineering S. 2019;26:651-663. DOI: 10.1515/eces-2019-0047

[13] Korzeniowska J, Panek E, Baran A. Testing of selected phytoindicators for assessment of trace metals (Cd, Cr, Cu,

Ni, Pb and Zn) traffic pollution along the road Cracow – Zakopane. Logistyka. 2014;4:4486-4495

[14] Markert B, Weckert V. Use of *Polytrichum formosum* (moss) as a passive biomonitor for heavy metal pollution (cadmium, copper, lead and zinc). Sci. Total Environ. 1989;86: 289-294

[15] Solon J, Borzyszkowski J, Bidłasik M, Richling A, Badora K, Balon J, Brzezińska-Wójcik T, Chabudziński Ł, Dobrowolski R, Grzegorczyk I, et al. Physico-geographical mesoregions of Poland: Verifi cation and adjustment of boundaries on the basis of contemporary spatial data. Geogr. Pol. 2018;91:143–170

[16] Bac-Moszaszwili M, Burchart J, Głazek J, Iwanow A, Jaroszewski W, Kotański Z, Lefeld J, Mastella L, Ziomkowski W, Roniewicz P, et al. Mapa geologiczna Tatr Polskich [Geological map of the Polish Tatras] 1: 30 000 1979, Wydawnictwo Geologiczne, Warszawa

[17] Piotrowska K, Kotański Z, Gawęda A, Piotrowski J, Rączkowski W. Szczegółowa Mapa Geologiczna Polski. Arkusz Tatry Zachodnie [Detailed Geological Map of Poland. The Western Tatras sheet]; NAG, PGI-NRI: Warsaw, Poland, 2008

[18] Piotrowska K, Michalik M, Rączkowski W, Iwanow A, Wójcik A, Derkacz M, Wasiluk R. Szczegółowa Mapa Geologiczna Polski. Arkusz Tatry Wysokie [Detailed Geological Map of Poland. The High Tatras sheet]; NAG, PIG-PIB: Warszawa, Poland, 2013

[19] Kotarba A, Kaszowski L, Krzemień K. High-mountain denudational system in the Polish Tatra Mountains: Geographical Studies; Ossolineum: Wrocław, Poland; 1987. 106p.

[20] Klimaszewski M. Rzeźba Tatr Polskich 1988 [Geomorphologic relief in the Polish part of Tatras]; Wydawnictwo PWN: Warszawa, Poland, 1988; pp. 1–707

[21] Hreško J, Boltizar M, Bugar G. The present-day development of landforms and landcover in alpine environment – Tatra Mts (Slovakia). Studia Geomorphol. Carpatho-Balc. 2005;39: 23–48

[22] Niedźwiedź, T. Sytuacje synoptyczne i ich wpływ na zróżnicowanie przestrzenne wybranych elementów klimatu w dorzeczu górnej Wisły [Synoptic situations and their impact on the spatial variability of selected climate elements in the upper Vistula basin], a Habilitation Dissertation at the Jagiellonian University; Jagiellonian University: Kazimierz, Poland, 1981; Volume 58, 165p.

[23] Niedźwiedź T. Climate of the Tatra Mountains. Mt. Res. Dev. 1992;12: 131–146

[24] Drewnik M. Geomorfologiczne uwarunkowania rozwoju pokrywy glebowej w obszarach górskich na przykładzie Tatr [Geomorphological conditions of the development of the soil cover in mountain areas on the example of the Tatra Mountains]; Wydawnictwo UJ: Kraków, Poland, 2008

[25] Troll C. High mountain belts between the polar caps and the equator: their definition and lower limit. Arctic and Alpine Research. 1973;5:19 – 28. https://www.jstor.org/stable/1550149

[26] Balon J, Jodłowski M, Krąż P. The Tatra Mountains: physico-geographical regions. In Atlas of The Tatr Mountains – Abiotic Nature; Dąbrowska K, Guzik M. Eds. plate I.4; Tatra National Park: Tatra County, Poland, 2015

[27] Żmudzka E, Nejedlik P, Mikulova K. Temperature, thermal indices. In Atlas

of The Tatr Mountains – Abiotic Nature; Dąbrowska K, Guzik M. Eds.; plate III.2; Tatra National Park: Tatra County, Poland, 2015.

[28] Ustrnul Z, Walawender E, Czekierda D, Štasny P, Lapin M, Mikulova K. Precipitation and snow cover. In Atlas of The Tatr Mountains – Abiotic Nature; Dąbrowska K, Guzik M. Eds. plate III.3; Tatra National Park: Tatra County, Poland, 2015.

[29] Skiba S, Koreň M, Drewnik M, Kukla J. Soils, In Atlas of The Tatr Mountains – Abiotic Nature; Dąbrowska K, Guzik M. Eds. plate V; Tatra National Park: Tatra County, Poland, 2015.

[30] Maňkovska B, Godzik B, Badea O, Shiparyk Y, Moravcik P. Chemical and morphological characteristic of key tree species in the Carpathian Mountains. Environ Pollut. 2004;130:41-54. DOI: 10.1016/j.envpol.2003.10.020

[31] Sawidis T, Breuste J, Mitrovic M, Pavlovic P, Tsigaridas K. Trees as bio indicator of heavy metal pollution in three European cities. Environ Pollut. 2011;159: 3560-70. DOI: 10.1016/j.envpol.2011.08.008

[32] Zechmeister H. Correlation between altitude and heavy metal deposition in the Alps. Environmental Pollution. 1995; 89:73-80. https://doi.org/10.1016/0269-7491(94)00042-C

[33] Gerdol R, Bragazza L. Effects of altitude on element accumulation in alpine moss. Chemosphere. 2006;64: 810–816. https://doi.org/10.1016/j.chemosphere.2005.10.053

[34] Kovář P. Conceptions of landscape-ecological relevance emerged in the Czech botany during the second half of twentieth century. Journal of Landscape Ecology. 2015;8:40–50. DOI: 10.1515/jlecol-2015-0013

[35] Shetekauri S, Shetekauri T, Kvlividze A, Chaligava O, Kalabegishvili T, Kirkesali E, Frontasyeva M, Chepurchenko O. Preliminary results of atmospheric deposition of major and trace elements in the greater and lesser Caucasus Mountains studied by the moss technique and neutron activation analysis. Annali di Botanica. 2015;5:89–95. https://doi.org/10.4462/annbotrm-13066

[36] Sahin I, Akcicek E, Guner O, Dogan Y, Ugulu I. An investigation on determining heavy metal accumulation in plants growing at Kumalar Mountain in Turkey. Eurasia J Biosci. 2016;10: 22-29. http://dx.doi.Org/10.5053/ejobios.2016.10.0.3

[37] Šoltés R. Correlation between altitude and heavy metal deposition in the Tatra Mts (Slovakia). Biologia. 1998; 53:85–90

[38] Samecka-Cymerman A, Stankiewicz A, Kolon K, Kempers A, Musiał M. Athyrium distentifolium used for bioindication at different altitudes in the Tatra National Park (South Poland). Ecotoxicology and Environmental Safety. 2012;79:184–188. https://doi.org/10.1016/j.ecoenv.2011.12.021

[39] Panek E. Trace metals in soils and in selected plant species in Poland's Carpathian region, a Habilitation Dissertation at the AGH University of Science and Technology; AGH University of Science and Technology: Kraków, Poland, 2000; Volume 79, 106p.

[40] Kuklová M, Hniličková H, Hnilička F, Pivková I, Kukla J. Toxic elements and energy accumulation in topsoil and plants of spruce ecosystems. Plant Soil Environ. 2017;63:402–408. DOI: 10.17221/364/2017-PSE

Chapter 19

Undesirable Neighbours: Eucalyptus and Protected Areas

Sandra Daniela Manzano Guzmán, Jose Augusto Drummond and Cristiane Gomes Barreto

Abstract

Eucalyptus is the common name of a set of exotic species present in the Brazilian territory. They have a strong invasive potential which is detrimental to the preservation of native floral formations, particularly in protected areas. This research seeks to (i) understand the stage of eucalyptus invasion in the Brasilia National Park; (ii) identify the main vectors of the invasive populations and (iii) verify the possible role of the adjoining Brasília National Forest in the invasion and (iv) consider possible conflicts between the roles of these two different categories of protected areas. A set of phytosociological sample areas were defined inside the park to pinpoint different eucalyptus populations. Findings indicate that eucalyptus populations inside the park behave invasively, having advanced 186.30 meters from their point of origin over the span of 45 years. Among the possible contamination vectors are a neighbouring nursery run by the local government and eucalyptus plantations in the adjoining Brasília National Forest. Results indicate the need for management actions to avoid continual seed dispersal by examined populations. They also indicate that the distinct conservation goals of national forests and national parks must be considered, especially when they are neighbours.

Keywords: Biological invasions, Brasília National Park, Cerrado, vectors of contamination, Brasilia National Forest

1. Introduction

Eucalyptus is the common name of hundreds of tree species of the Myrtaceae Family, currently classified in the genera Eucalyptus, Corymbia, and Angophora, native to Australia, a part of Papua New Guinea, and Indonesia. However, because they have several features that are useful for humans, eucalyptus species have been intentionally introduced in dozens of countries, including tropical and subtropical areas of the entire planet [1].

In 1823, Chile was the first South American country to introduce eucalyptus species. In Brazil, Federico de Albuquerque first introduced them in 1868 in the state of Rio Grande do Sul. From 1903 on, the Brazilian soil scientist Edmundo Navarro de Andrade (1881–1941) conducted systematic experiments to select tree species that could supply firewood as fuel for locomotives operating in the state of São Paulo [2]. He concluded that eucalyptus had greater energy efficiency than native Brazilian trees. He convinced railroad managers that the species should be planted in large scale in swaths along the railways in order to meet energy demands of locomotives.

This led to the establishment of numerous eucalyptus plantations along railways of the state of São Paulo [3].

Andrade studied the behaviour of 250 different species of eucalyptus, their physiology and their response to different types of soil and temperature. His experiments were conducted in the municipalities of Jundiaí and Rio Claro, in the interior of São Paulo [1]. Andrade later directed the planting of approximately 24 million trees in several plots belonging to the Companhia Paulista de Ferrocarril (a powerful railroad company). In 1940, as locomotives fuelled by wood were substituted by more modern ones equipped with diesel engines, those tree plantations started to be used for other purposes, such as manufacturing of pulp and paper [4].

Currently, eucalyptus represents 75% of all specimens found in Brazil's rather extensive commercially planted forests. In 2015, with almost 7.5 million hectares planted with eucalyptus, Brazil was one of the three largest eucalyptus growers in the world. 55.8% of all Brazilian plantations are concentrated in the Southeast region, especially in the states of Minas Gerais and São Paulo [5]. Globally, over the last 180 years, more than 200 eucalyptus species have been planted for several purposes outside their natural range [6].

Despite its contribution to the history of industrial and railway development, eucalyptus currently stands out in scientific research as a major "villain" identified by conservation biology. In recent years, several studies have highlighted the effects of exotic and invasive species on the loss of biodiversity [7–9]. Eucalyptus has become, for many scholars, a noteworthy enemy of biodiversity conservation. As an exotic species with a strong degree of invasiveness, eucalyptus jeopardizes conservation goals, mainly in protected areas [10].

In Brazil, the lack of control over the introduction of exotic species led to the dispersion of many species and to invasive processes in native formations. An example of this lack of control was the creation of fiscal incentives by the federal government, through Law 5,106 / 1966 and Decree-Law No. 1,376 / 1974, both of which stimulated commercial forest planting. One legal requirement was that supported projects should plant an annual minimum of 10,000 trees, but the species to be planted were not defined. As a result, extensive areas of Cerrado vegetation had their native flora replaced by extensive monocultures of eucalyptus and pinus [11]. Controversies about this type of afforestation gained importance when these species began to disperse spontaneously into natural areas, competing with native species [12, 13].

Despite this, on a global scale only a few eucalyptus populations have become invasive in the numerous areas to which they were transferred [14]. It is thus necessary to distinguish three concepts. An exotic species is a "species that would not naturally occur in a given geographic region without human (intentional or accidental) transport to the new region" [15]. An alien invasive species (AIS), on the other hand, is one that, once introduced (intentionally or not), has the ability (i) to reproduce autonomously and generate viable populations outside its natural range and (ii) to disperse a considerable distance from its point of introduction, in a short period of time, [16] without necessarily requiring additional human help. An exotic species is one that is found outside its natural distribution, with or without human assistance, but only those that can disperse at great distances in a short period of time are considered invasive.

Environmental problems emerge precisely when this invasive potential develops. Introduced populations become a more acute problem in areas designed to protect native ecosystems and biodiversity, such as officially protected areas (PAs). In these cases, invasive populations harm native biodiversity and hinder the achievement of PA objectives. The following eucalyptus species are currently present in several federal Brazilian PAs: *Eucalyptus angulosa* Schauer, *Corymbia citriodora*

Hook, *Eucalyptus crebra* F. Muell, *E. dunnii* Maiden, *E. grandis* W. Hill, *C. maculata* Hook, *E. paniculata* Sm., *Eucalyptus camaldulensis* Dhnh., *Eucalyptus saligna* Sm., *E. viminalis* Labill, and *E. robusta* Sm. [17]. Invasive populations of eucalyptus have been identified in PAs located in several Brazilian biomes and locations: Saltinho Biological Reserve (Pernambuco); Itapeva State Park (Rio Grande do Sul), Aratinga Ecological Station (Rio Grande do Sul), União Biological Reserve (Rio de Janeiro), and Vila Velha State Park (Paraná), among others [18–20]. Horowitz [20], surveying Brasília National Park (BNP), in the nation's Federal District, found eucalyptus populations that could behave as exotic invaders (in the initial stage of dispersion and colonization), while other populations were casual, that is, they have established themselves, but did not form long-term, viable populations.

The invasive potential of populations varies depending on the source of introduction, management practices, and the types affected ecosystems. Populations can behave invasively. This will influence management choices adopted by PA managers. These choices depend on how the invasive potential of the population is classified, on the identification of the source of contamination and, if it persists, on how the source can continue to influence dispersal and colonization of invading specimens.

The BNP is a highly restrictive type of PA - meaning that it is not open to any productive activities. Among its objectives is the maintenance of the integrity of native landscapes of the Cerrado biome. Despite this, activities carried out historically in areas adjacent to the park (human settlements, public works, road building, government buildings, depots, farms) created and left threats that affect the unit's native floral composition. Currently, in the immediate vicinity of the BNP there are industries, semi-rural housing complexes, agricultural and urban settlements. Other neighbours are two PAs - the Contagem Biological Reserve (also a strictly protected PA) and the Brasília National Forest - BNF (a multiple-use PA, which has plantations of exotic trees). The dominant landscape of the adjoining BNF is stands of planted eucalyptus and pine trees [21]. There is also a tree and shrub nursery run by nursery the local government (Novacap's Nursery) and associated with Brasília's urban landscaping policies. The nursery is used for growing a wide variety of exotic and native plants. A stretch of the area's major interstate highway (BR-040) and its heavy traffic pass quite close to the BNP (field observations). On account of all this, the BNP has been susceptible to the effects of many vectors that defy its conservation mandate, including invasions of exotic species. According to its managers, these invasions are the main challenge for the management of the BNP's native ecosystems.

Among the questions raised in this article, we seek to (i) understand the stage of eucalyptus invasion in the BNP; (ii) identify the main vectors of arboreal invasive populations; and (iii) verify the possible role of eucalyptus and pine plantations in the adjoining BNF in the invasions and thus (iv) consider the possible conflicts between the roles of these two different categories of PAs in view of the potential environmental risks generated by the proximity between them.

2. Materials and methods

In order to measure the invasive potential of a windbreaker planted with eucalyptus in 1960s along the boundary between the BNP and Novacap's nursery, 10 plots of 100 m2 (10 x 10 m) were delimited inside the BNP area, within a sampling area of 5,589 ha. All plots of the sampling area were identified as belonging to the gallery forest physiognomy, corridors along rivers in Cerrado landscapes. Reconnaissance trails were marked out to define the sampling area. In order to encompass the entire eucalyptus population, the area was demarcated by using the outermost individuals of the group (**Figure 1**). We used the stratified systematic

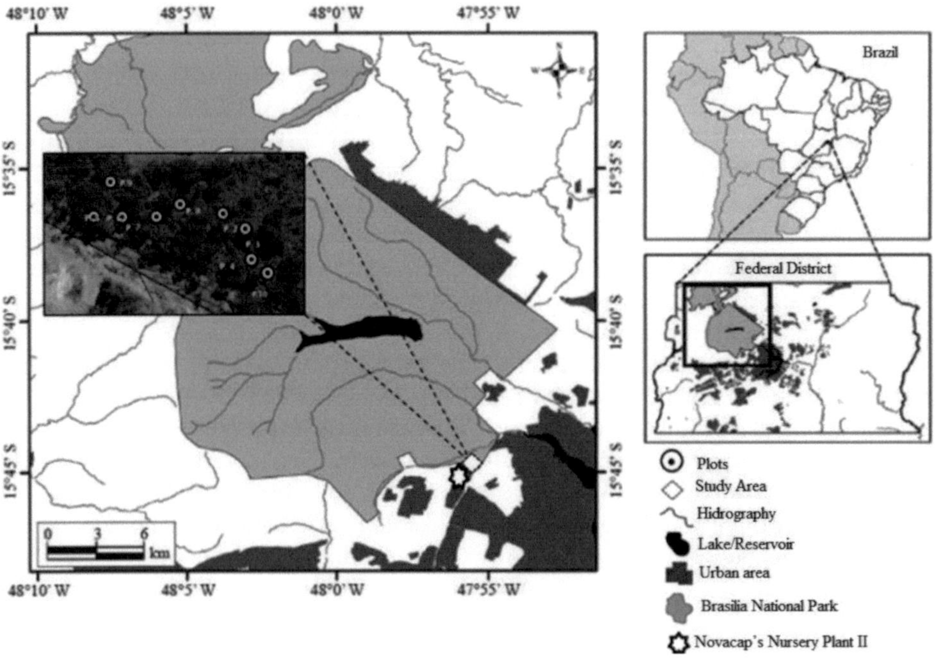

Figure 1.
Distribution of sampled areas in the Brasília National Park, close to Novacap's Nursery Plant II. Figure produced by the Authors. Source: Image adapted by the authors from [22].

sampling method, keeping a distance of 10 m between the plots. Plots were georeferenced with the GPS Garmin eTrex 10, version 3.1; the measuring point was placed on the central position of the plot. Eucalyptus individuals inside each plot were counted. Subsequently, phytosociological parameters of absolute taxon frequency, absolute taxon density, and absolute taxon dominance were computed.

Estimation of the rate of invasion as defined by [16] (more than 100 m covered in less than 50 years for plants with seed dispersal) required that we measure the distance between the windbreaker and the farthest eucalyptus within the polygon of the sampling area. Dispersion time was estimated from the age of the windbreaker. This age was determined with the help of aerial photographs. Local wind parameters (velocity, direction, and gust) were also considered for the period of August 2015 to August 2016. This was done to determine the period of the year in which wind favours dispersal of eucalyptus seed towards the BNP.

In our effort to determine the possible flow of exotic propagules from the BNF to the BNP, we marked out a trail inside the BNP, close to its border with the BNF. The location of eucalyptus or other trees found in the area was georeferenced; later the distances between each of them and the BNF were measured.

Data about wind speed, direction, and gust from August 2015 to August 2016 were provided by Brazil's National Institute of Meteorology. They were used to determine the period of the year that would favour seed dispersal in the BNP.

3. Results

We sampled 49 eucalyptus individuals near the Novacap nursery. The absolute frequency of the taxon (FAt) was 90%, which means that the species occurs in 9 of the 10 plots (there were no occurrences in plot 4). From the absolute density (DAt), the number of individuals present in the BNP area was estimated at 2.394

(density of 0.049 ind /m^2). In order to compute the absolute dominance (DoAt), the basal area (Gi) of each individual was estimated from the perimeter at breast height (1.30 m). The total measured basal area (GT) (equal to the sum of Gi) was 53,765.08 cm2. Therefore, 53.76% of the sampled area was occupied by eucalyptus.

Invasive status of the studied populations.

To determine whether the eucalyptus population in the sampling area meets the invasion rate defined by [16], we measured the distance between the windbreaker (a parental eucalyptus individual) and the farthest individual located at one of the points of the polygon in the sampling area. This distance was 186.30 m. The Google Earth distance measurement tool was used, based on the geographical coordinates of the selected eucalyptus individual and the windbreak. As a second measurement of the invasiveness parameter, parental age was determined by analysing aerial photographs belonging to the Cartographic System of the Federal District [23]. We analysed a 1965 photograph (before the nursery was created) and a 1975 one from (after its creation). The 1965 photograph recorded several rural roads and vegetation in the current area of the BNP. However, the windbreak did not exist; no disturbance of its area is perceptible. In the 1975 photograph, the windbreak appears as a thickened fringe located on the limits between the nursery and the BNP. As the nursery itself was created only in 1971, the planting of the windbreak must have occurred between 1971 and 1975.

To determine the type of exotic species (naturalized, casual, or invasive) in which the group of eucalyptus is classified, we used the results of the computation of the rate of invasion, as per [16], and the results obtained from phytosociological parameters. Measurement of the invasion status, together with observation of individuals, allowed an estimate of eucalyptus in its first stages of invasion. Both the phytosociological parameters and the dispersal capacity (up to 186.30 m in less than 50 years) indicate that this is an expanding group. In addition, no type of control of the studied individuals was recorded; the trees do not display hacking or girdling marks. Absence of control also favours invasion.

Eucalyptus found in this location represent a problem for individuals of native species peculiar to the Cerrado and to gallery forest phytophysiognomies. On the other hand, eucalyptus control measures, such as chopping and girdling, were recorded in other places inside the park, farther away from the windbreak. However, it is a group with high population density (5 ind./100m^2), when compared, for example, to similar values in Eucalyptus urophylla plantations in the Cerrado region of Minas Gerais (6.6 ind./100m^2) [24]. It is also reproductively active [25]. The data allow us to deduce that eucalyptus near the nursery is invasive and has the capacity to spread widely.

Since the nursery was created in 1971, it is assumed that the possible age of the eucalyptus in the windbreak is approximately 45 years. Therefore, eucalyptus trees in the park covered the distance of 186.30 m from the nursery in less than 50 years, which is substantially more than the 100 m in 50 years defined by the invasion index [16]. Concluding that the studied populations are invasive, we proceeded to investigate the possible vectors of invasion to be found in the area around the BNP.

3.1 Vectors of invasion

The BNF is composed by four neighbouring but disconnected areas. Area I is close to the BNP, at a distance of approximately 303.37 m; Area II is currently occupied by a rural settlement, created in 1996, and is located at a distance of approximately 46.51 m from the BNP; Areas III and IV are relatively distant from the BNP, at the approximate distances of 12,803 m and 3,403 m, respectively [25].

The reconnaissance trail, leading from the area of the BNP near the BNF, was 64.5 km long; it stretches from the park's entrance and uses the unit's main and secondary roads. In the area near the BNF the following trees were identified: A single eucalyptus individual (coordinates 15°45′9.69″S/48°2′7.46″W) and, also exotic, a juvenile and isolated pine tree (coordinates 15°44′20.23″-48°3′30.94″W). The first location corresponds to a distance of 42.44 m from Area II of the BNF, which was originally planted with eucalyptus. This area has been degazetted and is currently occupied by a rural settlement with a population of approximately 400 families. The pine tree is located 303 m from the BNF's Area I, which cultivates eucalyptus and pine. Other two pine trees were located in the park, but far from BNF areas.

Given the proximity of these individuals to the BNF area, historically planted with eucalyptus and pine, we can suppose, as park managers do, that it is a possible invasion vector. In order to corroborate this possibility, data on the region's wind characteristics (speed, direction and gust) provided by INMET (Brazil's National Meteorological Institute) were analysed for the period of September 2015 to August 2016. We sought to determine the possible role of wind in the dispersion of eucalyptus seeds from the BNF in the direction of the BNP [26]. Brazilian climatological data 1961–1990 for the study area point to a predominantly eastern wind direction, especially in the months of April to September [26]. This information was verified specifically for the period of September 2015 to August 2016. The record of monthly predominant winds can be seen in **Figure 2**.

Average wind speed was 1.57 m/s. The month with the highest gusts and the highest prevailing speed is January (2.2 m/s). January also stands out due to an inversion of the wind direction, which moves toward the NNW. January winds coincide with the peak of eucalyptus seed release. This may favour seed dispersion from the Viveiro II and Area II of BNF to the BNP (**Figure 3**).

Seed release occurs about six months after flowering, which in turn occurs in the dry season. Considering that the dry season occurs between April and September, seed release can go on from about November to March, the months in which the propagules can be dispersed to the BNP with the aid of winds. Based on the average predominance of the winds - to the East - and on the predominance of the winds in the month of greater speed and greater probability of dispersion - North-Northwest -, **Figure 3** illustrates the relation between wind directions and identifies the three potential sources of dispersion of eucalyptus seeds within the BNP: Area I of the BNF; Area II of the BNF; and Novacap's nursery.

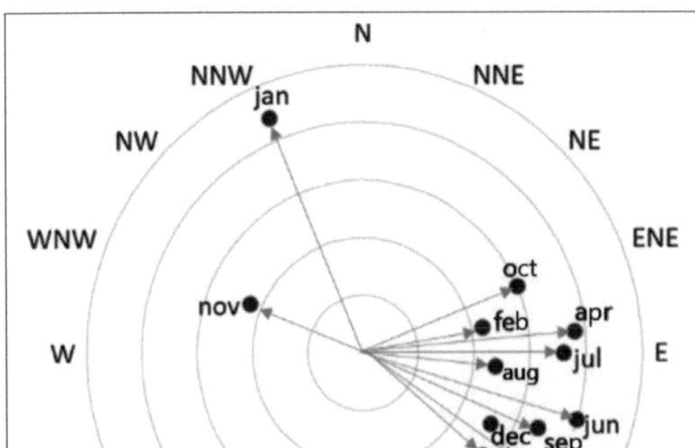

month/year	Predominant velocity	Predominant direction	
sep/15	1,7 m/s	104°	ESE
oct/15	1,5 m/s	75°	ENE
nov/15	1,1 m/s	290°	WNW
dec/15	1,3 m/s	109°	ESE
jan/16	2,2 m/s	327°	NNW
feb/16	1,1 m/s	82°	E
mar/16	1,4 m/s	116°	ESE
apr/16	1,9 m/s	88°	E
may/16	1,7 m/s	115°	ESE
jun/16	2 m/s	100°	E
jul/16	1,8 m/s	95°	E
aug/16	1,2 m/s	99°	E

Figure 2.
Wind speeds and predominant directions during one year in the Brasília National Park, from September (2015) to August (2016). Diagram produced by the Authors. Source: [26].

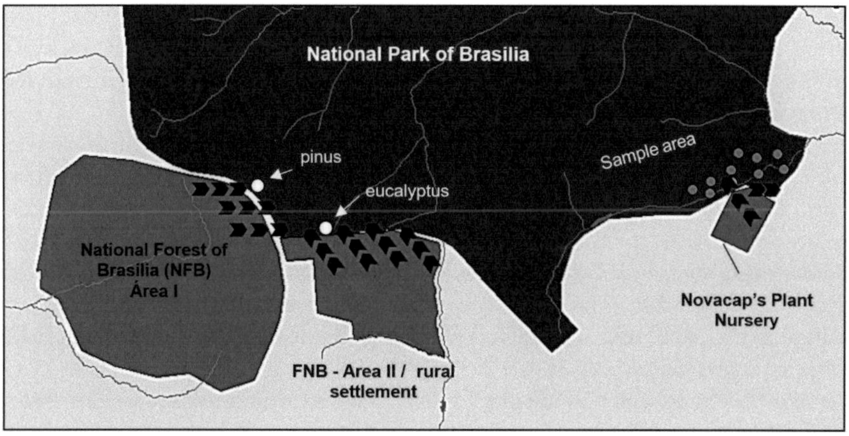

Figure 3.
Incidence of predominant winds as seed dispersion vectors into the Brasília National Park from eucalyptus sources located in the areas of the Brasília National Forest and the Novacap nursery. Figure produced by the Authors.

The predominant wind direction during the year, to the East, also directs the dispersion flow from the BNF's Area I to the BNP, which could favour the invasion of species from the first to the second. The predominantly Eastern direction also guides the dispersion flow of Novacap's nursery, reinforcing the invasive characteristics of eucalyptus in the BNP.

Our use of the reconnaissance trail near the BNF allowed us to evaluate its role in the dispersal of eucalyptus trees into the BNP. It was found that there is currently no group of eucalyptus in the BNP near its border with Area I of the BNF. Area I has pine and eucalyptus plantations [21]. The inference is that the relationship between wind speed, its direction, and the time of seed release, associated with the presence of a narrow trail between the two PAs, hinders or prevents the transfer of propagules from Area I of the BNF into the interior of the park.

However, a single eucalyptus individual was found at coordinates 15°45′9.69″S/48°2′7.46″W, inside the park (**Figure 3**). As this location does not show remnants of recent dwellings or old farms, we assumed that this individual was introduced from some external source. Area II of the BNF is close by. As informed above, since 1996 Area II has been occupied by a rural settlement, although it still displays remnants of pine and eucalyptus plantations.

According to **Figure 3**, the NNW wind direction, predominant at the time of seed release, is also the direction that favours dispersion of Area II of the BNF to the BNP. This may have favoured specifically the establishment of the single identified individual within the park (**Figure 3**).

4. Discussion

The BNP was established in an area of the Cerrado biome, with native ecosystems of grasslands, savannas, and forests. However, this area was previously occupied by farms, dwellings; exotic species were cultivated in it. In addition, its proximity to (Novacap's nursery, created in 1971 to produce seedlings of native and exotic species) aided, and continues to aid, colonization of the park by propagules of exotic species used mostly for urban landscaping. This is the case of the eucalyptus planted to form the windbreak located on the boundary between the park and the nursery [27].

Data generated by this research showed that eucalyptus populations in the BNP grew and dispersed since 1970s and, therefore, behaved as invasive. The lack of control and management reinforced this process. Most of the identified individuals are reproductively active and population density is high.

Regarding management options employed by the BNP staff to reduce or neutralize the risk and the influence of exotic species (the groups of eucalyptus within the park and in the windbreak), the most suitable one is eradication, because of the size of the groups [25]. According to observations of other groups of eucalyptus located inside the park, we found that the cutting method used to control the species does not result in eradication. The girdling method, observed in some individuals located elsewhere in the park, did manage to kill them. Therefore, we assume that girdling may work for the individuals studied.

On the other hand, the BNF displays a dominant anthropic landscape composed of stands of planted eucalyptus and pine trees – two exotic species with strong invasive potential. Although the BNF is managed under the same general regulations as national parks (Law 9,985, of July 18, 2000, and Decree 4,340, of August 22, 2002), its conservation goals differ from those of the BNP. While national parks, as "fully protected" areas, seek the preservation of nature, biodiversity, and ecosystems, national forests, as "sustainable use" areas, seek to combine conservation with the use of natural resources. To do so, national forests can plant and harvest commercially valuable exotic species, to the detriment of native species. In the case of the BNF, the area formerly occupied by Cerrado (open, savanna-like vegetation) was occupied by dense lines of tall pine and eucalyptus forests. This happened in the 1980s. Native flora was removed, and fauna was dispersed from the BNF area. Losses of biodiversity and native landscapes are obvious consequences. There are no gains in terms of environmental conservation.

Areas I and II of the BNF may be vectors of contamination of native vegetation inside the BNP, given their mutual proximity (**Figure 3**). According to the BNF's management plan, its dominant vegetation consists of pines and eucalyptus trees that lack proper cultivation practices is, a fact that increases the risk of invasion of these species into the BNP area [21]. Most of the BNF's Area I, which is closest to the park, is covered by eucalyptus plantations.

Although the park has followed its objectives consistently, the neighbouring BNF is a potential threat to its biodiversity by acting as a vector of contamination by exotic species. From the viewpoint of regional conservation, the BNF is a potential agent of environmental degradation than a protected area. The proximity of the BNF to the BNP generates no synergy or gains in terms of biodiversity protection. Brazilian PA law does allow the formation of "mosaics". Mosaics are management models, predicted by Law 9,985 of July 18, 2000, and Decree 4,340 of August 22, 2002, which seek to integrate the management of different types of neighbouring protected areas. However, we consider them to be unfeasible in the cases of national forests covered by plantations of exotic species which are located next to fully protected PAs. Environmental planning should always consider species with strong dispersion potentials as a risk and, therefore, should define containment measures in the same way as it does in the case of private commercial plantations located in the buffer zones of PAs, as predicted by law.

5. Conclusion

Field surveys conducted in the BNP led to the conclusion that the studied groups of eucalyptus have invasive characteristics. The BNP area adjacent to Novacap's nursery hosts a well-established and dispersing group of eucalyptus. If not controlled, it may become a very serious problem for the park. The populations of eucalyptus originated from windbreak propagules, located near the nursery, should be eliminated. Eradication is feasible due to the size of the groups and to their initial stage of dispersion. Managing these populations is essential for the BNP to fulfil its objectives, among which the preservation of biodiversity is paramount.

Area I of the BNF, close to BNP, with its plantations of eucalyptus and pines, is an obvious contamination risk, even though the risk so far has not been strong. This may be due to the combination of wind velocity and wind direction with the timing of seed release. This combination does not favour dispersion from Area I of the BNF to the BNP. However, this section of the BNP, on account of its proximity to the BNF, should become a prime target for the monitoring and control of exotic species in the park. On the other hand, eucalyptus propagules from the nursery windbreak, as well as remaining populations of Area II of BNF, have their dispersion towards the BNP favoured by wind speed and direction in the month of January, coinciding with peak seed release. This mode of dispersion is corroborated by the data obtained in this study.

In terms of control and management of eucalyptus in the BNP, this work points to the necessity of strengthening (i) measures of species eradication inside the unit and (ii) the management and planning of the landscape surrounding the park. Concerning the BNF, its proximity to the park jeopardizes the park's preservation goals, even though the BNF is an officially created and managed PA. Therefore, regardless of the formal categories of land use and occupation, vectors of contamination by exotic species can come from the most varied and even unexpected sources, such as a neighbouring PA.

Acknowledgements

The authors thank Capes and CNPq Foundations for financial support received during this research. We also thank Marília Teresinha de Souza Machado, for her participation in field work and data organization. Field research in the Brasília National Park was authorized by ICMBio (# 49670). Chapter financed by the Academic Excellence Program of the Brazilian Coordination for the Improvement of Higher Education Personnel (PROEX/CAPES).

Conflict of interest

The authors declare no conflict of interest.

Author details

Sandra Daniela Manzano Guzmán, Jose Augusto Drummond
and Cristiane Gomes Barreto*
Observatory of Conservation Units and Related Social Policies, University of Brasilia, Brasilia, Brazil

*Address all correspondence to: crisbarreto@unb.br

References

[1] Rejmánek M, Richardson D, Pysek P. 2012. Plant invasions and invasibility of plant communities, p. 332-355. In: Maarel EV, Franklin J. (eds.). Vegetation Ecology. 2nd ed. Wiley-Blackwell. 572p.

[2] Andrade, EN. 1909. A cultura dos Eucalyptus. São Paulo: Typographia Brazil de Rothschild & Comp.

[3] Lima, WP (org.). 1993. Impacto ambiental do Eucalipto. São Paulo: EDUSP.

[4] Machado, R. 2000. Navarro de Andrade, o Pai da Silvicultura Brasileira, p. 210-213. In: Machado R. A Floresta e o Homem. São Paulo: EDUSP. 428p.

[5] IBGE – Instituto Brasileiro de Geografia e Estatística. Produção da Extração Vegetal e da Silvicultura. Brasília, 2015. Available at: <https://sidra.ibge.gov.br/pesquisa/pevs/quadros/brasil/2015>. Accessed on: 9 Jul 2017

[6] Booth, TH. Eucalyptus and their potential for invasiveness particularly in frost-prone regions. International Journal of Forestry Research, 2012, 2012. doi:10.1155/2012/837165

[7] Simberloff, D et al. Impacts of biological invasions: what's what and the way forward. Trends in Ecology & Evolution, 28(1): 58-66, 2013.

[8] Ellis E, Antill EC, Kreft H. All is not loss: plant biodiversity in the anthropocene. Plos, 7(1): 1-9, 2012.

[9] Wardle D, Bardgett RD, Callaway RM, Van der PUTTEN WH. Terrestrial ecosystem responses to species gains and losses. Science, v.332: 1273-1277, 2011.

[10] Deus E, Silva JS, Castro-d'Iez P, Lomba A, Ortiz ML, Vicente J. Current and future conflicts between eucalypt plantations and high biodiversity areas in the Iberian Peninsula. Journal for Nature Conservation, 45: 107-117, 2018. https://doi.org/10.1016/j.jnc.2018.06.003

[11] Fidelis AC, LIMA JD. Impactos Ambientais da Silvicultura no Bioma Cerrado. V Simpósio Nacional de Geografia Agrária. 2009. Available at: http://www.uff.br/vsinga/trabalhos/Trabalhos%20Completos/Alex%20Camargo%20Fidelis.pdf Accessed on 10 Jul 2017.

[12] Buscardo, E et al. 2008. The early effects of afforestation on biodiversity of grasslands in Ireland, p. 133-148. In: Brockerhoff, EG et al. (orgs.). Plantation Forests and Biodiversity: Oxymoron or Opportunity?. Springer, Dordrecht. 296p.

[13] Dodet M, Collet C. When should exotic forest plantation tree species be considered as an invasive threat and how should we treat them?. Biological Invasions, 14(9): 1765-1778, 2012.

[14] Gordon D, Flory SL, Cooper AL, Morris SK. Assessing the Invasion Risk of Eucalyptus in the United States Using the Australian Weed Risk Assessment. International Journal of Forestry Research., 2012: 1-7, 2012. doi:10.1155/2012/203768

[15] Moro M. et al. Alienígenas na sala: o que fazer com espécies exóticas em trabalhos de taxonomia, florística e fitossociologia?. Acta Botanica Brasilica, 26(4): 991-999, 2012. Moro M. et al. Alienígenas na sala: o que fazer com espécies exóticas em trabalhos de taxonomia, florística e fitossociologia?. Acta Botanica Brasilica, 26(4): 991-999, 2012. (p. 993-994)

[16] Richardson DM et al. Naturalization and invasion of alien plants: concepts and definitions. Diversity and Distributions, 6(2): 93-107, 2000.

[17] Sampaio AB, Schmidt IB. Espécies Exóticas Invasoras em Unidades de Conservação Federais do Brasil. Biodiversidade Brasileira, 3(2): 32-49, 2013.

[18] Rocha FL, Lima SG, Martins VS, Torres PF, Reis RC. Avaliação da presença de espécies exóticas em unidades de conservação estaduais de Minas Gerais, Brasil. Revista de Ciências Agroambientais, 15(2): 242-246, 2017.

[19] Carvalho, F, Nascimento, MT, Braga, JMA. Estrutura e composição florística do estrato arbóreo de um remanescente de Mata Atlântica submontana no município de Rio Bonito, RJ, Brasil (Mata Rio Vermelho). Revista Árvore, 31(4): 717-730, 2007.

[20] Pellizzaro L, Souza T, Caminschi RM. Estudo de espécies arbóreas exóticas em unidade de conservação em Francisco Beltrão, Paraná, Brasil. EDUCERE - Revista da Educação, 13(1): 141-161, 2013.

[21] ICMBio. Plano de Manejo Floresta Nacional de Brasília. Brasília, 2016. Available at <http://www.icmbio.gov.br/portal/images/stories/plano-de-manejo/DCOM_plano_de_manejo_Flona_de_Brasilia_Diagnostico.pdf >. Accessed on: 22 Aug 2017.

[22] Martins, C. et al. Impacto da invasão e do manejo do capim-gordura (*Melinis minutiflora*) sobre a riqueza e biomassa da flora nativa do Cerrado sentido restrito. Revista Brasileira de Botânica. São Paulo, v. 34, n. 1, p. 73-90. 2011.

[23] Secretaria de Estado de Gestão do Território e Habitação - Segeth. Mapa Índice, articulação SICAD. Available at: <http://www.sedhab.df.gov.br/mapas_sicad/index2.htm>. Accessed on: 28 Apr 2016.

[24] Assis RL, Ferreira MM, Morais EJ, Fernandez LA. Produção de Biomassa de Eucalyptus urophylla ST. Blake sob diferentes espaçamentos na região de Cerrado de Minas Gerais. Revista Árvore, 23(2): 151-156, 1999.

[25] Guzmán SDM. 2016. Introdução de Espécies Exóticas Invasoras no Parque Nacional de Brasília: Contexto histórico e atual das influências do Viveiro II da Novacap e da Floresta Nacional de Brasília. Thesis (Master in Sustainable Development). Universidade de Brasília. 125p. Available at https://repositorio.unb.br/handle/10482/22171 Accessed on: 10 Jun 2021.

[26] INMET – Instituto Nacional de Meteorologia. Planilhas de Vento e Direção Predominante (1961-1990). Available at: fttp://www.inmet.gov.br/webcdp/climatologia/normais/imagens/normais/planilhas/Vento-Direcao-Predominante_NCB_1961-1990.xls Accessed in: 10 Jul 2017.

[27] Horowitz C, Martins CR, Machado T. 2007. Diagnóstico das Espécies Exóticas Arbóreas, Arbustivas e Herbáceas que Ocorrem nas Zonas de Uso Especial e de Uso Intensivo do Parque Nacional de Brasília. Brasília: Edições IBAMA.